PELICAN BOOKS

A 816

THE TREASURY OF MATHEMATICS

VOLUME 2

THE TREASURY OF
MATHEMATICS
VOLUME TWO

A COLLECTION OF
SOURCE MATERIAL IN MATHEMATICS
EDITED AND PRESENTED WITH
INTRODUCTORY BIOGRAPHICAL AND
HISTORICAL SKETCHES BY
HENRIETTA MIDONICK

Revised by
Minetta and Reginald Vesselo

PENGUIN BOOKS

Penguin Books Ltd, Harmondsworth, Middlesex, England
Penguin Books Pty Ltd, Ringwood, Victoria, Australia

—

Published in the U.S.A. 1965
First published in Britain by Peter Owen Ltd 1965
Published in Pelican Books 1968

—

Copyright © Philosophical Library, New York, 1965

—

Made and printed in Great Britain
by Butler & Tanner Ltd, Frome and London
Set in Monotype Times

CONTENTS

CONTENTS

CONTENTS

ACKNOWLEDGEMENTS

GRATEFUL acknowledgement and appreciation is extended to the publishers below for their 'gracious consent to the use of the material listed:

ISAAC BARROW
Selection from *Lecture X* of the *Geometrical Lectures of Isaac Barrow*, translated with notes by J. M. Child (Chicago and London: Open Court Press, 1916).

GEORGE BOOLE
Selection from Boole's *The Mathematical Analysis of Logic* (New York: Philosophical Library, 1948, reprint of 1847 edition). Selections from Boole's *An Investigation of the Laws of Thought* (London: Walton and Maberley; Cambridge: Macmillan, 1854).

GEORG CANTOR
Selections from '*Beiträge zur Begründüng der transfiniten Mengenlehre*', *Mathematische Annalen*, vol. xlvi, pp. 481–512, 1895, translated by Philip E. B. Jourdain and appearing in *Contributions to the Founding of the Theory of Transfinite Numbers* (New York: Dover).

ARTHUR CAYLEY
Selections from 'A Memoir on the Theory of Matrices' from *Collected Mathematical Papers of Arthur Cayley*, vol. II (Cambridge: C.U.P., 1889).

GEOFFREY CHAUCER
Chaucer and Messahalla on the Astrolabe, vol. 5 of *Early Science in Oxford*, pp. 1–28, 114–15, 118–19, by R. T. Gunther (Oxford: O.U.P., 1929), by courtesy of Mr A. E. Gunther.

GABRIEL CRAMER
Selections from *Introduction à l'Analyse des Lignes Courbes Algébriques*

ACKNOWLEDGEMENTS

(Geneva: Cramer; Philibert, 1750) specially translated for this edition by Henrietta Midonick.

AUGUSTUS DE MORGAN

Selections from 'On the Syllogism, No. III and On Logic in General', *Transactions of the Cambridge Philosophical Society*, vol. X, 1864.

RENÉ DESCARTES

Selections from *Discourse on Method*, translated by John Veitch (London and Washington: L. Walter Dunne, 1901). Selections from *The Geometry of René Descartes*, translated by D. E. Smith and M. L. Latham (New York: Dover, 1954).

DIOPHANTUS

Selections from the *Arithmetica*, taken from *Diophantus of Alexandria*, Thomas L. Heath, 2nd edition (Cambridge: C.U.P., 1910).

ALBRECHT DÜRER

Selections from *Four Books on Human Proportions* and *The Art of Mensuration* in *The Writings of Albrecht Dürer* (London: Peter Owen). Selections from *The Mathematics of Great Amateurs*, Julian Lowell Coolidge (Oxford: Clarendon Press, 1949).

GOTTLOB FREGE

Selections from *Die Grundlagen der Arithmetik*, translated by J. L. Austin (New York: Philosophical Library, 1953).

CARL FRIEDRICH GAUSS

Selections from: 1. *Disquisitiones Arithmeticae*, translated by Ralph G. Archibald, in *A Source Book in Mathematics*, ed. D. E. Smith (New York: McGraw Hill, 1929; Harvard University Press); 2. *Theoria Motus*, translated by Charles Henry Davis (New York: Little, Brown, 1857); 3. *The Foundation of Mathematics*, translated by G. Waldo Dunnington (Baton Rouge: Louisiana University Press, 1937); 4. *Superficies Curvas*, translated by J. C. Morehead and A. M. Hiltebeite, (Princeton: Princeton University Library, 1902).

GOTTFRIED WILHELM V. LEIBNIZ

Selections from *Non Inelegans Specimen Demonstrandi In Abstractis*, translated by C. I. Lewis, and *Scientia Generalis* and *Characteristica* from *A Survey of Symbolic Logic*, C. I. Lewis (Berkeley: University of California Press, 1918).

ACKNOWLEDGEMENTS

METRODORUS
Selections from *Greek Anthology*, translated by W. R. Paton, vol. 5, bk XIV (Cambridge, Mass.: Harvard University Press, 1953).

ISAAC NEWTON
Selections from: 1. *Two Treatises* (*Of the Quadrature of Curves* and *Analysis by Equations of an Infinite Number of Terms*), translated by John Stewart (London: James Bettenham, 1745); 2. *Mathematical Principles of Natural Philosophy and His System of the World*, translated by Andrew Motte and revised by Florian Cajori (Berkeley: University of California Press, 1946).

NICOMACHUS OF GERASA
Selections from *Introduction to Arithmetic*, translated by Martin Luther D'Ooge (New York: Macmillan, 1926).

OMAR KHAYYÁM
Selections from *The Algebra of Omar Khayyám*, translated by Daoud S. Kasir (New York: Columbia University Press, 1931).

BENJAMIN PEIRCE
Selections from 'Linear Associative Algebra', *American Journal of Mathematics*, vol. IV, 1881. (Published under the auspices of Johns Hopkins University, Baltimore.)

CHARLES SANDERS PEIRCE
Selections from *Exact Logic* and *The Simplest Mathematics*, vols. III and IV of *Collected Papers of Charles Sanders Peirce*, eds. Charles Hartshorne and Paul Weiss (Cambridge, Mass.: Harvard University Press, 1933).

ROBERT RECORDE
Selections from *The Whetstone of Witte* (London, 1557).

SIMON STEVIN
Selections from *De Thiende* in *The Principal Works of Simon Stevin*, vol. II, ed. D. J. Struik (Amsterdam: Swets and Zeitlinger, 1958).

JAMES JOSEPH SYLVESTER
Selections from: 1. Presidential Address to Section A of British Association, 1869, in vol. II of *Collected Mathematical Papers of James Joseph Sylvester* (Cambridge: C.U.P., 1908); 2. Address on Commemoration Day at Johns Hopkins University, 1877, vol. III.

ACKNOWLEDGEMENTS

JOHN VENN

Selection from *Symbolic Logic*, 2nd edition (London, New York: Macmillan, 1894).

CASPAR WESSEL

Selection from *On the Analytic Representation of Direction*, translated by Martin A. Nordgaard, in *A Source Book in Mathematics*, ed. D. E. Smith (New York: McGraw Hill, 1929; Harvard University Press).

FURTHER DEVELOPMENT

NICOMACHUS OF GERASA

(c. A.D. 100)

NICOMACHUS of Gerasa, a celebrated Neo-Pythagorean, was the first mathematician to write a book dealing specifically and systematically with arithmetic as the science of numbers. This work, the *Introduction to Arithmetic*, exerted a powerful influence over the study of arithmetic for more than a millennium. In the middle ages and later, the authority of Nichomachus in the field of arithmetic was comparable to that of Euclid in geometry. Nicomachus is said to have written eleven works. Two of these have been preserved in complete form, the *Introduction to Arithmetic* containing an elementary theory of numbers and the *Manuale Harmonicum*, a treatise on music. Parts of a third work, the *Theologumena Arithmeticae*, written by him or wholly based on his work, are also extant. The remaining treatises ascribed to him, dealing with astronomy, mathematics, music, philosophy and biography, are known only through allusions to them by other authors and through references made to them by Nicomachus himself.

For the details of his life, historians have had as their sources only the extant writings of Nicomachus and the evidence given by occasional quotations from his works or the mention of his name by other writers. The date, or period in which he flourished, has therefore been deduced by considering, in the first place, that in his *Manuale Harmonicum*, Nicomachus mentions a certain Thrasyllus, a Platonist, who is known to have lived in the reign of Tiberius (A.D. 14–37). Hence, the *Manuale* could not have been written much earlier than A.D. 14. Secondly, the *Introduction to Arithmetic* was translated into Latin by Apuleius who lived under Antoninus Pius (A.D. 138–61). Hence, the *Introduction* could not have been written after A.D. 161, and we arrive at the conclusion that Nicomachus lived most likely c. A.D. 100. As for his personal life, many inferences may be drawn from the preface written by Nicomachus to his *Manuale Harmonicum*. This was in the form of a letter addressed to an unnamed lady of noble birth, at whose request and for whose benefit Nicomachus wrote the treatise. The cultured tone of the letter, the reference to a more detailed and more advanced treatise to be written for her in the future and the remarks concerning journeys undertaken by him, leaving him short of time, are all significant. Nicomachus may well be regarded as a highly learned scientist and philosopher possessing considerable grace of personality. The elementary nature of the material in his extant works may by no means be a true measure of the extent of his achievements. That he was sought after and highly respected for his knowledge by persons of high rank points up the great reputation for learning which he enjoyed in his own day. His interests were obviously not confined to Gerasa (a city in ancient Palestine) which, however was probably the principal scene of his activity.

The commentaries of Iamblichus (fourth century) and the versions of the *Introduction* contained in the *De Institutione Arithmetica* by Boethius (d. 524) made Nicomachus's work available to scholars familiar with Greek and Latin. Through the translation by Thabit Qor'ah (836–901) the *Introduction* became known in the East. Even if Proclus (d. A.D. 485) had not listed Nicomachus as one of the 'golden chain' of true philosophers, the *Theologumena Arithmeticae* and the *Introduction to Arithmetic* would have clearly revealed the Pythagorean philosophy underlying Nicomachus's arithmetical system. Nicomachus distinguishes between the wholly conceptual, immaterial number, which he regards as the 'divine' number, and the number which measures material things, the 'scientific' number. The *Introduction* implements the study of 'scientific' numbers. Here, in a setting influenced by Pythagorean philosophy of numbers, he explains, defines and classifies numbers, and sets forth details of the principles governing their relations.

INTRODUCTION TO ARITHMETIC

translated by Martin Luther D'Ooge

BOOK I

The ancients, who under the leadership of Pythagoras first made science systematic, defined philosophy as the love of wisdom. Indeed the name itself means this, and before Pythagoras all who had knowledge were called 'wise' indiscriminately – a carpenter, for example, a cobbler, a helmsman, and in a word anyone who was versed in any art or handicraft. Pythagoras, however, restricting the title so as to apply to the knowledge and comprehension of reality, and calling the knowledge of the truth in this the only wisdom, naturally designated the desire and pursuit of this knowledge philosophy, as being desire for wisdom.

*

Therefore, if we crave for the goal that is worthy and fitting for man, namely, happiness of life – and this is accomplished by philosophy alone and by nothing else, and philosophy, as I said, means for us desire for wisdom, and wisdom the science of the truth in things, and of things some are properly so called, others, merely share the name – it is reasonable and most necessary to distinguish and systematize the accidental qualities of things.

Things, then, both those properly so called and those that simply have the name, are some of them unified and continuous, for example, an animal, the universe, a tree and the like, which are properly and peculiarly called 'magnitudes'; others are discontinuous, in a side-by-side arrangement, and, as it were, in heaps, which are called 'multitudes', a flock, for instance, a people, a heap, a chorus and the like.

Wisdom, then, must be considered to be the knowledge of these two forms. Since, however, all multitude and magnitude are by their own nature of necessity infinite – for multitude starts from a definite root and never ceases increasing; and magnitude, when division beginning with a limited whole is carried on, cannot bring the dividing process to an end, but proceeds therefore to infinity – and since sciences are always sciences of limited things, and never of infinites, it is accordingly evident that a science dealing either with magnitude, *per se*, or with multitude, *per se*, could never be formulated, for each of them is limitless in itself, multitude in the direction of the more, and magnitude in the direction of the less. A science, however, would arise to deal with something separated from each of them, with quantity, set off from multitude, and size, set off from magnitude.

Again, to start afresh, since of quantity one kind is viewed by itself, having no relation to anything else, as 'even', 'odd', 'perfect', and the like, and the other is relative to something else and is conceived of together with its relationship to another thing, like 'double', 'greater', 'smaller', 'half', 'one and one-half times', 'one and one-third times', and so forth, it is clear that two scientific methods will lay hold of and deal with the whole investigation[1] of quantity; arithmetic, absolute quantity, and music, relative quantity.

And once more, inasmuch as part of 'size' is in a state of rest and stability, and another part in motion and revolution, two other sciences in the same way will accurately treat of 'size', geometry the part that abides and is at rest, astronomy that which moves and revolves.

Without the aid of these, then, it is not possible to deal accurately with the forms of being nor to discover the truth in things, knowledge of which is wisdom, and evidently not even to philosophize properly.

*

Number is limited multitude or a combination of units or a flow of quantity made up of units; and the first division of number is even and odd.

The even is that which can be divided into two equal parts without a unit intervening in the middle; and the odd is that which cannot be divided into two equal parts because of the aforesaid intervention of a unit.

Now this is the definition after the ordinary conception; by the Pythagorean doctrine, however, the even number is that which admits of division into the greatest and the smallest parts at the same operation, greatest in size and smallest in quantity,

1. Nicomachus thus subdivides the subject matter and assigns the special fields of the four mathematical sciences: I, treating number (1) as such, absolutely, Arithmetic; and (2) relative number, Music; II, treating quantity (1) at rest, Geometry; (2) in motion, Astronomy. Proclus, op. cit., *Prol.*, p. 35, 21 ff., Friedl., gives the same division of the field of the mathematical sciences, using the same terms, in his report of the Pythagorean mathematics, probably drawing upon this work. It is to be noted that Nicomachus does not in fact adhere strictly to his classification, for he treats in this work of relative number, which falls in the domain of Music, and in the discussion of linear, plane and solid numbers he comes close to Geometry.

in accordance with the natural contrariety[1] of these two genera; and the odd is that which does not allow this to be done to it, but is divided into two unequal parts.

In still another way, by the ancient definition, the even is that which can be divided[2] alike into two equal and two unequal parts, except that the dyad,[3] which is its elementary form, admits but one division, that into equal parts; and in any division whatsoever it brings to light only one species of number, however it may be divided, independent of the other. The odd[4] is a number which in any division whatsoever, which necessarily is a division into unequal parts, shows both the two species of number together, never without intermixture one with another, but always in one another's company.

By the definition in terms of each other, the odd is that which differs by a unit from the even in either direction, that is, toward the greater or the less, and the even is that which differs by a unit in either direction from the odd, that is, is greater by a unit or less by a unit.

Every number is at once half the sum of the two on either side of itself,[5] and similarly half the sum of those next but one in either direction, and of those next beyond them, and so on as far as it is possible to go. Unity alone, because it does not have two numbers on either side of it, is half merely of the adjoining number; hence unity is the natural starting point of all number.

---

1. That is, halves are the greatest possible parts of a term in magnitude; and there is a smaller number of them than of any other fractional part. Thus greater magnitude of factors is associated with a smaller number of them; this is the 'natural contrariety' of magnitude and quantity.

2. When an even number is divided into two parts, whether equal or unequal, these parts are always either both odd or both even ('only one species of number,' as Nicomachus says), Iamblichus, (p. 12, 14 ff. Pistelli). See Heath, *History*, vol. I, p. 70.

3. Iamblichus (p. 13, 7 ff. Pistelli) notes that the monad is distinguished from all the odd numbers by not even admitting division into unequal parts, and the dyad from the even numbers by admitting division into equal parts only.

4. If an odd number is divided into two parts these will always be unequal and one odd, the other even ('the two species of number').

5. Thus 5 is half the sum of $4+6$, $3+7$, $2+8$, etc.

By subdivision of the even,[1] there are the even-times even, the odd-times even, and the even-times odd. The even-times even and the even-times odd are opposite to one another, like extremes, and the odd-times even is common to them both like a mean term.

Now the even-times even is a number which is itself capable of being divided into two equal parts, in accordance with the properties of its genus, and with each of its parts similarly capable of division, and again in the same way each of their parts divisible into two equals until the division of the successive subdivisions reaches the naturally indivisible unit. Take for example 64; one half of this is 32, and of this 16, and of this the half is 8, and of this 4, and of this 2, and then finally unity is half of the latter, and this is naturally indivisible and will not admit of a half.

It is a property of the even-times even that, whatever part of it be taken, it is always even-times even in designation, and at the same time, by the quantity of the units in it, even-times even in value; and that neither of these two things will ever share in the other class. Doubtless it is because of this that it is called even-times even, because it is itself even and always has its parts, and the parts of its parts down to unity, even both in name and in value; in other words, every part that it has is even-times even in name and even-times even in value.

*

Let us then set forth the odd numbers from 3 by themselves in due order in one series:

$$3, 5, 7, 9, 11, 13, 15, 17, 19, \ldots$$

and the even-times even, beginning with 4, again one after another in a second series after their own order:

$$4, 8, 16, 32, 64, 128, 256, \ldots$$

as far as you please. Now multiply by the first number of either series – it makes no difference which – from the beginning and in

1. Euclid, among the definitions of *Elem.*, VII, defines the even-times even, even-times odd, odd-times even and odd-times odd (the latter is 'one which is measured by an odd number an odd number of times'). Nicomachus confines himself to a tripartite division of the even only; Euclid's classification applies to all natural numbers.

order all those in the remaining series and note down the resulting numbers; then again multiply by the second number of the same series the same numbers once more, as far as you can, and write down the results; then with the third number again multiply the same terms anew, and however far you go you will get nothing but the odd-times even numbers.

For the sake of illustration let us use the first term of the series of odd numbers and multiply by it all the terms in the second series in order, thus: 3×4, 3×8, 3×16, 3×32, and so on to infinity. The results will be 12, 24, 48, 96, which we must note down in one line. Then taking a new start do the same thing with the second number, 5×4, 5×8, 5×16, 5×32. The results will be 20, 40, 80, 160. Then do the same thing once more with 7, the third number, 7×4, 7×8, 7×16, 7×32. The results are 28, 56, 112, 224; and in the same way as far as you care to go, you will get similar results.

Odd numbers		3	5	7	9	11	13	15
Even-times even		4	8	16	32	64	128	256
Odd-times even numbers	Breadth	16*	24	48	96	192	384	768
		20	40	80	160	320	640	1280
		28	56	112	224	448	896	1792
		36	72	144	288	576	1152	2304
		44	88	176	352	704	1408	2816
					Length			

* Ed. note. This number should be 12.

Now when you arrange the products of multiplication by each term in its proper line, making the lines parallel, in marvellous fashion there will appear along the breadth of the table the peculiar property of the even-times odd, that the mean term is always half the sum of the extremes, if there should be one mean, and the sum of the means equals the sum of the extremes if two. But along the length of the table the property of the even-times even will appear; for the product of the extremes is equal to the square of the mean, should there be one mean term, or their product, should there be two. Thus this one species has the peculiar properties of them both, because it is a natural mixture of them both.

Again, while the odd is distinguished over against the even in classification and has nothing in common with it, since the latter is divisible into equal halves and the former is not thus divisible, nevertheless there are found three species of the odd, differing from one another, of which the first is called the prime and incomposite, that which is opposed to it the secondary and composite, and that which is midway between both of these and is viewed as a mean among extremes, namely, the variety which, in itself, is secondary and composite, but relatively is prime and incomposite.

Now the first species, the prime and incomposite, is found whenever an odd number admits of no other factor save the one with the number itself as denominator,[1] which is always unity; for example, 3, 5, 7, 11, 13, 17, 19, 23, 29, 31. None of these numbers will by any chance be found to have a fractional part with a denominator different from the number itself, but only the one with this as denominator, and this part will be unity in each case; for 3 has only a third part, which has the same denominator as the number and is of course unity, 5 a fifth, 7 a seventh, and 11 only an eleventh part, and in all of them these parts are unity.

*

As an illustration, let 9 be compared with 25. Each in itself is secondary and composite, but relatively to each other they have only unity as a common measure, and no factors in them have the same denominator, for the third part in the former does not exist in the latter nor is the fifth part in the latter found in the former.

The production of these numbers is called by Eratosthenes the 'sieve', because we take the odd numbers mingled together and indiscriminate and out of them by this method of production separate, as by a kind of instrument or sieve, the prime and incomposite by themselves, and the secondary and composite by themselves, and find the mixed class by themselves.

1. As 1/3 in the case of 3.

The method of the 'sieve' is as follows. I set forth all the odd numbers in order, beginning with 3, in as long a series as possible, and then starting with the first I observe what ones it can measure, and I find that it can measure the terms two places apart, as far as we care to proceed. And I find that it measures not as it chances and at random, but that it will measure the first one, that is, the one two places removed, by the quantity of the one that stands first in the series, that is, by its own quantity, for it measures it 3 times; and the one two places from this by the quantity of the second in order, for this it will measure 5 times; and again the one two places further on by the quantity of the third in order, or 7 times, and the one two places still farther on by the quantity of the fourth in order, or 9 times, and so *ad infinitum* in the same way.

Then taking a fresh start I come to the second number and observe what it can measure, and find that it measures all the terms four places apart, the first by the quantity of the first in order, or 3 times; the second by that of the second, or 5 times; the third by that of the third, or 7 times; and in this order *ad infinitum*.

Again, as before, the third term 7, taking over the measuring function, will measure terms six places apart, and the first by the quantity of 3, the first of the series, the second by that of 5, for this is the second number, and the third by that of 7, for this has the third position in the series.

And analogously throughout, this process will go on without interruption, so that the numbers[1] will succeed to the measuring function in accordance with their fixed position in the series; the interval separating terms measured is determined by the orderly progress of the even numbers from 2 to infinity, or by the doubling of the position in the series occupied by the measuring term, and the number of times a term is measured is fixed by the orderly advance of the odd numbers in series from 3.

1. It is generally assumed (as by Heath, *History*, vol. I, p. 100) that in the 'sieve of Eratosthenes' only the odd prime numbers take on successively the measuring function, and indeed this is all that is necessary, for, e.g. 9 is a multiple of 3 and all its multiples are likewise multiples of 3. The text, however, seems to imply that all the odd numbers should be used, although perhaps Nicomachus did not intend that he should be so strictly interpreted.

BOOK II

Now a triangular number is one which, when it is analysed into units, shapes into triangular form the equilateral placement of its parts in a plane. 3, 6, 10, 15, 21, 28, and so on, are examples of it; for their regular formations, expressed graphically, will be at once triangular and equilateral. As you advance you will find that such a numerical series as far as you like takes the triangular form, if you put as the most elementary form the one that arises from unity, so that unity may appear to be potentially a triangle, and 3 the first actually.

Their sides will increase by the successive numbers, for the side of the one potentially first is unity; that of the one actually first, that is, 3, is 2; that of 6, which is actually second, 3; that of the third, 4; the fourth, 5; the fifth, 6; and so on.

The triangular number is produced from the natural series of numbers set forth in a line, and by the continued addition of successive terms, one by one, from the beginning; for by the successive combinations and additions of another term to the sum, the triangular numbers in regular order are completed. For example, from this natural series, 1, 2, 3, 4, 5, 6, 7, 8, 9, 10, 11, 12, 13, 14, 15, I take the first term and have the triangular number which is potentially first, 1, △ ; then adding the next term I get the triangle actually first, for 2 plus 1 equals 3. In its graphic representation it is thus made up: Two units, side by side, are set beneath one unit, and the 'number' three 'is made a triangle: △ Then when next after these the following number, 3, is added, simplified unto units, and joined to the former, it gives 6, the second triangle in actuality, and furthermore, it graphically represents this number: △ Again, the number that naturally follows, 4, added in and set down below the former, reduced to units, gives the one in order next after the aforesaid, 10, and takes a triangular form: △ 5, after this, then 6, then 7, and all the numbers in order, are added, so that regularly the sides of each triangle will consist of as many numbers as have been added from the natural series to produce it:

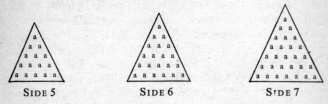

SIDE 5   SIDE 6   SIDE 7

The square is the next number after this, which shows us no longer 3, like the former, but 4, angles in its graphic representation, but is none the less equilateral. Take, for example, 1, 4, 9, 16, 25, 36, 49, 64, 81, 100; for the representations of these numbers are equilateral, square figures, as here shown; and it will be similar as far as you wish to go:

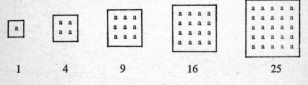

1   4   9   16   25

It is true of these numbers, as it was also of the preceding, that the advance in their sides progresses with the natural series. The side of the square potentially first, 1, is 1; that of 4, the first in actuality, 2; that of 9, actually the second, 3; that of 16, the next, actually the third, 4; that of the fourth, 5; of the fifth, 6, and so on in general with all that follow.

This number also is produced if the natural series is extended in a line, increasing by 1, and no longer the successive numbers are added to the numbers in order, as was shown before, but rather all those in alternate places, that is, the odd numbers. For the first, 1, is potentially the first square; the second, 1 plus 3, is the first in actuality; the third, 1 plus 3 plus 5, is the second in actuality; the fourth, 1 plus 3 plus 5 plus 7, is the third in actuality; the next is produced by adding 9 to the former numbers, the next by the addition of 11, and so on.

In these cases, also, it is a fact that the side of each consists of as many units as there are numbers taken into the sum to produce it.[1]

The pentagonal number is one which likewise upon its resolution into units and depiction as a plane figure assumes the form of an equilateral pentagon. 1, 5, 12, 22, 35, 51, 70, and analogous numbers are examples. Each side of the first actual pentagon, 5, is 2, for 1 is the side of the pentagon potentially first, 1; 3 is the side of 12, the second of those listed; 4, that of the next, 22; 5, that of the next in order, 35, and 6 of the succeeding one, 51, and so on. In general the side contains as many units as are the numbers that have been added together to produce the pentagon, chosen out of the natural arithmetical series set forth in a row. For in a like and similar manner, there are added together to produce the pentagonal numbers the terms beginning with 1 to any extent whatever that are two places apart, that is, those that have a difference of 3.

Unity is the first pentagon, potentially, and is thus depicted:

5, made up of 1 plus 4, is the second, similarly represented:

12, the third, is made up out of the two former numbers with 7 added to them, so that it may have 3 as a side, as three numbers have been added to make it. Similarly the preceding pentagon, 5,

---

1. So in the first square, 1, the side is 1 and only one term is taken to produce it. In the second, 4, the side is 2 and two terms are taken to produce it $(1+3)$. Generally, the algebraic sum of 1, 3, 5 . . . to n terms is n^2.

was the combination of two numbers and had 2 as its side. The graphic representation of 12 is this:

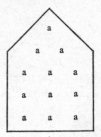

The other pentagonal numbers will be produced by adding together one after another in due order the terms after 7 that have the difference 3, as for example, 10, 13, 16, 19, 22, 25, and so on. The pentagons will be 22, 35, 51, 70, 92, 117, and so forth.

The hexagonal, heptagonal and succeeding numbers will be set forth in their series by following the same process, if from the natural series of number there be set forth series with their differences increasing by 1. For as the triangular number was produced by admitting into the summation the terms that differ by 1 and do not pass over any in the series; as the square was made by adding the terms that differ by 2 and are one place apart, and the pentagon similarly by adding terms with a difference of 3 and two places apart (and we have demonstrated these, by setting forth examples both of them and of the polygonal numbers made from them), so likewise the hexagons will have as their root-numbers[1] those which differ by 4 and are three places apart in the series, which added together in succession will produce[2] the hexagons. For example, 1, 5, 9, 13, 17, 21, and so on; so that the hexagonal numbers produced will be 1, 6, 15, 28, 45, 66, and so so on, as far as one wishes to go.

The heptagonals, which follow these, have as their root-numbers terms differing by 5 and four places apart in the series, like 1, 6, 11, 16, 21, 26, 31, 36, and so on. The heptagons that thus arise are 1, 7, 18, 34, 55, 81, 112, 148, and so forth.

The octagonals[3] increase after the same fashion, with a

1. That is, gnomons; the term being used in the broader sense.

27

2. MS. G gives the following diagram of the hexagonal number 15:

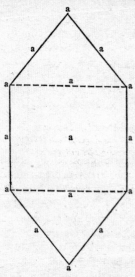

3. The following illustrations are from the same MS.:

Derivation of heptagonals:

$$\overbrace{1, 2, 3, 4, 5, 6,}^{7} \overbrace{7, 8, 9, 10, 11,}^{18} \overbrace{12, 13, 14, 15, 16,}^{34} \overbrace{17, 18, 19, 20, 21}^{55}$$

Heptagonal   *Octagonal*

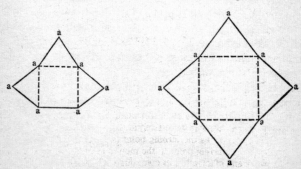

difference of 6 in their root-numbers and corresponding variation in their total constitution.

In order that, as you survey all cases, you may have a rule generally applicable,[1] note that the root-numbers of any polygonal differ by 2 less than the number of the angles shown by the name of the polygonal – that is, by 1 in the triangle, 2 in the square, 3 in the pentagon, 4 in the hexagon, 5 in the heptagon, and so on, with similar increase.

Any square figure[2] diagonally divided is resolved into two

1. Cf. also Theon, pp. 34, 6, and p. 40, 11 ff. The principle here stated by Nicomachus had already been given by Hypsicles (*c.* 180 B.C.), whose theorem is cited by Diophantus (*De Polygonis Numeris, Prop.* IV) as follows: 'If as many numbers as you please be set out at equal interval from 1, and the interval is 1, their sum is a triangular number; if the interval is 2, a square; if 3, a pentagonal; and generally the number of angles is greater by 2 than the interval.'

2. MS. G gives the following figure as an illustration. The principle may be proved from the formulas of arithmetic progression,

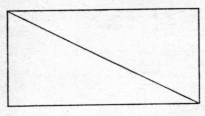

$$S = \frac{n}{2}(a+l), \; l = a + (n-1)d.$$

Two successive triangular numbers, formed according to definition by the summation of n and $n+1$ terms respectively, will therefore be $\frac{n^2+n}{2}$ and $\frac{n^2+3n+2}{2}$, and their sum is n^2+2n+1, which is $(n+1)^2$, a perfect square.

The Neo-Pythagoreans employed an interesting development of this principle to display the relative characters of the monad and the dyad (cf. *Theol. Arith.*, p. 9 Ast, and Iamblichus *In Nic.*, p. 75, 20 ff.). The matter is stated in *Theol. Arith.*, loc. cit., as follows: The monad is the cause of squares not only because the odd numbers successively arranged about it give squares, but also 'because each side, as the turning point (*sc.* of a double race course) from the monad as starting point to the monad as finish line has the sum of its going forth and of its return its own square'. That is, to take the side 5, triangles and every square number is resolved into two consecutive triangular numbers, and hence is made up of two successive triangular numbers. For example, 1, 3, 6, 10, 15, 21, 28, 36, 45, 55, and so on, are triangular numbers and 1, 4, 9, 16, 25, 36, 49, 64, 81, 100, squares. If you add any two consecutive triangles that you please, you will always make a square, and hence, whatever square you resolve, you will be able to make two triangles of it.

Again, any triangle joined to any square figure makes a pentagon, for example, the triangle 1 joined with the square 4 makes the pentagon 5; the next triangle, 3 of course, with 9, the next square, makes the pentagon 12; the next, 6, with the next square, 16, gives the next pentagon, 22; 10 and 25 give 35; and so on.

Similarly,[1] if the triangles are added to the pentagons, following the same order, they will produce the hexagonals in due order, and again the same triangles with the latter will make the heptagonals in order, the octagonals after the heptagonals, and so on to infinity.

To remind us, let us set forth rows of the polygonals, written in parallel lines, as follows: The first row, triangles, the next

---

when the successive numbers up to 5 are set out as one side of the race-track, 5 is made the turning point and the other side is made up of the descending numbers to 1, e.g.

$$\begin{array}{ccccc} 1 & 2 & 3 & 4 & \\ & & & & 5 \\ 1 & 2 & 3 & 4 & \end{array}$$

the sum of the whole series is 25, or 5^2. The series 1 ... 5, of course, is one triangular number, and the descending series 4 ... 1 the immediately preceding one. From its resemblance to the double race course of the Greek games this proposition was apparently recognized under the name 'diaulos' (cf. Iamblichus, p. 75, 25). Its further application to the heteromecic numbers is not pertinent to the present subject.

1. This proposition and the preceding are special cases of the theorem that the polygonal number of r sides with side n, plus the triangular number with side $n-1$, makes the polygonal number with $r+1$ sides and side n. Algebraically

$$\frac{n+1}{2}(2+nd) + \frac{n(n+1)}{2} = \frac{n+1}{2}[2+n(d+1)].$$

squares, after them pentagonals, then hexagonals, then heptagonals, then if one wishes the succeeding polygonals.

Triangles	1	3	6	10	15	21	28	36	45	55
Squares	1	4	9	16	25	36	49	64	81	100
Pentagonals	1	5	12	22	35	51	70	92	117	145
Hexagonals	1	6	15	28	45	66	91	120	153	190
Heptagonals	1	7	18	34	55	81	112	148	189	235

You can also set forth the succeeding polygonals in similar parallel lines.

In general, you will find that the squares are the sum of the triangles above those that occupy the same place in the series, plus the numbers of that same class in the next place back;[1] for example, 4 equals 3 plus 1, 9 equals 6 plus 3, 16 equals 10 plus 6, 25 equals 15 plus 10, 36 equals 21 plus 15, and so on.

The pentagons are the sum of the squares above them in the same place in the series, plus the elementary triangles that are one place further back in the series; for example, 5 equals 4 plus 1, 12 equals 9 plus 3, 22 equals 16 plus 6, 35 equals 25 plus 10, and so on.

Again, the hexagonals are similarly the sums of the pentagons above them in the same place in the series plus the triangles one place back; for instance, 6 equals 5 plus 1, 15 equals 12 plus 3, 28 equals 22 plus 6, 45 equals 35 plus 10, and as far as you like.

The same applies to the heptagonals, for 7 is the sum of 6 and 1, 18 equals 15 plus 3, 34 equals 28 plus 6, and so on. Thus each polygonal number is the sum of the polygonal in the same place in the series with one less angle, plus the triangle, in the highest row, one place back in the series.

Naturally, then, the triangle is the element of the polygon[2] both in figures and in numbers, and we say this because in the table, reading either up or down or across, the successive numbers in the rows are discovered to have as differences the triangles in regular order.

1. That is, in the column next to the left.
2. *Theol. Arith.*, p. 8 Ast, states that the triangle is the element of both magnitudes and numbers and is made by the congress of the monad and the dyad.

Still further, every square plus its own side becomes hetero-mecic, or by Zeus, if its side is subtracted from it. Thus, 'the other' is conceived of as being both greater and smaller than 'the same', since it is produced, both by addition and by subtraction, in the same way that the two kinds of inequality[1] also, the greater and the less, have their origin from the application of addition or subtraction to equality.

1. The results obtained by adding to or subtracting their sides from the square numbers are as follows:

$$4+2 = \ 6 = 2 \times 3$$
$$9+3 = 12 = 3 \times 4$$
$$16+4 = 20 = 4 \times 5$$
$$25+5 = 30 = 5 \times 6$$

or

$$m^2 + m = (m+1)m$$

$$4-2 = \ 2 = 1 \times 2$$
$$9-3 = \ 6 = 2 \times 3$$
$$16-4 = 12 = 3 \times 4$$
$$25-5 = 20 = 4 \times 5$$

or

$$m^2 - m = (m-1)m$$

32

DIOPHANTUS
(c. A.D. 250)

DIOPHANTUS OF ALEXANDRIA, an ancient Greek mathematician, was the author of one of the greatest mathematical treatises of ancient times, the *Arithmetica*, a masterly exposition of algebraic analysis, so thorough and complete in its time, that all previous works in its field ceased to be of interest and passed into oblivion. In the *Arithmetica*, moreover algebraic methods advanced to a peak of achievement which was not to be surpassed before the sixteenth century. The pronounced influence exerted by Diophantus upon the development of algebra in the Late Renaissance began in a flurry of activity centring around the rediscovery of a manuscript of his work in 1570. In 1572, Bombelli published his *Algebra* containing many problems taken from the *Arithmetica*. While Bombelli's translation of the problems was excellent, he was not careful to state which problems in his *Algebra* were his own and which had come from Diophantus. A Latin translation of the *Arithmetica* written by Xylander (William Holzmann) appeared in 1574. In 1585, Simon Stevin published a French version of the first four books of the *Arithmetica*. In 1621, Bachet de Meziriac published an edition of the *Arithmetica* which contained the Greek text as well as his Latin translation. Bachet's edition was made famous by the notes written in the margin of a copy of his book by Fermat (publ. 1670). Since that day, Fermat's notes have stirred the speculative instincts of mathematicians, have stimulated a prodigious output in the theory of numbers, and nevertheless still offer problems which continue to be completely baffling.

The few details of the personal life of Diophantus available to us are all contained in his epitaph as quoted in the *Anthologia Graecia*. It is said to have been composed shortly after his death by a close friend who shared Diophantus's love of an algebraic problem. The epitaph states that a sixth of Diophantus's life was spent in childhood, that after a twelfth more had elapsed he grew a beard, that when a seventh more had passed he married, and that five years later his son was born;

the son lived to half his father's age and four years after the son's death the father died. We learn in this way that Diophantus lived to the age of eighty-four. It is generally believed that he lived in the third century A.D., since neither Nicomachus (c. A.D. 100) nor Theon of Smyrna (c. A.D. 130) make any mention of him, and on the other hand he is quoted by Theon of Alexandria (c. 365). The commentary on his work written by Hypatia (d. 415), daughter of Theon of Alexandria, is the ultimate source of all extant manuscripts and translations of the *Arithmetica*.

The *Arithmetica* was originally written in thirteen books. Hypatia's commentary extends only to the first six books. The remainder of the work was probably lost before the tenth century. Mathematical notation clearly had its beginnings in the *Arithmetica* where Diophantus employed symbols to represent operations or quantities occurring repeatedly in his solutions. No substantial improvement over his essentially abbreviative symbolism was invented until Vieta's time (c. 1580). In the *Arithmetica*, Diophantus dealt with a wide variety of problems. Some were solved by determinate equations, others by indeterminate equations, the latter type being by far the more numerous. His clear mastery of this area of mathematics, the impressive variety of his clever devices for effecting his solutions, and his comprehensive presentation have made his name descriptive of this type of analysis. Diophantine analysis is the name of a branch of the theory of numbers which is concerned with the rational solutions of indeterminate problems (Diophantine problems) involving one or more indeterminate equations (Diophantine equations) of the first, second or higher degrees.

THE ARITHMETICA

From Diophantus of Alexandria *by Sir Thomas L. Heath*

BOOK I

Preliminary

DEDICATION

'Knowing, my most esteemed friend Dionysius, that you are anxious to learn how to investigate problems in numbers, I have tried, beginning from the foundations on which the science is built up, to set forth to you the nature and power subsisting in numbers.

'Perhaps the subject will appear rather difficult, inasmuch as it is not yet familiar (beginners are, as a rule, too ready to despair of success); but you, with the impulse of your enthusiasm and the benefit of my teaching, will find it easy to master; for eagerness to learn, when seconded by instruction, ensures rapid progress.'

'All numbers are made up of some multitude of units, so that it is manifest that their formation is subject to no limit.'

*

DEFINITIONS

A *square* ($=x^2$) is δύναμις ('power'), and its sign is a Δ with Y superposed, thus Δ^Y.

A *cube* ($=x^3$) is κύβος, and its sign is K^Y.

A *square-square* ($=x^4$) is δυναμοδύναμις, and its sign is $\Delta^Y\Delta$.

A *square-cube* ($=x^5$) is δυναμόκυβος, and its sign is ΔK^Y.

A *cube-cube* ($=x^6$) is κυβόκυβος, and its sign is $K^Y K$.

'It is, from the addition, subtraction or multiplication of these numbers or from the ratios which they bear to one another or to their own sides respectively that most arithmetical problems are formed' ... 'each of these numbers ... is recognized as an element in arithmetical inquiry.'

'But the number which has none of these characteristics, but *merely has in it an indeterminate multitude of units* (πλῆθος μονάδων ἀόριστον) *is called* ἀριθμός, "*number*", *and its sign is* ς[$=x$].'

'And there is also another sign denoting that which is invariable in determinate numbers, namely the unit, the sign being M with o superimposed, thus $\overset{o}{M}$.'

Thus

from ἀριθμός [x]	we derive the term	ἀριθμοστόν [..$1/x$]
,, δύναμις [x^2]	,, ,, ,,	δυναμοστόν [..$1/x^2$]
,, κύβος [x^3]	,, ,, ,,	κυβοστόν [..$1/x^3$]
,, δυναμοδυναμις [x^4]	,, ,, ,,	δυναμοδυναμοστόν [..$1/x^4$]
,, δυναμόκυβος [x^5]	,, ,, ,,	δυναμοκυβοστόν [..$1/x^5$]
,, κυβόκυβος [x^6]	,, ,, ,,	κυβοκυβοστόν [..$1/x^6$],

and each of these has the same sign as the corresponding original species, but with a distinguishing mark.

Thus $\Delta^{\gamma\chi} = 1/x^2$, just as $\gamma^\chi = \frac{1}{3}$.

Sign of Subtraction (*minus*).

'*A minus multiplied by a minus makes a plus;*[1] *a minus multiplied by a plus makes a minus; and the sign of a minus is a truncated* Ψ *turned upside down thus* ⋔'.

'It is well that one who is beginning this study should have acquired practice in the addition, subtraction and multiplication of the various species. He should know how to add positive and negative terms with different coefficients to other terms,[2] themselves either positive or likewise partly positive and partly negative, and how to subtract from a combination of positive and negative terms other terms either positive or likewise partly positive and partly negative.

'Next, if a problem leads to an equation in which certain terms are equal to terms of the same species but with different coefficients, it will be necessary to subtract like from like on both sides, until one term is found equal to one term. If by chance there are on either side or on both sides any negative terms, it will be necessary to add the negative terms on both sides, until the terms on both sides are positive, and then again to subtract like from like until one term only is left on each side.

'This should be the object aimed at in framing the hypotheses of propositions, that is to say, to reduce the equations, if possible, until one term is left equal to one term; *but I will show you later how, in the case also where two terms are left equal to one term, such a problem is solved.*'

Problems

1. To divide a given number into two having a given difference.

Given number 100, given difference 40.

1. The literal rendering would be 'A wanting multiplied by a wanting makes a forthcoming'. The word corresponding to *minus* is λεῦψις ('wanting'): when it is used exactly as our *minus* is, it is in the dative λεύψει, but there is some doubt whether Diophantus himself used this form.

2. εἶδος, 'species', is the word used by Diophantus throughout.

Lesser number required x. Therefore

$$2x+40=100,$$
$$x=30.$$

The required numbers are 70, 30.

2. To divide a given number into two having a given ratio.

Given number 60, given ratio 3 : 1.
Two numbers x, $3x$. Therefore $x=15$.
The numbers are 45, 15.

26. Given two numbers, to find a third number which, when multiplied into the given numbers respectively, makes one product a square and the other the side of that square.

Given numbers 200, 5; required number x.
Therefore $200x=(5x)^2$, and
$$x=8.$$

27. To find two numbers such that their sum and product are given numbers.

Necessary condition. The square of half the sum must exceed the product by a square number.

Given sum 20, given product 96.
$2x$ the difference of the required numbers.
Therefore the numbers are $10+x$, $10-x$.
Hence $100-x^2=96$.
Therefore $x=2$, and
the required numbers are 12, 8.

28. To find two numbers such that their sum and the sum of their squares are given numbers.

Necessary condition. Double the sum of their squares must exceed the square of their sum by a square. ἔστι δὲ καὶ τοῦτο πλασματικόν.[1]

---

1. There has been controversy as to the meaning of this difficult phrase. Xylander, Bachet, Cossali, Schulz, Nesselmann, all discuss it. Xylander translated it by '*effictum aliunde*'. Bachet, of course, rejects this, and, while leaving the word untranslated, maintains that it has an active rather than a passive signification; it is, he says, not something 'made up' (*effictum*) but something '*a quo aliud quippiam effingi et plasmari potest*', 'from which something else can be made up', and this he interprets as meaning that from the conditions to which the term is applied, combined with the solutions

Given sum 20, given sum of squares 208.

Difference $2x$.

Therefore the numbers are $10+x$, $10-x$.

Thus $200+2x^2=208$, and $x=2$.

The required numbers are 12, 8.

29. To find two numbers such that their sum and the difference of their squares are given numbers.

Given sum 20, given difference of squares 80.

Difference $2x$.

The numbers are therefore $10+x$, $10-x$.

Hence $(10+x)^2-(10-x)^2=80$,

or $40x=80$, and $x=2$.

The required numbers are 12, 8.

30. To find two numbers such that their difference and product are given numbers.

Necessary condition. Four times the product together with the square of the difference must give a square. ἔστι δὲ καὶ τοῦτο πλασματικόν.

Given difference 4, given product 96.

$2x$ the sum of the required numbers.

Therefore the numbers are $x+2$, $x-2$; accordingly $x^2-4=96$, and $x=10$.

The required numbers are 12, 8.

31. To find two numbers in a given ratio and such that the sum of their squares also has to their sum a given ratio.

Given ratios 3 : 1 and 5 : 1 respectively.

Lesser number x.

Therefore $10x^2=5.4x$, whence $x=2$, and the numbers are 2, 6.

---

of the respective problems in which it occurs, the rules for solving mixed quadratics can be evolved. Of the two views I think Xylander's is nearer the mark. πλασματικόν should apparently mean 'of the nature of a πλάσμα', just as δραματικόν means something connected with or suitable for a drama; and πλάσμα means something 'formed' or 'moulded'. Hence the expression would seem to mean 'this is of the nature of a formula', with the implication that the formula is not difficult to make up or discover. Nesselmann, like Xylander, gives it much this meaning, translating it '*das lässt sich aber berwerkstelligen*'. Tannery translates πλασματικόν by '*formativum*'.

32. To find two numbers in a given ratio and such that the sum of their squares also has to their difference a given ratio.

Given ratios 3 : 1 and 10 : 1.

Lesser number x, which is then found from the equation $10x^2 = 10.2x$.

Hence $x = 2$, and
    the numbers are 2, 6.

BOOK II

6. To find two numbers having a given difference and such that the difference of their squares exceeds their difference by a given number.

Necessary condition. The square of their difference must be less than the sum of the said difference and the given excess of the difference of the squares over the difference of the numbers.

Difference of numbers 2, the other given number 20.

Lesser number x. Therefore $x + 2$ is the greater, and $4x + 4 = 22$.

Therefore $x = 4\frac{1}{2}$, and
    the numbers are $4\frac{1}{2}$, $6\frac{1}{2}$.

7. To find two numbers such that the difference of their squares is greater by a given number than a given ratio of their difference.[1] [*Difference assumed.*]

Necessary condition. The ratio being 3 : 1, the square of the difference of the numbers must be less than the sum of three times that difference and the given number.

Given number 10, difference of required numbers 2.

Lesser number x. Therefore the greater is $x + 2$, and $4x + 4 = 3.2 + 10$.

Therefore $x = 3$, and
    the numbers are 3, 5.

---

1. Here we have the identical phrase used in Euclid's *Data*: The difference of the squares is τῆς ὑπεροχῆς αὐτῶν δοθέντι ἀριθμῷ μείζων ἢ ἐν λόγῳ literally 'greater than their difference by a given number (more) than in a (given) ratio', by which is meant 'greater by a given number than a given proportion or fraction of their difference'.

8. To divide a given square number into two squares.[1]

Given square number 16.

x^2 one of the required squares. Therefore $16 - x^2$ must be equal to a square.

Take a square of the form[2] $(mx - 4)^2$, m being any integer and 4 the number which is the square root of 16, e.g. take $(2x - 4)^2$, and equate it to $16 - x^2$.

Therefore $4x^2 - 16x + 16 = 16 - x^2$,

or $5x^2 = 16x$, and $x = 16/5$.

The required squares are therefore 256/25, 144/25.

1. It is to this proposition that Fermat appended his famous note in which he enunciates what is known as the 'great theorem' of Fermat. The text of the note is as follows:

'On the other hand it is impossible to separate a cube into two cubes, or a biquadrate into two biquadrates, or generally *any power except a square into two powers with the same exponent*. I have discovered a truly marvellous proof of this, which however the margin is not large enough to contain.'

Did Fermat really possess a proof of the general proposition that $x^m + y^m = z^m$ cannot be solved in rational numbers where m is any number > 2? As Wertheim says, one is tempted to doubt this, seeing that, in spite of the labours of Euler, Lejeune-Dirichlet, Kummer and others, a general proof has not even yet been discovered. Euler proved the theorem for $m = 3$ and $m = 4$, Dirichlet for $m = 5$, and Kummer, by means of the higher theory of numbers, produced a proof which only excludes certain particular values of m, which values are rare, at all events along the smaller values of m; thus there is no value of m below 100 for which Kummer's proof does not serve. (I take these facts from Weber and Wellstein's *Encyclopädie der Elementar-Mathematik*, I_2, p. 284, where a proof of the formula for $m = 4$ is given.)

It appears that the Göttingen Academy of Sciences has recently awarded a prize to Dr A. Wieferich, of Münster, for a proof that the equation $x^p + y^p = z^p$ cannot be solved in terms of positive integers not multiples of p, if $2^p - 2$ is not divisible by p^2. 'This surprisingly simple result represents the first advance, since the time of Kummer, in the proof of the last Fermat theorem' (*Bulletin of the American Mathematical Society*, February 1910).

Fermat says ('*Relation des nouvelles découvertes en la science des nombres*', August 1659, *Oeuvres*, II., p. 433) that he proved that *no cube is divisible into two cubes* by a variety of his method of *infinite diminution* (*descente infinie* or *indéfinie*) different from that which he employed for other negative or positive theorems.

2. Diophantus's words are: 'I form the square from any number of ἀριθμοί minus as many units as there are in the side of 16.' It is implied throughout that m must be so chosen that the result may be *rational* in Diophantus's sense, i.e. rational and positive.

9. To divide a given number which is the sum of two squares into two other squares.[1]

Given number $13 = 2^2 + 3^2$.

As the roots of these squares are 2, 3, take $(x+2)^2$ as the first square and $(mx-3)^2$ as the second (where m is an integer), say $(2x-3)^2$.
Therefore $(x^2+4x+4)+(4x^2+9-12x)=13$,
or $5x^2+13-8x=13$.
Therefore $x=8/5$, and
the required squares are 324/25, 1/25.

10. To find two square numbers having a given difference.

Given difference 60.
Side of one number x, side of the other x *plus* any number the square of which is not greater than 60, say 3.
Therefore $(x+3)^2-x^2=60$;
$x=8\frac{1}{2}$, and
the required squares are $72\frac{1}{4}$, $132\frac{1}{4}$.

11. To add the same (required) number to two given numbers so as to make each of them a square.

(1) Given numbers 2, 3; required number x.
Therefore

$$\left. \begin{array}{r} x+2 \\ x+3 \end{array} \right\} \text{ must both be squares.}$$

1. Diophantus's solution is substantially the same as Euler's (*Algebra*, tr. Hewlett, Part II, Art. 219), though the latter is expressed more generally.
Required to find xy, such that
$$x^2+y^2=f^2+g^2.$$

If $x \gtrless f$, then $y \lessgtr g$.

Put therefore $\qquad x=f+pz, \quad y=g-qz$:
hence $\qquad 2fpz+p^2z^2-2gqz+q^2z^2=0$,

and
$$z=\frac{2gq-2fp}{p^2+q^2},$$

so that
$$x=\frac{2gpq+f(q^2-p^2)}{p^2+q^2}, \quad y=\frac{2fpq+g(p^2-q^2)}{p^2+q^2},$$

in which we may substitute all possible numbers for p, q.

This is called a double-equation (διπλοϊσότης).

To solve it, *take the difference between the two expressions and resolve it into two factors;*[1] in this case let us say 4, ¼.

Then *take either*

(*a*) *the square of half the difference between these factors and equate it to the lesser expression,*

or (*b*) *the square of half the sum and equate it to the greater.*

In this case (*a*) the square of half the difference is 225/64.

Therefore, $x+2=225/64$, and $x=97/64$, the squares being 225/64, 289/64.

Taking (*b*) the square of half the sum, we have $x+3=289/64$, which gives the same result.

(2) To avoid a double-equation,

first find a number which when added to 2, or to 3, gives a square.

Take e.g. the number x^2-2, which when added to 2 gives a square.

Therefore, since this same number added to 3 gives a square,

$x^2+1=$ a square $=(x-4)^2$, say,

the number of units in the expression (in this case 4) being so taken that the solution may give $x^2>2$.

Therefore $x=15/8$, and

the required number is 97/64, as before.

12. To subtract the same (required) number from two given numbers so as to make both remainders squares.

Given numbers 9, 21.

Assuming $9-x^2$ as the required number, we satisfy one condition, and the other requires that $12+x^2$ shall be a square.

Assume as the side of this square x *minus* some number the square of which >12, say 4.

Therefore $(x-4)^2=12+x^2$,

and $\qquad x=\frac{1}{2}$.

The required number is then $8\frac{3}{4}$.

1. Here, as always, the factors chosen must be suitable factors, i.e. such as will lead to a 'rational' result, in Diophantus's sense.

[Diophantus does not reduce to lowest terms, but says $x=4/8$ and then subtracts 16/64 from 9 or 576/64.]

BOOK III

6. To find three numbers such that their sum is a square and the sum of any pair is a square.

> Let the sum of all three be x^2+2x+1, sum of first and second x^2, and therefore the third $2x+1$; let sum of second and third be $(x-1)^2$.
>
> Therefore the first $= 4x$, and the second $= x^2-4x$.
>
> But first $+$ third $=$ square,
>
> that is, $6x+1 =$ square $= 121$, say.
>
> Therefore $x=20$, and
>
> > the numbers are 80, 320, 41.

[An alternative solution, obviously interpolated, is practically identical with the above except that it takes the square 36 as the value of $6x+1$, so that $x=35/6$, and the numbers are $140/6 = 840/36$, 385/36, 456/36.]

7. To find three numbers in A.P. such that the sum of any pair gives a square.

> First find three square numbers in A.P. and such that half their sum is greater than any one of them. Let x^2, $(x+1)^2$ be the first and second of these; therefore the third is $x^2+4x+2 = (x-8)^2$, say.
>
> Therefore $x=62/20$ or $31/10$;
>
> > and we may take as the numbers 961, 1681, 2401. We have now to find three numbers such that the sums of pairs are the numbers just found.
>
> The sum of the three $=5043/2=2521\frac{1}{2}$, and
>
> > the three numbers are $120\frac{1}{2}$, $840\frac{1}{2}$, $1560\frac{1}{2}$.

10. To find three numbers such that the product of any pair of them added to a given number gives a square.

> Let the given number be 12. Take a square (say 25) and subtract 12. Take the difference (13) for the product of the first and second numbers, and let these numbers be $13x$, $1/x$ respectively.

Again subtract 12 from another square, say 16, and let the difference (4) be the product of the second and third numbers.

Therefore the third number $=4x$.

The third condition gives $52x^2 + 12 =$ a square; now $52 = 4.13$, and 13 is not a square; but, if it were a square, the equation could easily be solved.[1]

Thus we must find two numbers to replace 13 and 4 such that their product is a square, while either $+12$ is also a square.

Now the product is a square if both are squares; hence we must find two squares such that either $+12 =$ a square.

'This is easy[2] and, as we said, it makes the equation easy to solve.'

The squares 4, $\frac{1}{4}$ satisfy the condition.

19. To find four numbers such that the square of their sum *plus* or *minus* any one singly gives a square.

Since, in any right-angled triangle,

(sq. on hypotenuse) $\pm$ (twice product of perps.) = a square, we must seek four right-angled triangles [in rational numbers] having the same hypotenuse,

or we must find a square which is divisible into two squares in four different ways; and 'we saw how to divide a square into two squares in an infinite number of ways'. [II, 8]

Take right-angled triangles in the smallest numbers, (3, 4, 5) and (5, 12, 13); and multiply the sides of the first by the hypotenuse of the second and *vice versa*.

This gives the triangles (39, 52, 65) and (25, 60, 65); thus 65^2 is split up into two squares in *two* ways.

Again, 65 is 'naturally' divided into two squares in two ways,

---

1. The equation $52x^2 + 12 = u^2$ can in reality be solved as it stands, by virtue of the fact that it has one obvious solution, namely $x = 1$. Another solution is found by substituting $y + 1$ for x, and so on. The value $x = 1$ itself gives (13, 1, 4) as a solution of the problem.

2. We have to find two pairs of squares differing by 12. (*a*) If we put $12 = 6.2$, we have $\{\frac{1}{2}(6-2)^2 + 12 = \{\frac{1}{2}(6+2)\}^2$, and 16, 4 are squares differing by 12, or 4 is a square which when added to 12 gives a square. (*b*) If we put $12 = 4.3$, we find $\{\frac{1}{2}(4-3)\}^2$ or $\frac{1}{4}$ to be a square which when added to 12 gives a square.

namely into 7^2+4^2 and 8^2+1^2, 'which is due to the fact that 65 is the product of 13 and 5, each of which numbers is the sum of two squares'.

Form now a right-angled triangle[1] from 7, 4. The sides are $(7^2-4^2, 2.7.4, 7^2+4^2)$ or (33, 56, 65).

Similarly, forming a right-angled triangle from 8, 1, we obtain $(2.8.1, 8^2-1^2, 8^2+1^2)$ or 16, 63, 65.

Thus 65^2 is split into two squares in *four* ways.

Assume now as the sum of the numbers $65x$ and

as first number $2.39.52x^2=4056x^2$,
,, second ,, $2.25.60x^2=3000x^2$,
,, third ,, $2.33.56x^2=3696x^2$,
,, fourth ,, $2.16.63x^2=2016x^2$,

the coefficients of x^2 being four times the areas of the four right-angled triangles respectively.

The sum $12768x^2=65x$, and $x=\frac{65}{12768}$.

The numbers are

$\frac{17136600}{163021824}$, $\frac{12675000}{163021824}$, $\frac{15615600}{163021824}$, $\frac{8517600}{163021824}$.

BOOK IV

29. To find four square numbers such that their sum added to the sum of their sides makes a given number.[2]

1. If there are two numbers p, q to 'form a right-angled triangle' from them means to take the numbers p^2+q^2, p^2-q^2, $2pq$. These are the sides of a right-angled triangle, since

$$(p^2+q^2)^2=(p^2-q^2)^2+(2pq)^2.$$

2. On this problem Bachet observes that Diophantus appears to assume, here and in some problems of Book V, that any number not itself a square is the sum of two or three or four squares. He adds that he has verified this statement for all numbers up to 325, but would like to see a scientific proof of the theorem. These remarks of Bachet's are the occasion for another of Fermat's famous notes: 'I have been the first to discover a most beautiful theorem of the greatest generality, namely this: Every number is either a triangular number or the sum of two or three triangular numbers; every number is a square or the sum of two, three, or four squares; every number is a pentagonal number or the sum of two, three, four or five pentagonal numbers; and so on *ad infinitum*, for hexagons, heptagons and any polygons whatever, the enunciation of this general and wonderful theorem being

Given number 12.

Now $x^2 + x + \frac{1}{4} =$ a square.

Therefore the sum of four squares + the sum of their sides + 1 = the sum of four other squares = 13, by hypothesis.

Therefore we have to divide 13 into four squares; then, if we subtract $\frac{1}{2}$ from each of their sides, we shall have the sides of the required squares.

Now $13 = 4 + 9 = (\frac{63}{25} + \frac{36}{25}) + (\frac{144}{25} + \frac{81}{25})$,

and the sides of the required squares are

  11/10, 7/10, 19/10, 13/10,

the squares themselves being

  121/100, 49/100, 361/100, 169/100.

30. To find four squares such that their sum *minus* the sum of their sides is a given number.

Given number 4.

Now $x^2 - x + \frac{1}{4} =$ a square.

Therefore (the sum of four squares) − (sum of their sides) + 1 = the sum of four other squares = 5, by hypothesis.

Divide 5 into four squares, as

  9/25, 16/25, 64/25, 36/25.

The sides of these squares *plus* $\frac{1}{2}$ in each case are the sides of the required squares.

Therefore sides of required squares are

  11/10, 13/10, 21/10, 17/10,

and the squares themselves

  121/100, 169/100, 441/100, 289/100.

---

varied according to the number of the angles. The proof of it, which depends on many various and abstruse mysteries of numbers, I cannot give here; for I have decided to devote a separate and complete work to this matter and thereby to advance arithmetic in this region of inquiry to an extraordinary extent beyond its ancient and known limits.'

Unfortunately the promised separate work did not appear. The theorem so far as it relates to squares was first proved by Lagrange (*Nouv. Mémoires de l'Acad. de Berlin*, année 1770, Berlin 1772, pp. 123–33; *Oeuvres*, III, pp. 189–201), who followed up results obtained by Euler. Cf. also Legendre, *Zahlentheorie*, tr. Maser, I, pp. 212, et seq. Lagrange's proof is set out as shortly as possible in Wertheim's *Diophantus*, pp. 324–30. The theorem of Fermat in all its generality was proved by Cauchy (*Oeuvres*, IIe série, vol. VI, pp. 320–53); cf. Legendre, *Zahlentheorie*, tr. Maser, II, pp. 332 et seq.

31. To divide unity into two parts such that, if given numbers are added to them respectively, the product of the two sums gives a square.

Let 3, 5 be the numbers to be added; x, $1-x$ the parts of 1.
Therefore $(x+3)(6-x)=18+3x-x^2=$ a square $=4x^2$, say;
thus $18+3x=5x^2$, *which does not give a rational result.*

Now 5 comes from a square $+1$; and, in order that the equation may have a rational solution, we must substitute for the square taken (4) a square such that

(the square $+1$). $18+(3/2)^2=$ a square.

Put $(m^2+1)18+2\frac{1}{4}=$ a square,
or $72m^2+81=$ a square $=(8m+9)^2$, say,
and $m=18$, $m^2=324$.
Hence we must put
$(x+3)(6-x)=18+3x-x^2=324x^2$.
Therefore[1] $325x^2-3x-18=0$.

$$x=78/325=6/25,$$

and $(\frac{6}{25}, \frac{19}{25})$ is a solution.

BOOK V

29. To find three squares such that the sum of their squares is a square.

Let the squares be x^2, 4, 9 respectively.[2]
Therefore $x^4+97=$ a square $=(x^2-10)^2$, say; whence $x^2=3/20$.
If the ratio of 3 to 20 were the ratio of a square to a square, the problem would be solved; but it is not.

Therefore *I have to find two squares (p^2, q^2, say) and a number (m, say) such that $m^2-p^4-q^4$ has to $2m$ the ratio of a square to a square.*

1. Observe the solution of a mixed quadratic equation.
2. 'Why', says Fermat, 'does not Diophantus seek *two* fourth powers such that their sum is a square? This problem is in fact impossible, as by my method I am in a position to prove with all rigour.' It is probable that Diophantus knew the fact without being able to prove it generally. That neither the sum nor the difference of two fourth powers can be a square was proved by Euler (*Commentationes Arithmeticae*, I, pp. 24 et seq., and *Algebra*, part II, c. XIII).

Let $p^2 = z^2$, $q^2 = 4$ and $m = z^2 + 4$.

Therefore $m^2 - p^4 - q^4 = (z^2 + 4)^2 - z^4 - 16 = 8z^2$.

Hence $8z^2/(2z^2 + 8)$, or $4z^2/(z^2 + 4)$, must be the ratio of a square to a square.

Put $z^2 + 4 = (z + 1)^2$, say;

therefore $z = 1\frac{1}{2}$, and the squares are $p^2 = 2\frac{1}{4}$, $q^2 = 4$, while $m = 6\frac{1}{4}$;

or, if we take 4 times each, $p^2 = 9$, $q^2 = 16$, $m = 25$.

Starting again, we put for the squares x^2, 9, 16;

then the sum of the squares $= x^4 + 337 = (x^2 - 25)^2$, and $x = 12/5$.

The required squares are 144/25, 9, 16.

30. [The enunciation of this problem is in the form of an epigram, the meaning of which is as follows.]

A man buys a certain number of measures of wine, (χόες) some at 8 drachmas, some at 5 drachmas each. He pays for them a *square* number of drachmas; and if we add 60 to this number, the result is a square, the side of which is equal to the whole number of measures. Find how many he bought at each price.

Let $x =$ the whole number of measures; therefore $x^2 - 60$ was the price paid, which is a square $= (x - m)^2$, say.

Now $\frac{1}{5}$ of the price of the five-drachma measures $+ \frac{1}{8}$ of the price of the eight-drachma measures $= x$;

so that $x^2 - 60$, the total price, has to be divided into two parts such that $\frac{1}{5}$ of one $+ \frac{1}{8}$ of the other $= x$.

We cannot have a real solution of this unless $x > \frac{1}{8}(x^2 - 60)$ and $< \frac{1}{5}(x^2 - 60)$.

Therefore $5x < x^2 - 60 < 8x$.

(1) Since $x^2 > 5x + 60$,

$x^2 = 5x +$ a number greater than 60,

whence x is *not less than* 11.

(2) $x^2 < 8x + 60$

or $x^2 = 8x +$ some number less than 60,

whence x is *not greater than* 12.

Therefore $11 < x < 12$.

Now (from above) $x = (m^2 + 60)/2m$;

therefore $22m < m^2 + 60 < 24m$.

Thus (1) $22m = m^2 + $(some number less than 60),
and therefore m is *not less than* 19.
(2) $24m = m^2 + $(some number greater than 60),
and therefore m is *less than* 21.
Hence we put $m = 20$, and
$$x^2 - 60 = (x-20)^2,$$
so that $x = 11\frac{1}{2}$, $x^2 = 132\frac{1}{4}$, and $x^2 - 60 = 72\frac{1}{4}$.
Thus we have to divide $72\frac{1}{4}$ into two parts such that $\frac{1}{5}$ of one
part *plus* $\frac{1}{8}$ of the other $= 11\frac{1}{2}$.
Let the first part be $5z$.
Therefore $\frac{1}{8}$ (second part) $= 11\frac{1}{2} - z$,
or second part $= 92 - 8z$;
therefore $5z + 92 - 8z = 72\frac{1}{4}$;
and $z = 79/12$.

Therefore the number of five-drachma $\chi\acute{o}\varepsilon\varsigma = 79/12$.
Therefore the number of eight-drachma $\chi\acute{o}\varepsilon\varsigma = 59/12$.

BOOK VI

18. To find a right-angled triangle such that the area added to the
hypotenuse gives a cube, while the perimeter is a square.

Area x, hypotenuse some cube *minus* x, perpendiculars x, 2.
Therefore we have to find a cube which, when 2 is added to it,
becomes a square.
Let the side of the cube be $m - 1$.
Therefore $m^3 - 3m^2 + 3m + 1 = $a square $= (1\frac{1}{2}m + 1)^2$, say.
Thus $m = 21/4$, and the cube $= (17/4)^3 = 4913/64$.
Put now x for the area, x, 2 for the perpendiculars, and
$4913/64 - x$ for the hypotenuse;
and x is found from the equation $(4913/64 - x)^2 = x^2 + 4$.
[$x = 24121185/628864$, and the triangle is
(2, 24121185/628864, 24153953/628864).]

19. To find a right-angled triangle such that its area added to
one of the perpendiculars gives a square, while the perimeter is a
cube.

Make a right-angled triangle from some indeterminate odd number,[1] say $2x+1$;

then the altitude $=2x+1$, the base $=2x^2+2x$, and the hypotenuse $=2x^2+2x+1$.

Since the perimeter $=$ a cube,

$$4x^2+6x+2=(4x+2)(x+1)=\text{a cube};$$

and, if we divide all the sides by $x+1$, we have to make $4x+2$ a cube.

Again, the area $+$ one perpendicular $=$ a square.

Therefore $\quad \dfrac{2x^3+3x^2+x}{(x+1)^2}+\dfrac{2x+1}{x+1}=$ a square;

that is, $\quad \dfrac{2x^3+5x^2+4x+1}{x^2+2x+1}=2x+1=$ a square.

But $4x+2=$ a cube;

therefore we must find a cube which is double of a square; this is of course 8.

Therefore $4x+2=8$, and $x=1\frac{1}{2}$.

The required triangle is (8/5, 15/5, 17/5).

20. To find a right-angled triangle such that the sum of its area and one perpendicular is a cube, while its perimeter is a square.

Proceeding as in the last problem, we have to make

$$\left.\begin{array}{l}4x+2 \text{ a square}\\ 2x+1 \text{ a cube}\end{array}\right\}.$$

We have therefore to seek a square which is double of a cube; this is 16, which is double of 8.

Therefore $4x+2=16$, and $x=3\frac{1}{2}$.

The triangle is (16/9, 63/9, 65/9).

1. This is the method of formation of right-angled triangles attributed to Pythagoras. If m is any odd number, the sides of the right-angled triangle formed therefrom are m, $\frac{1}{2}(m^2-1)$, $\frac{1}{2}(m^2+1)$,

$$\text{for } m^2+[\tfrac{1}{2}(m^2-1)]^2=[\tfrac{1}{2}(m^2+1)]^2.$$

Cf. Proclus, *Comment. on Eucl.* 1 (ed. Friedlein), p. 428, 7 et seq., etc.

33

METRODORUS

(*c.* A.D. 500)

METRODORUS, a Greek grammarian and mathematician, is credited
with the compilation of the mathematical portion of the Greek An-
thology. He is said by some to have lived as early as the third century
A.D., but it is more likely that he flourished during the reigns of the
Emperor Anastasius I (A.D. 491–518) and of his successor, the Emperor
Justin (A.D. 518–27).

The modern version of the Greek Anthology is based on a tenth-
century manuscript found in the library of the Elector Palatine at
Heidelberg, in 1606. The manuscript, known from the place of its
discovery as the Palatine MS., consisted of fifteen books of short
poems or epigrams, organized according to subject matter. After
enthusiastic and unusually extended study, it was translated, edited
and published in thirteen volumes (1794–1814). Selections found in
previous anthologies, but not included in the Palatine MS. were added
to the Anthology as a sixteenth book. The Palatine MS. and the
Planudean MS. (Florence 1494), which it replaced in popularity, were
both based on previous Greek anthologies. Of these, the *Garland* of
Meleagar of Gadara (first century B.C.), was the first to be compiled.
Meleagar's *Garland* was greatly admired in his time and the basic
value of the Greek Anthology today rests in a large measure upon the
material taken from his collection. In the *Garland*, Meleagar added
some epigrams of his own to selections from the works of contem-
porary poets and earlier writers, going as far back as 700 B.C. Similar
anthologies were compiled in the first century A.D. and later. In the
sixth century the format was changed from an alphabetical order
according to the initial letter of the first word of the selection to an
arrangement according to subject matter. As each of the great collections added some selections written subsequently to those already
compiled, the ninth- and tenth-century manuscripts of the Greek
anthologies contained representative extracts of the most highly valued
writings composed over a span of about seventeen hundred years.

Book XIV of the Greek Anthology is devoted to mathematical problems, riddles and oracles. The problems are written in epigrammatic form, and on the basis of their style they are believed to have been written by Metrodorus. Although Metrodorus may have added some problems of his own, he is generally credited, not with devising the problems, but with collecting them. Judging by their content, many of the problems may have originated in the fifth century B.C. or earlier. There are simple problems of the type found in the Rhind Papyrus (seventeenth century B.C.). Problem 49 is illustrative of the application of the rule given by Thymaridas, an early Pythagorean (sixth century B.C.). The rule, known as the 'flower' or 'bloom' of Thymaridas, was used to solve simultaneous equations when the number of unknowns was the same as the number of equations. Problems 12 and 50 are reminiscent of references by Plato (428–347 B.C.) to the use of bowls made of the same or of different metals as aids in the education of Egyptian children (*Laws* VII, 819). Problems are included which lead to types of determinate and indeterminate equations dealt with by Diophantus (*c*. A.D. 250). Problem 126 is the famous source of our information concerning the details of the life of Diophantus.

THE GREEK ANTHOLOGY

BOOK XIV

translated by William R. Paton

Problem 6

'Best of clocks, how much of the day is past?' There remain twice two-thirds of what is gone.

Solution: $5\frac{1}{7}$ hours are past and $6\frac{6}{7}$ remain.

Problem 7

I am a brazen lion; my spouts are my two eyes, my mouth, and the flat of my right foot. My right eye fills a jar in two days, my left eye in three, and my foot in four. My mouth is capable of filling it in six hours; tell me how long all four together will take to fill it.

Solution: The scholia propose several, two of which, by not counting fractions, reach the result of four hours; but the strict sum is $3\frac{33}{37}$ hours.[1]

Problem 11

I desire my two sons to receive the thousand staters of which I am possessed, but let the fifth part of the legitimate one's share exceed by ten the fourth part of what falls to the illegitimate one.
Solution: $577\frac{7}{9}$ and $422\frac{2}{9}$.

Problem 12

Croesus the king dedicated six bowls weighing six minae,[2] each one drachm heavier than the other.
Solution: The weight of the first is $97\frac{1}{2}$ drachmae, and so on.

Problem 48

The Graces were carrying baskets of apples, and in each was the same number. The nine Muses met them and asked them for apples, and they gave the same number to each Muse, and the nine and three had each of them the same number. Tell me how many they gave and how they all had the same number.
Solution: The three Graces had three baskets with four apples in each, i.e. twelve in all, and they each gave three to the Muses. Any multiple of twelve does equally well.

Problem 49

Make me a crown weighing sixty minae, mixing gold and brass, and with them tin and much-wrought iron. Let the gold and bronze together form two-thirds, the gold and tin together three-fourths, and the gold and iron three-fifths. Tell me how much gold you must put in, how much brass, how much tin, and how much iron, so as to make the whole crown weigh sixty minae.
Solution: Gold $30\frac{1}{2}$, brass $9\frac{1}{2}$, tin $14\frac{1}{2}$, iron $5\frac{1}{2}$.

1. In this and subsequent problems, one day is taken as equal to 12 hours.
2. One mina = 100 drachmae.

Problem 50

Throw me in, silversmith, besides the bowl itself, the third of its weight, and the fourth, and the twelfth; and casting them into the furnace stir them, and mixing them all up take out, please, the mass, and let it weigh one mina.

Solution: The bowl weighs $\frac{3}{5}$ of a mina, or 60 drachmae.

Problem 51

A. I have what the second has and the third of what the third has.
B. I have what the third has and the third of what the first has.
C. And I have ten minae and the third of what the second has.
Solution: *A* has 45 minae, *B* has $37\frac{1}{2}$, and *C* has $22\frac{1}{2}$.

Problem 116

Mother, why dost thou pursue me with blows on account of the walnuts? Pretty girls divided them all among themselves. For Melission took two-sevenths of them from me, and Titane took the twelfth. Playful Astyoche and Philinna have the sixth and third. Thetis seized and carried off twenty, and Thisbe twelve, and look there at Glauce smiling sweetly with eleven in her hand. This one nut is all that is left to me.

Solution: There were 336; $(96+28+56+112+20+12+11+1)$.

Problem 117

A. Where are thy apples gone, my child? *B*. Ino has two-sixths and Semele one-eighth, and Autonoe went off with one-fourth, while Agave snatched from my bosom and carried away a fifth. For thee ten apples are left, but I, yes I swear it by dear Cypris, have only this one.

Solution: There were 120; $(40+15+30+24+11)$.

Problem 118

Myrto once picked apples and divided them among her friends; she gave the fifth part to Chrysis, the fourth to Hero, the nineteenth to Psamathe, and the tenth to Cleopatra, but she presented the twentieth part to Parthenope and gave only twelve to Evadne. Of the whole number a hundred and twenty fell to herself.

Solution: 380; (76+95+20+38+19+12+120).

Problem 119

Ino and Semele once divided apples among twelve girl friends who begged for them. Semele gave them each an even number and her sister an odd number, but the latter had more apples. Ino gave to three of her friends three-sevenths, and to two of them one-fifth of the whole number. Astynome took eleven away from her and left her only two apples to take to the sisters. Semele gave two quarters of the apples to four girls, and to the fifth one sixth part, to Eurychore she made a gift of four; she remained herself rejoicing in the possession of the four other apples.

Solution: Ino distributed 35; (15+7+11+2) and Semele 24; (12+4+4+4).

Problem 120

The walnut-tree was loaded with many nuts, but now someone has suddenly stripped it. But what does he say? 'Parthenopea had from me the fifth part of the nuts, to Philinna fell the eighth part, Aganippe had the fourth, and Orithyia rejoices in the seventh, while Eurynome plucked the sixth part of the nuts. The three Graces divided a hundred and six, and the Muses got nine times nine from me. The remaining seven you will find still attached to the farthest branches.'

Solution: There were 1680 nuts.

Problem 121

From Cadiz to the city of the seven hills the sixth of the road is to the banks of Baetis, loud with the lowing of herds, and hence a fifth to the Phocian soil of Pylades – the land is Vaccaean, its name derived from the abundance of cows. Thence to the precipitous Pyrenees is one-eighth and the twelfth part of one-tenth. Between the Pyrenees and the lofty Alps lies one-fourth of the road. Now begins Italy and straight after one-twelfth appears the amber of the Po. O blessed am I who have accomplished two thousand and five hundred stades journeying from thence! For the Palace on the Tarpeian rock is my journey's object.

Solution: The total distance is 15,000 stades (say 1,500 miles); from Cadiz to the Guadalquivir, i.e. to its upper waters, 2,500, thence to the Vaccaei (south of the Ebro) 3,000, thence to the Pyrenees 2,000, thence to the Alps 3,750, thence to the Po 1,250, thence to Rome 2,500.

Problem 123

Take, my son, the fifth part of my inheritance, and thou, wife, receive the twelfth; and ye four sons of my departed son and my two brothers, and thou my grieving mother, take each an eleventh part of the property. But ye, my cousins, receive twelve talents, and let my friend Eubulus have five talents. To my most faithful servants I give their freedom and these recompenses in payment of their service. Let them receive as follows. Let Onesimus have twenty-five minae and Davus twenty minae, Syrus fifty, Synete ten and Tibius eight, and I give seven minae to the son of Syrus, Synetus. Spend thirty talents on adorning my tomb and sacrifice to Infernal Zeus. From two talents let the expense be met of my funeral pyre, the funeral cakes, and grave-clothes, and from two let my corpse receive a gift.[1]

Solution: The whole sum is 660 talents; (132+55+420+12+5+2+34).

Problem 124

The sun, the moon, and the planets of the revolving zodiac spun such a nativity for thee; for a sixth part of thy life to remain an orphan with thy dear mother, for an eighth part to perform forced labour for thy enemies. For a third part the gods shall grant thee home-coming, and likewise a wife and a late-born son by her. Then thy son and wife shall perish by the spears of

1. Probably precious ointment.

the Scythians, and then having shed tears for them thou shalt reach the end of thy life in twenty-seven years.

Solution: He lived 72 years; (12+9+24+27).

Problem 126

This tomb holds Diophantus. Ah, how great a marvel! the tomb tells scientifically the measure of his life. God granted him to be a boy for the sixth part of his life, and adding a twelfth part to this, he clothed his cheeks with down; He lit him the light of wedlock after a seventh part, and five years after his marriage He granted him a son. Alas! late-born wretched child; after attaining the measure of half his father's life, chill Fate took him. After consoling his grief by this science of numbers for four years he ended his life.

Solution: He was a boy for 14 years, a youth for 7, at 33 he married, at 38 he had a son born to him who died at the age of 42. The father survived him for 4 years, dying at the age of 84.

Problem 129

A traveller, ploughing with his ship the broad gulf of the Adriatic, said to the captain, 'How much sea have we still to traverse?' And he answered him, 'Voyager, between Cretan Ram's Head and Sicilian Peloris are six thousand stades, and twice two-fifths of the distance we have traversed remains till the Sicilian strait.'

Solution: They had travelled $3,333\frac{1}{3}$ stades and had still $2,666\frac{2}{3}$ to travel.

Problem 130

Of the four spouts one filled the whole tank in a day, the second in two days, the third in three days, and the fourth in four days. What time will all four take to fill it?

Solution: $\frac{12}{25}$ of a day.

Problem 131

Open me and I, a spout with abundant flow, will fill the present cistern in four hours; the one on my right requires four more hours to fill it, and the third twice as much. But if you bid them both join me in pouring forth a stream of water, we will fill it in a small part of the day.

Solution: In $2\frac{2}{11}$ hours.

Problem 132

This is Polyphemus the brazen Cyclops, and as if on him someone made an eye, a mouth, and a hand, connecting them with pipes. He looks quite as if he were dripping water and seems also to be spouting it from his mouth. None of the spouts are irregular; that from his hand when running will fill the cistern in three days only, that from his eye in one day, and his mouth in two-fifths of a day. Who will tell me the time it takes when all three are running?

Solution: $\frac{6}{23}$ of a day.

Problem 133

What a fine stream do these two river-gods and beautiful Bacchus pour into the bowl. The current of the streams of all is not the same. Nile flowing alone will fill it up in a day, so much water does he spout from his paps, and the thyrsus of Bacchus, sending forth wine, will fill it in three days, and thy horn, Achelous, in two days. Now run all together and you will fill it in a few hours.

Solution: $\frac{6}{11}$ of a day.

Problem 134

O woman, how hast thou forgotten Poverty? But she presses hard on thee, goading thee ever by force to labour. Thou didst use to spin a mina's weight of wool in a day, but thy eldest daughter spun a mina and one-third of thread, while thy younger daughter contributed a half-mina's weight. Now thou providest them all with supper, weighing out one mina only of wool.

Solution: The mother in a day $\frac{6}{17}$, the daughters respectively $\frac{8}{17}$ and $\frac{3}{17}$.

Problem 135

We three Loves stand here pouring out water for the bath, sending streams into the fair-flowing tank. I on the right, from my long-winged feet, fill it full in the sixth part of a day; I on the left, from my jar, fill it in four hours; and I in the middle, from my bow, in just half a day. Tell me in what a short time we should fill it, pouring water from wings, bow, and jar all at once.

Solution: $\frac{1}{11}$ of a day.

Problem 136

Brick-makers, I am in a great hurry to erect this house. Today is cloudless, and I do not require many more bricks, but I have all I want but three hundred. Thou alone in one day couldst make as many, but thy son left off working when he had finished two hundred, and thy son-in-law when he had made two hundred and fifty. Working all together, in how many hours can you make these?

Solution: $\frac{2}{5}$ of a day.

Problem 137

Let fall a tear as you pass by; for we are those guests of Antiochus whom his house slew when it fell, and God gave us in equal shares this place for a banquet and a tomb. Four of us from Tegea lie here, twelve from Messene, five from Argos, and half of the banqueters were from Sparta, and Antiochus himself. A fifth of the fifth part of those who perished were from Athens, and do thou, Corinth, weep for Hylas alone.

Solution: There were 50 guests.

Problem 138

Nicarete, playing with five companions of her own age, gave a third of the nuts she had to Cleis, the quarter to Sappho, and the fifth to Aristodice, the twentieth and again the twelfth to Theano, and the twenty-fourth to Philinnis. Fifty nuts were left for Nicarete herself.

Solution: She had 1,200 nuts; (400+300+240+160+50+50).

Problem 139

Diodorus, great glory of dial-makers, tell me the hour since when the golden wheels of the sun leapt up from the east to the pole.

Four times three-fifths of the distance he has traversed remain until he sinks to the western sea.

Solution: $3\frac{9}{17}$ hours had passed, $8\frac{8}{17}$ hours remained.

Problem 140

Blessed Zeus, are these deeds pleasing in thy sight that the Thessalian women[1] do in play? The eye of the moon is blighted by mortals; I saw it myself. The night still wanted till morning twice two-sixths and twice one-seventh of what was past.

Solution: $6\frac{6}{41}$ of the night had gone by and $5\frac{35}{41}$ remained.

Problem 141

Tell me the transits of the fixed stars and planets when my wife gave birth to a child yesterday. It was day, and till the sun set in the western sea it wanted six times two-sevenths of the time since dawn.

Solution: It was $4\frac{8}{19}$ hours from sunrise.

Problem 142

Arise, work-women, it is past dawn; a fifth part of three-eighths of what remains is gone by.

Solution: $\frac{36}{43}$ of an hour had gone by.

Problem 143

The father perished in the shoals of the Syrtis, and this, the eldest of the brothers, came back from that voyage with five talents. To me he gave twice two-thirds of his share, on our mother he bestowed two-eighths of my share, nor did he sin against divine justice.

Solution: The elder brother had $1\frac{5}{7}$ talents, the younger $2\frac{2}{7}$, the mother 1 talent.

Problem 144

A. How heavy is the base I stand on together with myself?

1. Witches.

B. And my base together with myself weighs the same number of talents. *A.* But I alone weigh twice as much as your base. *B.* And I alone weigh three times the weight of yours.

   Solution: From these data not the actual weights but the proportions alone can be determined. The statue *A* was a third part heavier than *B*, and *B* only weighed $\frac{3}{4}$ of the statue *A*. The base of *B* weighed thrice as much as the base of *A*.

Problem 145

A. Give me ten minas and I become three times as much as you.
B. And if I get the same from you I am five times as much as you.
   Solution: $A = 15\frac{5}{7}$, $B = 18\frac{4}{7}$.

Problem 146

A. Give me two minas and I become twice as much as you. *B.* And if I got the same from you I am four times as much as you.
   Solution: $A = 3\frac{5}{7}$, $B = 4\frac{6}{7}$.

Problem 147

Answer of Homer to Hesiod when he asked the Number of the Greeks who took part in the War against Troy. There were seven hearths of fierce fire, and in each were fifty spits and fifty joints on them. About each joint were nine hundred Achaeans.
   Solution: 315,000.

34

OMAR KHAYYÁM
(c. 1044–1123)

THE phenomenal rise of the fame of Omar Khayyám, the astronomer–poet of Persia, in all parts of the civilized world began in 1859 with the anonymous publication of a hundred of his four-line verses in a book entitled the *Rubaiyatt of Omar Khayyám*. The verses had been translated rather freely into English by Edward Fitzgerald, an English writer and student of Iranian philology. Within fifty years after the appearance of that edition, more than three hundred editions in English of the first and subsequent versions of the *Rubaiyatt* by Fitzgerald were published. Within seventy years after the first publication, more than thirteen hundred works connected with the *Rubaiyatt* had appeared. Of these almost two hundred were written in thirty-three languages other than English. In 1900 the Omar Khayyám Club of America was organized in Boston, Mass., and interest in his writing continued undiminished. The enormous vogue for his poetry everywhere evoked a correspondingly widespread desire to know more intimately the details of the poet's life and work. In this quest some searched in the great libraries of the world hoping to find manuscripts of other works written by Omar. Others closely scrutinized his poetry and interpreted it. Fact and fancy mingled on a wide scale, and there was no lack of controversy over the factual story of Omar Khayyám's life, his quality as a poet and as a philosopher, as well as the depth and sincerity of his devotion to religious principles. Despite the emergence of certain apparently irreconcilable points of view, continued research has narrowed the differences, and substantial agreement has been found in some matters.

Omar's full name is Omar ibn Ibrahim al-Khayyám, Giyat ed-din Abu'l Fath (Ghiyath ud Din Abu'l Fatah 'Omar bin Ibrahim Khayyám). The date of his birth may have been as early as 1042. His family name, al-Khayyám, means 'the tent-maker', but this occupation was most likely far back in his ancestry, his immediate forebears having been a literary family. Omar himself says, 'Khayyám who stitched in the tents of science'.

During his lifetime Omar was held in the highest esteem by princes, by scholars, and by the common man as the greatest scientist of his day. He was a teacher of scientific subjects, although his known pupils are few in number. By his own statement he preferred to learn rather than to teach. Omar was celebrated for his mastery of astrology and metaphysics, for his prodigious feats of memory, and for the mystical bent of his philosophy. His reform of the Persian calendar achieved extraordinary accuracy, greatly adding to Omar's stature in the field of science. His quatrains, too, were well known in his own lifetime. However, the same quatrains which were admired and enjoyed by many people, were found shocking by others, and some verses had the effect of bringing important enmities upon Omar. In his later years Omar gave up writing and went into seclusion for long periods.

It was the field of algebra which brought Omar the first complete success of his long career, and his contributions to this area of mathematics were of lasting scientific influence. His treatise on algebra became widely known through the translation (1931) of an Arabic manuscript in the library of Professor David Eugene Smith. This work, now known to us as *The Algebra of Omar Khayyám,* had been in use as a school text in Persia for hundreds of years. It continued in use until a very late date, especially in the more isolated regions of Persia. Far ahead of his time, in mathematical methods, Omar supported his algebraic solutions by geometrical constructions and proofs. Celebrated as the astronomer-poet of Persia, Omar Khayyám was also the first mathematician to study and classify cubic equations and to employ conic sections in their solution.

THE ALGEBRA OF OMAR KHAYYÁM

translated from the Arabic, with explanatory notes,
by Daoud S. Kasir

CHAPTER I: DEFINITIONS

Algebra. By the help of God and with His precious assistance, I say that Algebra is a scientific art. The objects with which it deals are absolute numbers and measurable quantities which, though themselves unknown, are related to 'things' which are known, whereby the determination of the unknown quantities is possible. Such a thing is either a quantity or a unique relation, which is only determined by careful examination. What one searches for in the algebraic art are the relations which lead from the known to the unknown to discover which is the object of Algebra[1] as stated above. The perfection of this art consists in knowledge of the scientific method by which one determines numerical and geometric unknowns.

Measurable Quantities. By measurable quantities I mean continuous quantities of which there are four kinds, viz., line, surface, solid and time, according to the customary terminology of the Categories[2] and what is expounded in metaphysics.[3] Some consider space a subdivision of surface, subordinate to the division of continuous quantities, but investigation has disproved this claim. The truth is that space is a surface only under circumstances the determination of which is outside the scope of the present field of investigation. It is not customary to include 'time' among the objects of our algebraic studies, but if it were mentioned it would be quite admissible.

The Unknown. It is a practice among algebraists in connexion with their art to call the unknown which is to be determined a 'thing',[4] the product obtained by multiplying it by itself a 'square',[5] and the product of the square and the 'thing' itself a 'cube'. The product of the square multiplied by itself is 'the square of the square', the product of its cube multiplied by its square 'the cube of the square', and the product of a cube into itself 'a cube of the cube, and so on, as far as the succession is

1. The author refers here to the algebraic relations existing between the known and the unknown quantities which the algebraist has to establish. For other Arabic definitions of algebra see *Haji Khalpha, Mohammed ibn Musa*, by Karpinski, p. 67; *Mukadamat ibn Khaldun*, p. 422 (Egypt).

2. *Category* of Aristotle, cap. 6; phys. lv, cap. 4 ult. According to Aristotle's definitions, point, line and surface are first principles and must be assumed. Heath, *Euclid*, vol. I, pp. 155–6, 158, 159, 165, 170.

3. Here the author is referring to his book on metaphysics. See A. Christensen: '*Un Traité de Metaphysique d'Omar*', *Le Monde Oriental*, 1908, vol. I, pp. 1–16.

4. The Arabic word *shai* (literally a 'thing') here means the 'unknown'. Latin translators used the word *res*.

5. *Mal* (literally, 'substances') is the word used by the author to indicate the second power of the unknown. Gherardo of Cremona (c. 1150) used *census*, which has the same meaning.

carried out.[1] It is known from Euclid's book, the *Elements*,[2] that all the steps are in continuous proportion; i.e. that the ratio of one to the root is as the ratio of the root to the square and as the ratio of the square to the cube.[3] Therefore, the ratio of a number to a root is as the ratio of roots to squares, and squares to cubes, and cubes to the squares of the squares, and so on after this manner.[4]

Sources. It should be understood that this treatise cannot be comprehended except by those who know thoroughly Euclid's books, the *Elements* and the *Data*, as well as the first two books from Apollonius' work on *Conics*. Whoever lacks knowledge of any one of these books cannot possibly understand my work, as I have taken pains to limit myself to these three books only.

Algebraic Solutions. Algebraic solutions are accomplished by the aid of equations; that is to say, by the well-known method of equating these degrees one with the other. If the algebraist were to use the square of the square in measuring areas, his result would be figurative[5] and not real, because it is impossible to consider the square of the square as a magnitude of a measurable nature. What we get in measurable quantities is first one dimension, which is the 'root'[6] or the 'side'[7] in relation to its square; then two dimensions, which represent the surface and the (algebraic) square representing the square surface; and finally, three dimensions, which represent the solid.[8] The cube in quanti-

1. $x.x = x^2$, $x.x.x = x^3$, $x^2.x^2 = x^4$, $x^2.x^3 = x^5$, $x^3.x^3 = x^6$
2. *Elements of Euclid*, Heath, vol. II, p. 390.
3. $1:x = x:x^2 = x^2:x^3$.
4. $a:ax = ax:ax^2 = ax^2:ax^3 = ax^3:ax^4$, etc.
5. The literal translation of the Arabic word *majaz* is 'path, way'. The author means that a quantity raised to the fourth power cannot be represented geometrically and therefore has no real geometric meaning, while algebraically it has. In other words, the author meant by *majaz*, 'hypothetical'. See *Akrabu'l-Mawarid* (Beirut, 1889).
6. The Arab writers used also the word 'root' for the first power of the unknown quantity in an equation.
7. *Dulu* (literally, side), meaning an unknown quantity, is represented geometrically by a line, while *mal* (square) is represented by a surface, and *muk'ab* (cube) by a solid.
8. The author's idea of dimension conforms fundamentally with that of Aristotle, Proclus and al-Nairizi. See Heath, *Euclid*, vol. I, pp. 157–9.

ties is the solid bounded by six squares, and since there is no other dimension, the square of the square does not fall under measurable quantities. This is even more true in the case of higher powers.[1] If it is said that the square of the square is among measurable quantities, this is said with reference to its reciprocal value in problems of measurement and not because it in itself is measurable. This is an important distinction to make.

The square of the square is, therefore, neither essentially nor accidentally a measurable quantity, and is not as even and odd numbers, which are accidentally included in measurable quantities, depending on the way in which they represent continuous measurable quantities as discontinuous.

What is found in the books of algebra relative to these four geometric quantities – namely, the absolute numbers, the 'sides', the squares, and the cubes – are three equations containing numbers, sides and squares. We, however, shall present methods by which one is able to determine the unknown quantities in equations including four degrees concerning which we have just said that they are the only ones that can be included in the category of measurable quantities, namely, the number, the thing, the square, and the cube.

The demonstration[2] (of solutions) depending on the properties of the circle – that is to say, as in the two works of Euclid, the *Elements* and the *Data* – is easily effected; but what we can demonstrate only by the properties of conic sections should be referred to the first two books on conics.[3] When, however, the object of the problem is an absolute number,[4] neither we, nor any

1. It was Descartes who first defeated this method of reasoning, which had been universally accepted before him.
2. The author refers here to the demonstration of processes which constitute the solution.
3. The author is referring here to the above-mentioned works of Apollonius.
4. Or 'if it is required to satisfy the proposed equation by a whole number'. The author means here that cubic equations can be solved geometrically, but that neither he nor his predecessors could solve them algebraically. It was to Archimedes, al-Mahani, Thabit ibn Qorra, and al-Khazin that he referred here. It was not until the sixteenth century that Cardan and Tartaglia succeeded in solving cubic equations algebraically. Smith, *History of Mathematics*, vol. II, pp. 455 (footnote 2) and 459.

of those who are concerned with algebra, have been able to prove this equation – perhaps others who follow us will be able to fill the gap – except when it contains only the three first degrees, namely, the number, the thing, and the square.[1] For the numerical demonstration given in cases that could also be proved by Euclid's book, one should know that the geometric proof of such procedure does not take the place of its demonstration by number, if the object of the problem is a number and not a measurable quantity. Do you not see how Euclid proved certain theorems relative to proportions of geometric quantities in his fifth book, and then in the seventh book gave a demonstration of the same theorems for the case when their object is a number?[2]

CHAPTER II: TABLE OF EQUATIONS

The equations among those four quantities are either simple or compound. The simple equations are of six species:

(i) A number equals a root.
(ii) A number equals a square.
(iii) A number equals a cube.
(iv) Roots equal a square.
(v) Squares equal a cube.
(vi) Roots equal a cube.

Three of these six species are mentioned in the books of the algebraists. The latter said a thing is to a square as a square is to a cube. Therefore, the equality between the square and the cube is equivalent to the equality of the thing to the square. Again they said that a number is to a square as a root is to a cube. And they did not prove by geometry. As for the number which is equal to a cube there is no way of determining its side when the problem is numerical except by previous knowledge of the order of cubic numbers. When the problem is geometrical it cannot be solved except by conic sections.

---

1. For example, as in $x^2 - 7x + 12 = 0$, where 12 is the number, x is the 'thing', and x^2 is the square.
2. Compare, for example, Prop. 5 of Book V with Prop. 8 of Book VII.

CHAPTER V: PRELIMINARY THEOREMS FOR THE CONSTRUCTION OF CUBIC EQUATIONS

After presenting those species of equations which it has been possible to prove by means of the properties of the circle, i.e. by means of Euclid's book, we take up now the discussion of the species which cannot be proved except by means of the properties of conics. These include fourteen species: one simple equation, namely, that in which *a number is equal to a cube*; six trinomial equations; and seven tetranomial equations.

Let us precede this discussion by some propositions based on the book on *Conics* so that they may serve as a sort of introduction to the student and so that our treatise will not require familiarity with more than the three books already mentioned, namely, the two books of Euclid on the *Elements* and the *Data*, and the first two parts of the book on *Conics*.

Between two given lines it is required to find two other lines such that all four will form a continuous proportion.[1]

Let there be two straight lines (given) *AB*, *BC* (Figure 34.1), and let them enclose the right angle *B*. Construct a parabola the vertex of which is the point *B*, the axis *BC*, and the parameter *BC*. Then the position of the conic *BDE* is known because the positions of its vertex and axis are known, and its parameter is given. It is tangent to the line *BA*, because the angle *B* is a right angle and it is equal to the angle of the ordinate, as was shown in the figure of the thirty-third proposition in the first book on *Conics*.[2] In the same manner construct another parabola, with vertex *B*, axis *AB*, and parameter *AB*. This will be the conic *BDZ*, as was shown also by Apollonius in the fifty-sixth proposition of the first book.[3] The conic *BDZ* is tangent to the line *BC*.

---

1. $AB:x = x:y = y:BC$.

In the parabola BDE $\overline{HD}^2 = BT^2 = BH.BC$, hence $BC:BT = BT:HB$.

In the parabola BDZ $\overline{DT}^2 = HB^2 = BA.BT$, hence $BT:HB = HB:BA$.

Consequently $BC:BT = BT:HB = HB:BA$ or $AB:BH = BH:BT = BT:BC$.

This is the second of the two constructions of the problem attributed to Menaechmus.

2. Apollonius, ed. Oxford, 1710, fol., p. 57. The author refers here to the 32nd proposition of this book.

3. Apollonius, ed. Oxford, book I, proposition 52.

Therefore, the two (parabolas) necessarily intersect. Let D be the point of intersection. Then the position of point D is known because the position of the two conics is known. Let fall from the point D two perpendiculars, DH and DT, on AB and BC

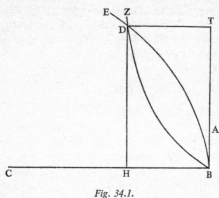

Fig. 34.1.

respectively. These are known in magnitude, as was shown in the *Data*.[1] I say that the four lines AB, BH, BT, BC are in continuous proportion.

Demonstration

The square of HD is equal to the product of BH and BC, because the line DH is the ordinate of the parabola BDE. Consequently BC is to HD, which is equal to BT, as BT to HB. The line DT is the ordinate of the parabola BDZ. The square of DT (which is equal to BH) is equal to the product of BA and BT. Consequently BT is to BH as BH to BA. Then the four lines are in continuous proportion and the line DH is of known magnitude, as it is drawn from the point the position of which is known, to a line whose position is known, at an angle whose magnitude is known. Similarly, the length of DT is known. Therefore, the two lines, BH and BT, are known and are the means of the proportion between the two lines, AB and BC; that is to say,

1. See propositions 30, 25, 26.

AB to *BH* is as *BH* to *BT* and is as *BT* to *BC*. That is what we wanted to demonstrate.

Given (Figure 34.2) *the rectangular parallelepiped ABCDE, whose base is the square AD, and the square MH, construct on MH a rectangular parallelepiped equal to ABCDE.*

Let *AB* to *MZ* be as *MZ* to *K* and let *AB* to *K* be as *ZT* to *ED*.[1] Then make *ZT* perpendicular to the plane *MH* at the point *Z* and complete the solid *MZTH*. Then I say that this solid is equal to the given solid.

Demonstration

The square *AC* to the square *MH* is as *AB* to *K*. Then the square *AC* to the square *MH* is as *ZT*, which is the height of the solid

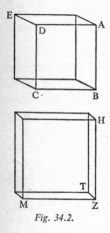

Fig. 34.2.

MTH to *ED*, which is the height of the solid *BE*. Therefore the two solids are equal, for their bases are reciprocally proportional to their heights, as it was demonstrated in the eleventh book of the *Elements*.

Whenever we speak of 'a solid' we mean a rectangular parallelepiped and whenever we say 'plane' we refer to the rectangle.

Given the solid ABCD (Figure 34.3), *whose base AC is a square, it is required to construct a solid whose base is a square, whose height is equivalent to a given line ET, and which is equal to the solid ABCD.*

Let *ET* be to *BD* as *AB* to *K* and take between *AB* and *K* a

---

1. That is, construct *K* from the proportion $AB:MZ = MZ:K$, in which AB^2, MZ are known and ZT as the third term of $AB:K = ZT:ED$. The volume of $ABCDE = \overline{AB}^2 . DE$. But $\overline{MZ}^2 . ZT$ = volume of the parallelepiped on $MH = \overline{AB}^2 . DE$, since the first proportion gives $\overline{MZ}^2 = AB . K$ and the second $ZT = \dfrac{AB . ED}{K}$. But $\overline{AB}^2 . DE$ = volume of parallelepiped *ABCDE*. Hence the two solids are equal.

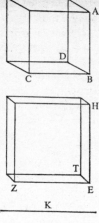

Fig. 34.3.

mean proportional line EZ.[1] Make EZ perpendicular to ET and complete ZT. Then make EH perpendicular to the plane TZ and equal to EZ, and complete the solid $HETZ$. Then I say the volume of solid T, whose base is the square HZ and height the given line ET, is equal to the volume of given solid D.

Demonstration

The square AC to the square HZ is as AB to K. Consequently, the square AC to the square HZ is as ET to BD. Therefore the bases of the two solids are also reciprocally proportional to their heights and the solids then are equal. That is what we wanted to demonstrate.

1. Construct K from the proportion
$$ET : BD = AB : K,$$
and EZ from
$$AB : EZ = EZ : K.$$
It follows that $\quad \overline{AB^2} : \overline{EZ^2} = ET : BD.$

Then $\quad\quad\quad \overline{AB^2} . BD = \overline{EZ^2} . ET.$

Or $\quad\quad\quad\quad$ solid D = solid T.

71

After these preliminary proofs we shall be able to give the solution of the third species of the simple equation, *a cube is equal to a number*.[1]

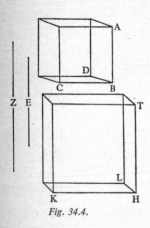

Fig. 34.4.

Let the number be equal to the solid *ABCD* (Figure 34.4), and its base *AC* the square of one, as we have said previously. Its length is equal to the given number. It is desired to construct a cube equal to this solid. Take between the two lines *AB* and *BD* two mean proportionals.[2] These are known in magnitude, as has been demonstrated.[3] They are *E* and *Z*. Then draw *HT* equal to the line *E* and construct on it the cube *THKL*. This cube and its side are known in magnitude. Then I say that this cube is equal to the solid *D*.

Demonstration

The square *AC* to the square *TK* is twice the ratio *AB* to *HK*,[4] and twice the ratio of *AB* to *HK* is equal to the ratio *AB* to *Z*. But the first is to the third of the four lines as *HK*, the second, is to *BD*, the fourth. The bases (*TK*, *AC*) of the cube *L* and the solid *D* are then reciprocally proportional to their height. Then the solids are equal, which is what we wanted to demonstrate.

1. $a = x^3$.

2. $AB = E^2 : Z$; $\overline{AB^2} = \dfrac{AB \cdot E^2}{Z}$ (*E*, *Z* mean proportional lines)

$$\overline{AB^2} : E^2 = AB : Z$$

But
$$AB : E = Z : BD \text{ gives}$$
$$AB : Z = E : BD$$

Hence
$$\overline{AB^2} : E^2 = E : BD$$

∴
$$E^2 = \overline{AB^2} : BD, \text{ etc.}$$

3. See footnote 2, p. 68.

4. That is $(AB : HK)(AB : HK)$ or $\overline{AB^2} : \overline{HK^2}$.

CHAPTER VI: TRINOMIAL EQUATIONS

Capable of Being Proved by Means of the Properties of the Conic Sections

After this we are to work on the remaining six trinomial equations.

I. The first species. *A cube and sides are equal to a number.*

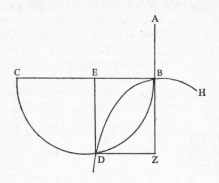

Fig. 34.5.

Let the line *AB* (Figure 34.5) be the side of a square equal to the given number of roots. Construct a solid whose base is equal to the square on *AB*, equal in volume to the given number. The construction has been shown previously. Let *BC* be the height of the solid. Let *BC* be perpendicular to *AB*. You know already what meaning is applied in this discussion to the phrase *solid number*. It is a solid whose base is the square of unity and whose height is equal to the given number; that is, the height is a line whose ratio to the side of the base of the solid is as the ratio of the given number to one. Produce *AB* to *Z* and construct a parabola whose vertex is the point *B*, axis *BZ*, and parameter *AB*. Then the position of the conic *HBD* will be known, as has been shown previously and it will be tangent to *BC*. Describe on *BC* a semi-circle. It necessarily intersects the conic. Let the point of intersection be *D*; drop from *D*, whose position is known, two

73

perpendiculars DZ and DE on BZ and BC. Both the position and the magnitude of these lines are known. The line DZ is an ordinate of the conic. Its square is then equal to the product of BZ and AB. Consequently, AB to DZ, which is equal to BE, is as BE to ED, which is equal to ZB. But BE to ED is as ED to EC. The four lines then are in continuous proportion, AB, BE, ED, EC, and consequently the square of the parameter AB, the first, is to the square of BE, the second, as BE, the second, is to EC, the fourth. The solid whose base is the square AB and whose height is EC is equal to the cube BE, because the heights of these figures are reciprocally equal to their bases. Let the solid whose base is the square of AB and height is EB be added to both. The cube BE plus the solid then is equal to the solid whose base is the square AB and whose height is BC, which solid we have assumed to be equal to the given number. But the solid whose base is the square of AB, which is equal to the number of roots, and whose height is EB, which is the side of the cube, is equal to the number of the given sides of the cube EB. The cube EB, then, plus the number of its given sides is equal to the given number, which was required.

This species does not present varieties of cases or impossible problems. It has been solved by means of the properties of the circle combined with those of the parabola.

II. The second species. *A cube and a number are equal to sides.*

Let the line AB (Figure 34.6) be the side of a square equal to the number of the roots, and construct a solid having as its base the square of AB and equal to the given number, and let its height BC be perpendicular to AB. Describe a parabola having as its vertex the point B and its axis along the direction AB and its parameter AB. This is, then, the curve DBE, whose position is known. Construct also a second conic, namely, a hyperbola whose vertex is the point C and whose axis is along the direction of BC. Each one of its two parameters, the perpendicular and the oblique, is equal to BC. It is the curve ECZ. This hyperbola also is known in position, as was shown by Apollonius in the 58th proposition of his first book. The two conics will either meet or will not meet. If they do not meet, the problem is impossible of solution. If they do meet, they do it tangentially at a point or by intersection at two points.

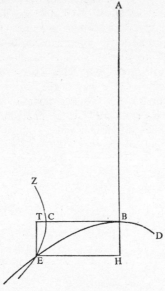

Fig, 34.6.

Suppose they meet at a point and let it be at *E*, whose position is known. Then drop from it two perpendiculars *ET* and *EH* on the two lines *BT* and *BH*. The two perpendiculars are known unerringly in position and magnitude. The line *ET* is an ordinate of the hyperbola. Consequently, the square of *ET* is to the product of *BT* and *TC* as the parameter is to the oblique, as was demonstrated by Apollonius in the twentieth proposition of the first book. The two sides, the perpendicular and the oblique, are equal. Then the square *ET* is equal to the product of *BT* and *TC*, and *BT* to *TE* is as *TE* to *TC*. But the square of *EH*, which is equivalent to *BT*, is equal to the product of *BH* and *BA*, as was demonstrated in the second proposition of the first book of the treatise on conics. Consequently, *AB* is to *BT* as *BT* is to *BH* and as *BH*, which is equal to *ET*, is to *TC*. The four lines, *AB*,

BT, *ET*, *TC*, then, are in continuous proportion, and the square of *AB*, the first, is to the square of *BT*, the second, as *BT*, the second, is to *TC*, the fourth. The cube of *BT*, then, is equal to the solid whose base is the square *AB* and whose height is *CT*. Let the solid whose base is the square of *AB* and whose height is *BC*, which was made equal to the given number, be added to both. Then the cube *BT* plus the given number is equal to the solid whose base is the square of *AB* and whose height is *BT*, which represents the number of the sides of the cube.

Thus it is shown that this species includes different cases and among its problems are some that are impossible. The species has been solved by means of the properties of the two conics, the parabola and the hyperbola.

35

GEOFFREY CHAUCER

(1340?–1400)

GEOFFREY CHAUCER, first great poet to write extensively in the English language, was also a superb master of English prose. His works, immediately popular, have been celebrated by all the generations of readers from his day to the present, for the dramatic clarity of the story they tell and for the breadth of human nature they portray. In addition to his acute observations on life and the vivid touches which make his characters spring alive Chaucer's poetry exhibits his magnificent power in poetical techniques and verse forms. Deeply learned in Latin, French and Italian literature, he introduced and embodied in his English compositions many technical features of style and form which had hitherto been known only in the older languages. His prose writings include only one scientific work, *A Treatise on the Astrolabe* (1391). Intended primarily as a text book of instruction for his 'litell son Lowys' whose attainments in Latin had not reached the proficiency required for current Latin expositions, the *Treatise* was written in 'plain English words' superbly accommodated to the scientific needs of clarity, directness and accuracy. The prologue of the *Treatise*, moreover, is an excellent example of Chaucer's graceful and flowing prose style, and of his incomparable felicity in arousing in his readers the pleasant sense of personal, sole discovery of the author's meaning and matchless charm.

Geoffrey Chaucer was born in London about 1340, and most of his life was spent in that city. His career was twofold, divided between writing and his service to the royal family, the periods of his greatest activity in the latter capacity being the least productive of literary works. His first employment in the service of the royal family was probably obtained through the intercession of his father, John Chaucer, a vintner, who is known to have been in attendance on King Edward III as early as 1338. The household account books (1357) of Princess Elizabeth, wife of Lionel, Duke of Clarence (third son of Edward III), contain items relating to the purchase of clothes and other articles for

Geoffrey Chaucer, serving in that household as a page. Toward the end of the first phase of the Hundred Years War, Chaucer took part in the one military adventure of his life. Serving in the English army that invaded France in 1359, he was unfortunately taken prisoner, but was released some months later on the payment by King Edward of £16 toward his ransom. When his patron Lionel died in 1368, Chaucer entered the service of the next brother, John of Gaunt, Duke of Lancaster, and his fortunes thereafter were closely connected to those of the House of Lancaster. Chaucer's wife, Philippa, was the sister of Katherine, widow of Sir Hugh de Swynford. Katherine was governess to John of Gaunt's children, later becoming his mistress, and in 1396, his wife. For more than twenty years Chaucer served in various civil and diplomatic offices. Diplomatic missions of the highest importance were entrusted to him and were the occasions of his journeys to the distant cities of Genoa, Pisa and Florence. Through his fees as the Comptroller of the Customs for the port of London, pensions and other income, Chaucer appears generally to have prospered, although there are records which indicate some periods of hardship.

Nothing more has been learned concerning Chaucer's son whose importunities at the age of ten led to the purchase of the portable Arabian astrolabe described in the *Treatise*. Chaucer states that his claim in this work is not for originality, but only for his selection of the most reliable authorities on the subject, for the accuracy of his instruction and for rendering it accessible in the English language. For the *Conclusions* of Part II, and for the methods of *Umbra Recta* for finding the height of an accessible object and of *Umbra Versa* for finding the height of an inaccessible object, Chaucer drew heavily upon the Latin manuscript of the *Compositio et Operatio Astrolabii*, by Messahalla, many copies of which are extant. The five parts which Chaucer projected for his work do not all seem to have been completed. The first two parts of Chaucer's *A Treatise on the Astrolabe* and the supplement soon became and still remain a standard work on the construction and the uses of the astrolabe.

GEOFFREY CHAUCER

TREATISE ON THE ASTROLABE

From Chaucer and Messahalla on the Astrolabe
by Robert T. Gunther

BREAD AND MILK FOR CHILDREN

Little Lewis my son, I have perceived well by certain signs thy
ability to learn sciences touching numbers and proportions; and
I also consider thy earnest prayer specially to learn the Treatise
of the Astrolabe. Then forasmuch as a philosopher saith, 'he
wrappeth him in his friend, who condescendeth to the rightful
prayers of his friend', therefore I have given thee an astrolabe
for our horizon, composed for the latitude of Oxford, upon
which, by means of this little treatise, I purpose to teach thee a
certain number of conclusions appertaining to the same instrument. I say certain conclusions, for three reasons.

The first is this: understand that all the conclusions that have
been found, or possibly might be found in so noble an instrument
as an astrolabe, are not known perfectly to any mortal man in
this region, as I suppose.

Another reason is this: that truly, in any treatise of the astrolabe that I have seen, there are some conclusions that will not
in all things perform their promises. And some of them are too
hard for thy tender age of ten years to understand.

I will show thee this treatise, divided into five parts, under full
easy rules and in plain English words; for Latin thou knowest as
yet but little, my little son. But nevertheless, these true conclusions are sufficient for thee in English, as they are in Greek for
noble Greek scholars, and in Arabic for Arabians, and in
Hebrew for Jews, and in Latin for the Latin folk, for they have
written them first out of other different languages, in their own
tongue, that is to say, in Latin. And God knows, that in all these
languages, and in many more, these conclusions have been
sufficiently learned and taught, though by diverse rules, just as
diverse paths lead diverse folk the right way to Rome. Now I will
meekly pray every discreet person that readeth or heareth this
little treatise, to have my rude inditing and my superfluity of

words excused, for two causes. The first, that curious inditing and hard sentences are at once too difficult for a child to learn. And the second, that indeed it seems better to me to write a good sentence twice unto a child, rather than that he forget it once. And Lewis, if I show thee in my easy English as true conclusions touching this matter, and not only as true but as many and as subtle conclusions as are shown in Latin in any common treatise of the astrolabe, grant me the more thanks; and pray God save the King, who is lord of this language, and all that are true to him and obey him, each in his degree, the more and the less. But consider well, that I do not claim to have found this work by my own labour or ingenuity, I am but an unlearned compiler of the labour of old astrologians. I have translated it into English only for thy instruction; and with this sword shall I slay envy.

The first part of this treatise will rehearse the figures and the parts of thy astrolabe, so that thou mayest have the greater knowledge of thy own instrument.

The second part will teach thee to work the exact practice of the aforesaid conclusions, as far and as exactly as may be showed in so small and portable an instrument. For every astrologian well knows that the smallest fractions are not shown in so small an instrument, as they are in subtle tables calculated on purpose.

The third part will contain diverse tables of longitudes and latitudes of fixed stars for the astrolabe, and tables of declinations of the sun, and tables of longitudes of cities and of towns; both for the regulation of a clock and to find the meridian altitude, and many another notable conclusion, in accordance with the calendars of the reverend scholars, friar John Somer and friar Nicholas Lenne.[1]

The fourth part will be a theory to explain the moving of the celestial bodies with the causes. In particular it will show a table of the exact moving of the moon from hour to hour, every day and in every sign, after thy almanac; after which table there follows an explanation, sufficient to teach both the manner of the working of that same conclusion, and to know in our horizon the degree of the zodiac with which the moon rises in any

1. Somer's Calendar was calculated for 140 years from 1367, Lynn's for 76 years from 1387.

latitude, and the arising of any planet in accordance with its latitude from the ecliptic line.

The fifth part will be an introduction according to the rules of
our doctors, in which thou mayest learn a great part of the general
rules of theory in astrology. In this fifth part thou wilt find
tables of equations of 'houses' for the latitude of Oxford; and
table of dignities of planets, and other useful things, if God and
his mother, the maid, will grant more than I promise.

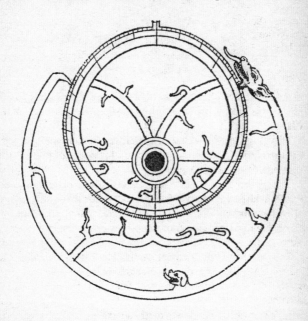

Fig. 35.1. The Rete of Chaucer's Astrolabe.

PART I

Here beginneth the Description of the Astrolabe

1. Thy astrolabe hath a *ring* to put on the thumb of thy right hand when taking the height of things.

Fig. 35.2.

And note that from henceforward, I will call the height of anything that is taken by thy 'rule', the *altitude*, without more words.

2. This ring runs in a kind of *eyelet*, fastened to the 'mother' of thy astrolabe, in so roomy a space that it does not prevent the instrument from hanging plumb.

3. The *mother* of thy astrolabe is a very thick plate, hollowed out with a large cavity, which receives within it the thin *plates* marked for different climates, and thy *rete*, shaped like a net or the web of a spider. For more explanation, lo here the figure (Figure 35.3):

4. This mother is divided on the backhalf with a line, which descends from the ring down to the lowest border. This line, from the aforesaid ring to the centre of the large cavity in the middle, is called the *south line*, or the *line meridional*. And the remainder of this line down to the border is called the *north line*, or the *line of midnight*. And for more explanation, lo here the figure (Figure 35.4):

5. At right angles to the meridional line, there crosses it another line of the same length from east to west. This, from a little cross + in the border to the centre of the large cavity, is called the *east line*, or *line oriental*; and the remainder of the line from the aforesaid + to the border, is called the *west line*, or *line occidental*. Now hast thou here the 4 quarters of thy astrolabe,

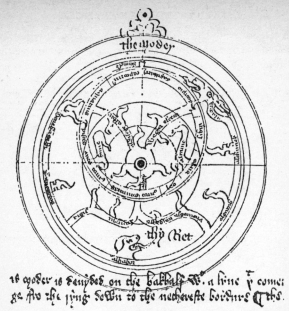

the Rete is deuyded on the bakhalf W^t a line y comes
ga fro the pyng down to the nethereste bordure of the.

Fig. 35.3. The Rete lying in the Mother. Like the other text-figures it has
been taken from MS. Cambridge Rd. 3.53, and is obviously from the same
source as the Rete, with the head of the Dogstar, Alhabor, reversed.

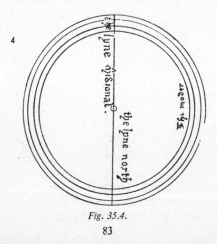

Fig. 35.4.

83

divided according to the 4 principal quarters of the compass, or quarters of the firmament. And for more explanation, lo here thy figure (Figure 35.5):

6. The east side of thy astrolabe is called the right side, and the west side is called the left side. Forget not this, little Lewis.

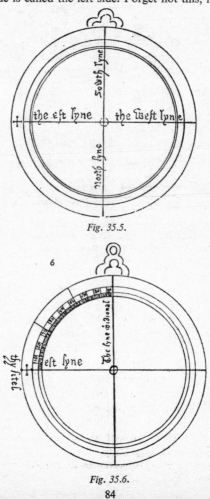

Fig. 35.5.

Fig. 35.6.

Put the ring of thy astrolabe upon the thumb of thy right hand, and then its right side will be towards thy left side, and its left side will be towards thy right side; take this as a general rule, as well on the back as on the hollow side. Upon the end of the east line, as I first said, is marked a little +, which is always regarded as the beginning of the first degree in which the sun rises. And for more explanation, lo here the figure (Figure 35.6):

7. From this little + up to the end of the meridional line, under the ring, thou wilt find the *border divided into* 90 *degrees*; and every quarter of thy astrolabe is divided in the same proportion. Over these degrees are numbers, and the degrees are divided into fives as shown by long lines between. The space between the long lines containeth a mile-way.[1] And every degree of the border contains 4 minutes, that is to say, minutes of an hour. And for more explanation, lo here the figure:

[*Ed. Note. The figure is similar to Figure 35.6.*]

8. Under the circle of these degrees are written the *names of the* 12 *Signs*, as Aries, Taurus, Gemini, Cancer, Leo, Virgo, Libra, Scorpio, Sagittarius, Capricornus, Aquarius, Pisces; and the numbers of the degrees of the Signs are written in Arabic numerals above, and with long divisions, from 5 to 5, divided from the time that the Sign entereth unto the last end. But understand well, that these degrees of Signs are each of them considered to be of 60 minutes, and every minute of 60 seconds, and so forth into small fractions infinite, as saith Alkabucius,[2] and therefore, know well, that a degree of the border containeth 4 minutes, and a degree of a Sign containeth 60 minutes; remember this. And for more explanation, lo here thy figure (Figure 35.7):

9. Next follows the *Circle of the Days*, in number 365 that are numbered as are the degrees, and divided also by long lines from 5 to 5; and the numbers under that circle are written in Arabic numerals. And for more explanation, lo here thy figure (Figure 35.8):

10. Next the circle of the days follows the *Circle of the Months*;

1. The time it takes to walk a mile.
2. Abdilazi Alchabitius, *Introductorium ad Scientiam Judicialem Astronomiae*, printed 1473.

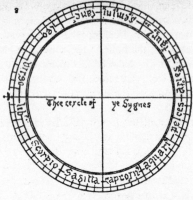

Fig. 35.7.

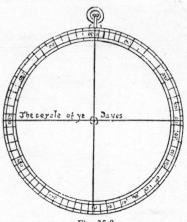

Fig. 35.8.

that is to say, January, February, March, April, May, June, July, August, September, October, November, December. These months were named amongst the Arabians, some for their prophets, and some by statutes of lords, some by other lords of Rome. Also, as it pleased Julius Caesar and Caesar Augustus, some months were composed of different numbers of days, as

86

July and August. Then hath January 31 days, February 28, March 31, April 30, May 31, June 30, July 31, August 31, September 30, October 31, November 30, December 31. Nevertheless, although Julius Caesar took 2 days out of February and put them in his month of July,[1] and Augustus Caesar called the month of August after his own name and ordained it of 31 days; yet trust well, that the sun never dwelleth on that account more or less in one sign than in another.

11. Then follow the names of the *Holydays* in the calendar, and next them the letters of the a, b, c, on which they fall. And for more explanation, lo here thy figure (Figure 35.9):[2]

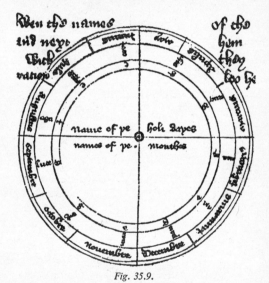

Fig. 35.9.

1. This is not right. Julius Caesar added two days to January, August and December, and one day to April, June, September and November.

2. The Festivals marked in the figure are those of St Paul (25 January), Purification (2 February), Annunciation (25 March), Invention of the Cross (3 May), St John Baptist (24 June), St James (25 July), St Lawrence (10 August)?, Nativity B.V.M. (8 September), St Luke (18 October), All Souls (2 November), Conception B.V.M. (8 December). But the scribe has put them in the wrong months.

87

12. Next to the aforesaid circle of the a, b, c, and under the crossline is marked a *scale*, like 2 measuring-rules or else like ladders, that serveth by its 12 points and its divisions for full many a subtle conclusion. Of this aforesaid scale, the part from the cross-line to the right angle, is called *umbra versa*, and the nether part is called *umbra recta*, or else *umbra extensa*.[1] And for more explanation, lo here the figure (Figure 35.10):

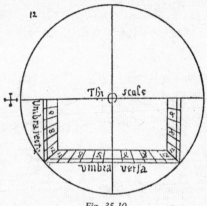

Fig. 35.10.

13. Then hast thou a broad *rule*, that hath on either end a square plate pierced with certain holes, to receive the streams of the sun by day, and also by means of thy eye, to know the altitude of stars by night. And for more explanation, lo here thy figure (Figure 35.11):

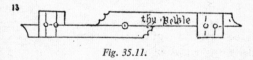

Fig. 35.11.

14. Then is there a large *pin* like an axle-tree, that goeth through the hole, and holdeth the tables of the climates and the rete in the cavity of the mother. Through this pin there goeth a little *wedge*, called the *horse*, which compresses all the parts in a heap;

1. The names are transposed in the original MS. and in the figure.

the pin which resembles an axle-tree, is imagined to be the pole arctic (north pole) in thy astrolabe. And for the more explanation, lo here thy figure (Figure 35.12):

Fig. 35.12.

15. The hollow side of thy astrolabe is also divided with a long cross into 4 *quarters* from east to west, from south to north, from right side to left side, as is the back-side. And for the more explanation, lo here thy figure (Figure 35.13):

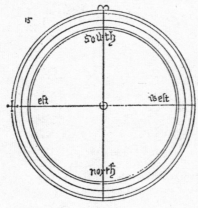

Fig. 35.13.

16. The border of the hollow-side is divided from the point of the east line to the point of the south line under the ring, into 90 *degrees*; and every quarter is divided by that same proportion. So too is the back-side divided, and that amounteth to 360 *degrees*. And understand well, that degrees of this border correspond with, and are concentric to, the degrees of the equinoctial, that is divided into the same number as is every other circle in the high heaven. This same border is also divided with 23 capital letters and a small cross + above the south line, so as to show the

24 *equal* hours of the clock; and, as I have said, 5 of these degrees make a mile-way, and 3 mile-ways make an hour. And every degree of this border contains 4 minutes of time, and every minute contains 60 seconds. Now have I told thee twice, and for more explanation, lo here the figure (Figure 35.14):

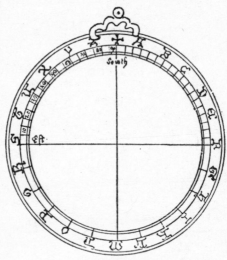

Fig. 35.14.

17. The *plate* under thy rete is marked with 3 principal circles; of which the least is called the *Circle of Cancer*, because that the head of Cancer, or the beginning of the Sign of Cancer in the rete, turneth evermore concentric upon this same circle. In this head of Cancer is the greatest declination northward of the sun. And therefore is it called the solstice of summer; which declination, according to Ptolemy is 23 degrees and 50 minutes, as well in Cancer as in Capricorn. This sign of Cancer is called the *tropic* of summer, from *tropos,* that is to say a turning, for then beginneth the sun to pass away from us; and for the more explanation, lo here the figure (Figure 35.15):

The middle circle in wideness, of these 3, is called the *Equinoctial Circle,* upon which turns evermore the heads of Aries and

Libra. And understand well, that evermore this equinoctial circle turns exactly from very east to very west; as I have shown thee in the solid sphere. This same circle is called also the weigher, *equator*, of the day, for when the sun is in the heads of Aries and Libra, then are the days and the nights equal in length in all the world. And therefore are these two signs called the *equinoxes*.

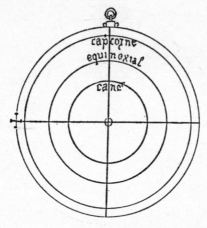

Fig. 35.15.

And all that moveth within the heads of these Aries and Libra, their moving is called northward, and all that moveth without these heads, their moving is called southward as from the equinoctial. Take heed of these latitudes north and south, and forget it not. By the equinoctial circle the 24 hours of the clock are considered; for [evermore] the arising of 15 degrees of the equinoctial maketh an equal hour of the clock. This equinoctial is called the girdle of the *first moving*, or else of the *angulus primi motus vel primi mobilis*. And note, that first moving is called 'moving' of the first movable of the 8th sphere, which motion is from east to west, and after again into east, also it is called 'girdle' of the first moving, for it divideth the first movable, that is to say, the sphere, in 2 equal parts, evenly distant from the poles of this world.

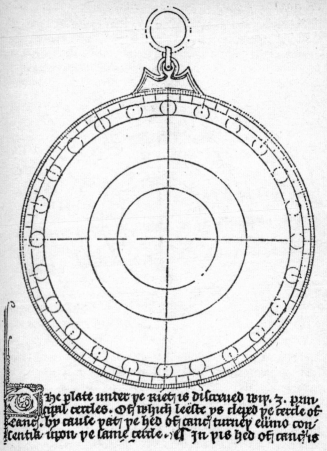

The plate under ye riet is discrived with .z. principal cercles. Of which leeste ys clepid ye cercle of cancr. by cause pat ye hed of cancr turney evmo concentrik vpon ye same cercle. & in pis hed of cancr is

Fig. 35.16. Alternative figure of the Mother of Chaucer's Astrolabe.

The widest of these 3 principal circles is called the *Circle of Capricorn*, because that the head of Capricorn turneth evermore concentric upon the same circle, in the head of Capricorn is the greatest declination southward of the sun, and therefore is it called the *solstice of winter*. This sign of Capricorn is also called the *tropic of winter* for then beginneth the sun to come again towards us. And for the more explanation, lo here thy figure (Figure 35.17):

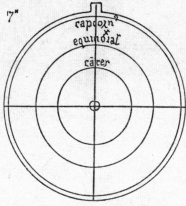

Fig. 35.17.

18. Upon this aforesaid plate are drawn certain circles [of altitude] that are called *Almicanteras*, some of which seem perfect circles, and some seem imperfect. The centre that standeth amidst the narrowest circle is called the *zenith*; and the lowest circle, or the first circle, is called the *horizon*, that is to say, the circle that divides the two hemispheres, i.e. the part of the heaven above the earth, and the part beneath. These almicanteras are compounded by 2 and 2 [or are two degrees apart], but some other astrolabes have the almicanteras divided by one degree, others by two, and others by three degrees according to the size of the astrolable. The aforesaid zenith is imagined to be the point exactly over the crown of thy head, and also the zenith is the exact pole of the horizon in every region. And for more explanation, lo here thy figure (Figure 35.18).

19. From this zenith, as it seemeth, there comes a kind of crooked lines like the claws of a spider, or else like the work of a woman's caul, crossing the almicanteras at right angles. These lines or divisions are called *azimuths*. They divide the horizon of thy astrolabe into 24 divisions. And serve to indicate the direc-

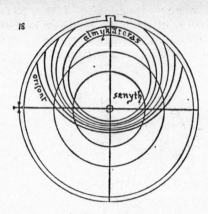

Fig. 35.18.

tions of the firmament, and to other conclusions, such as the position of the cenith[1] of the sun and of every star. And for more explanation, lo here thy figure (Figure 35.19):

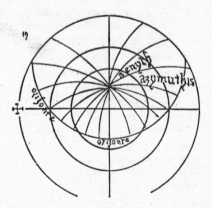

Fig. 35.19.

1. The cenith (*not* zenith) was the 'point of the horizon denoting the sun's place in azimuth' (Skeat). In the figure the 18 azimuth lines have been carelessly sketched: they should be symmetrical and 24 in number.

20. Next the azimuths, under the Circle of Cancer, there are 12 oblique divisions, much like to the shape of the azimuths; they show the spaces of the *hours of planets*. And for more explanation, lo here thy figure (Figure 35.20):

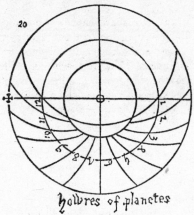

Fig. 35.20.

21. The *rete* of thy astrolabe with thy zodiac, shaped like a net or a spider's web, according to the old description, thou mayest turn up and down as thyself liketh. It contains a certain number of fixed stars, with their longitudes and latitudes properly ascertained, if the maker have not erred. The *names of the stars* are written in the margin of the rete where they are situate; and the small point of each star is called the centre. Understand also that all stars situated within the zodiac of thy astrolabe are called stars of the north, for they rise north of the east line. And all the rest of the fixed stars, out of the zodiac, are called stars of the south; but I say not that they all rise to the south of the east line; witness one, Aldebaran and Algomeisa. Understand generally this rule, that those stars that are called stars of the north rise sooner than the degree of their longitude; and all the stars of the south rise later than the degree of their longitude; that is to say, the fixed stars in thy astrolabe. The measure of this longitude of stars is taken in the *ecliptic line* of heaven, on[1] which line,

1. Chaucer wrote 'under'.

when the sun and moon are in an exact line, or else closely bordering upon it, then an eclipse of the sun or of the moon is possible, as I shall declare, and also the cause why. But truly the ecliptic line of the zodiac is the outermost border of thy zodiac, where the degrees are marked.

The *zodiac* of thy astrolabe is shaped like a circle that contains a large breadth, in proportion to the size of thy astrolabe, to signify that the zodiac in heaven is imagined to be a surface containing a latitude of 12 degrees, whereas all the rest of the circles in the heaven are imagined true lines without any latitude. Amidst this celestial zodiac is imagined a line, called the ecliptic line, on which line is evermore the way of the sun. Thus there are 6 degrees of the zodiac on one side of the line, and 6 degrees on the other. In the rete of an astrolabe the zodiac band represents the 6 degrees of the zodiac on the northern side of the ecliptic line. The zodiac is divided into 12 principal divisions, dividing the 12 signs. And, for the accuracy of thy astrolabe, every small division in a sign is divided by two degrees and two; I mean degrees containing 60 minutes. And this aforesaid heavenly zodiac is called the circle of the signs, or the circle of the beasts, for 'zodia' in the Greek language means 'beasts' in the Latin tongue. And in the zodiac are the 12 *signs* that have names of beasts; either because when the sun enters into any of the signs, he taketh the property of such beasts; or else because the stars that are fixed there are disposed in signs of beasts, or shaped like beasts; or else, when the planets are under these signs, they act upon us by their influence, operations and effects like to the operations of beasts. And understand also, that when a hot planet comes into a hot sign, then its heat increaseth; and if a planet be cold, then its coldness diminisheth, because of the hot sign. And by this conclusion thou mayest take example in all the signs, be they moist or dry, movable or fixed; reckoning the quality of the planet as I first said.

And each of these 12 signs hath respect to a certain part of the body of a man and hath it in subjection; as Aries hath thy head, and Taurus thy neck and thy throat, Gemini thy armholes and thy arms, and so forth; as shall be shown more plainly in the 5th part of this treatise. This zodiac, which is part of the 8th sphere, cuts across the Equinoctial; and crosses it again in equal parts, of which one half declineth southward, and the other northward, as the *Treatise of the Sphere* plainly declareth. And for more explanation, lo here thy figure (Figure 35.21):

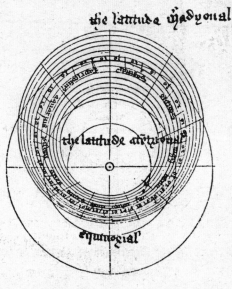

Fig. 35.21.

22. Then thou hast a *label*, that is shaped like a rule, save that it is straight and hath no plates with holes at the ends; but by the point of the aforesaid label, thou wilt calculate thy equations in the border of thy astrolabe, as by thy almury. And for more declaration, lo here thy figure (Figure 35.22):

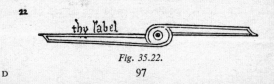

Fig. 35.22.

23. Thy *almury* is called the denticle of Capricorn or else the calculator. It is situate fixed in the head of Capricorn, and it serveth for many a necessary conclusion in equations of things, as shall be shown. And for the more declaration, lo here thy figure (Figure 35.23):

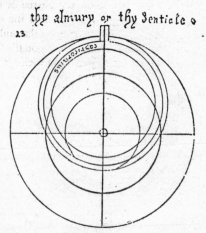

Her endith the Description of the Astrelabie.

Fig. 35.23.

PART II

Her bygynnen the Conclusions of the Astrelabie

1. To find the degree in which the sun is day by day, after her[1] course about.

Ascertain the day of thy month, and lay thy rule upon that day, then the true point of thy rule will sit in the border, on the degree of thy sun.

Example as thus: In the year of our Lord 1391, on the 12th day of March[2] at midday, I wished to know the degree of the sun. I

1. The sun was of the feminine gender in Anglo-Saxon.
2. Chaucer's dates are about 8 days behind ours, e.g. his 12 March is our 21 March.

sought in the back-half of my astrolabe, and found the circle of the days, which I know by the names of the months written under the circle. Then I laid my rule over the said day, and found the point of my rule in the border upon the first degree of Aries, a little within the degree; thus I know this conclusion.

On another day, I wanted to know the degree of my sun at midday on the 13th day of December; I found the day of the month as I have said, then I laid my rule upon the said 13th day, and found the point of my rule in the border upon the first degree of Capricorn, a little within the degree. And then I had the full experience of this conclusion; and for the more explanation, lo here thy figure (Figure 35.24):

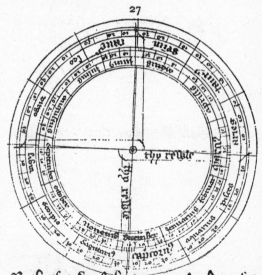

Fig. 35.24.

2. To find the altitude of the sun, or of other celestial bodies.
Put the ring of thy astrolabe upon thy right thumb and turn thy left side against the light of the sun. And move thy rule up

and down till the streams of the sun shine through both holes of thy rule. Then look to see how many degrees thy rule is raised from the little cross upon thy east line, and take there the altitude of thy sun. In this same wise thou mayest find by night the altitude of the moon, or of bright stars. This chapter is so general, that there needeth no more explanation; but forget it not. And for the more explanation, lo here thy figure (Figure 35.25):

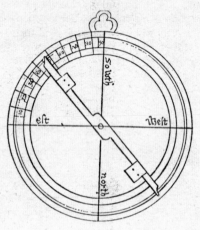

Fig. 35.25.

SUPPLEMENT

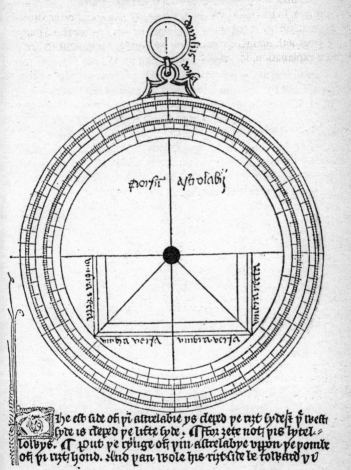

Fig. 35.26. Back of Chaucer's Astrolabe. The names of the shadow scales, *umbra*, *recta* and *umbra versa*, have been transposed in this figure.

41. *Umbra Recta*
[St John's College, Cambridge MS.]

If it so be that thou willt work by *umbra recta*, and canst get to the base of the tower, thou shalt work in this manner:

Take the altitude of the tower by both holes so that the rule lies even on a point. For example: [Suppose] I see him [the top of the tower] at the point of 4; I then measure the space between myself and the tower and find it 20 feet; then I say that as 4 is to 12, right so is the space between thee and the tower to the height of the tower. 4 is a third part of 12, so is the space between thee and the tower a third part of the height of the tower. Then thrice 20 feet is the height of the tower, with the addition of [the height of] thy person to thine eye. And this rule is general for *umbra recta* from the point of 1 to 12. And if thy rule fall upon 5, then the space between thee and the tower is five-twelfths of the height of the tower, with the addition of thine own height.

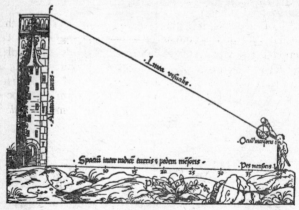

Fig. 35.27.

42. *Umbra Versa*

Another method of working is by *umbra versa*, if so be that thou canst not get to the base of the tower. I see him [the top of the tower] through the number of 1 [on the scale], and I set a mark at my foot. Then I go nearer to the tower and look at him through the point of 2, and there, I set another mark. I then consider the ratio of 1 to 12, and find that [1 goes into 12] 12 times; and I consider the ratio of 2 to 12, and thou wilt find it to be 6 times. Then thou willt find that as 12 is more than 6 by the number of 6, so is the space between the two marks the space of 6 times the height [you wish to measure]. And note that at the first altitude of 1, thou settest a mark, and again when thou sawest it at 2, thou settest another mark, and thou findest [the space] between the two marks to be 60 feet; then thou willt find that 10 is the sixth part of 60. Hence 10 feet is the height of the tower.

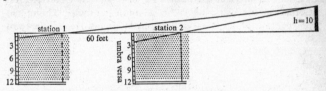

Fig. 35.28. Figure to illustrate Chaucer's method.

For other points, if it [the rule] fall in *umbra versa*, as thus: suppose it fell upon 2, and at the second [sighting] upon 3, then thou willt find that 2 is one-sixth of 12, and 3 is one-fourth of 12; then as 6 is greater than 4 by 2, so is the space between the two marks twice the height of the tower, and if the difference were thrice, then will it be 3 times; and thus thou mayest proceed from 2 to 12; and if it be 4, 4 times, or 5, 5 times and *sic de ceteris.*

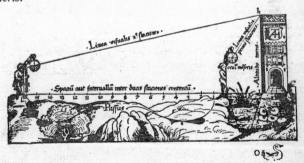

Fig. 35.29. Measuring height of inaccessible tower. (After Stoffler, 1524.)

36
ALBRECHT DÜRER
(1471–1528)

THE letters, manuscripts and printed works of Albrecht Dürer, greatest of German artists, reflect the spirit, philosophy, keen perception and theoretical interests of this true Renaissance genius. He was the third of eighteen children born to Albrecht Dürer, a goldsmith of Magyar stock, who settled in Nuremberg, a city of growing importance, to which, for decades, persons of outstanding ability in all fields had been drawn, through the cross currents in the religious, commercial and political life of Europe. Nuremberg of the fifteenth century possessed two great prerequisites for cultural development, a paper mill and numerous printing presses manned by skilled printers. In addition to the books and pamphlets issuing from her presses, the products of her master craftsmen, watchmakers, bell-founders, builders of organs, glass painters and so on, were in widespread demand. Most important of all were the goldsmiths, whose works of art, coins and seals were kept to a certain standard of purity in the precious metals of which they were fashioned. Direct commercial and banking ties connected Nuremberg to all the principal cities of Europe. In this hard-driving community, Albrecht Dürer rose from obscure and humble origins to gain acceptance at the highest level in intellectual and aristocratic circles, wholly on the strength of his extraordinary achievements.

'And when I had learned reading and writing,' Dürer wrote, 'my father took me from school and taught me the goldsmith's trade.' Nevertheless, his chief interest was in painting, and in 1486 he was apprenticed to the finest painter in Nuremberg, Michael Wolgemut, in whose studio numerous woodcuts for book illustrations were also produced. In 1490 the customary *wanderjahre* followed, during which Dürer learned the art of copper engraving in the Schongauer studio at Colmar. Recalled to Nuremberg in 1494 he married pretty Agnes Frey, fifteen years old. The marriage had previously been arranged by their parents. Nothing has been found in Dürer's writings to indicate that the marriage was a happy one. We learn that his wife kept her accounts efficiently, but portraits of her rather give an impression of petulance. Dürer's lifelong friend, Pirkheimer, wrote of her as avaricious, and the true cause of her husband's death through her incessant demands. This harsh judgement of her has been explained by some as the bitter outcry of a grief-stricken man, inconsolable in his loss of a deeply loved and vastly admired friend. As a member of the inner circle of Pirkheimer's friends, and as a friend of Lazarus Spengler, writer and leader in the local reform movement, Dürer came to enjoy the association and friendship of the leading political, religious and intellectual figures of his time. He was court painter and protégé of Emperor Maximilian I, and later he became court painter to Maximilian's successor, Charles V. Deeply religious, he was a friend and supporter of Martin Luther. Through his great friend Erasmus, Dürer made the acquaintance of many of the leading scientists and mathematicians of the world. Fragments of his correspondence with them, showing his keen interest in mathematics, survive. A prolific artist, Dürer netted a modest estate from the sale of his paintings, his engravings on wood and copper, his watercolours, his altar pieces, book illustrations and illuminations, even though, as his writings indicate, payments were frequently hard to collect.

In 1505, on the occasion of his second trip to Italy, Dürer came as a distinguished artist with an important commission. He was well received both as an artist and as a gentleman. This marked not only the emerging maturity of his art, but also the beginning of the vigorous development of his theoretical interest in measurement, perspective and proportion. Convinced that mastery of these subjects was fundamental to the improvement and advance of artistic achievement, he diligently pursued the study of these branches of mathematics, ingeniously applying the theoretical principles learned here to the practice of the artistic professions. Dürer also devoted a great deal of time to basic theoretical geometry. The culmination of many years of persistent effort in research was the publication (1525) of his first literary work, the *Unterweysung der Messung mit dem Zirkel und Richtscheyt* (' Instruction in the Art of Mensuration with Compass and Rule'). Durer's 'Art of Mensuration' is a treatise on descriptive geometry founded largely on Euclid. It contains numerous geometrical figures and unusual curves devised by Dürer, as well as his original paper-folding methods for the construction of geometrical solids. Dürer's *Etliche Unterricht zu Befestigung der Stett, Schloss und Flecken*, on the art of fortification, appeared in 1527, and his *Vier Bücher von menschlichen Proportion*, on human proportion, was published in 1528 (posthumously). A great teacher as well as an incomparable artist, Dürer wrote in German for the instruction of German youth. In accomplishing his immediate aim, the improvement of artistic methods and practices, he also contributed to the development of the German language and to the spread of scientific knowledge among the great numbers of people to whom Latin was not available. Dürer's books were immediately popular all over Europe and soon were translated into Latin, French, Italian, Portuguese and Dutch.

INSTRUCTION IN THE ART OF MENSURATION

From The Writings of Albrecht Dürer
translated by William M. Conway

The Teaching of Measurement with the rule and compass, in lines, and solids, put together and brought into print with accompanying figures by Albrecht Dürer, for the use of all lovers of art, in the year 1525.

To my very dear Master and friend, Herr Wilibald Pirkheimer, I, Albrecht Dürer, wish health and happiness.

Gracious Master and friend. Heretofore many talented scholars in our German land have been taught the art of painting, without any foundation and almost according to mere every-day rule-of-thumb. Thus they have grown up in ignorance, like a wild unpruned tree. And, though some of them have acquired a free hand by continuous practice, so that it cannot be denied that their work has been done skilfully, yet, instead of being grounded upon principle, it has merely been made according to their tastes. If, however, painters of understanding and artists worthy of the name were to see so rash a work, they would scorn the blindness of these fellows, and that not without justice. For, to one who really knows, nothing is more unpleasant to see in a picture than fundamental error, however carefully the details may be painted. That such painters have found satisfaction in their errors is only because they have not learnt the *Art of Measurement*, without which no one can either be or become a master of his craft. But that again has been the fault of their masters, who themselves were ignorant of this art.

Considering, however, that this is the true foundation for all painting, I have proposed to myself to propound the elements for the use of all eager students of Art, and to instruct them how they may employ a system of *Measurement with Rule and Compass*, and thereby learn to recognize the real Truth, seeing it before their eyes. Thus they will not only acquire a delight in and love towards art, but attain an increasingly correct understanding of it. And they will not be misled by those now amongst us who, in our own day, revile the Art of Painting and say that it is servant to Idolatory. For a Christian would no more be led to superstition by a picture or effigy than an honest man to commit murder because he carries a weapon by his side. He must indeed be an unthinking man who would worship picture, wood or stone. A picture therefore brings more good than harm, when it is honourably, artistically, and well made.

In what honour and respect these arts were held by the Greeks and Romans the old books sufficiently prove. And, although in the course of time the arts were lost, and remained lost for more than a thousand years, they were once more brought to light by the Italians, two centuries ago. For arts very quickly disappear, but only with difficulty and after a long time can they be rediscovered.

Therefore I hope that no man understanding will censure this project and teaching of mine, for it is well meant and will be useful to all who study art. It will not alone be serviceable to painters, but also to goldsmiths, sculptors, stone-masons, joiners and all who require measurements. No one indeed is obliged to avail himself of this doctrine of mine, but I am sure that whosoever does adopt it will not only thereby gain a firm grounding, but, arriving by daily practice at a better comprehension of it, will pursue the search and discover far more than I now point out.

Knowing, as I do, gracious Master and friend, that you are a lover of all the arts, and for the great affection and friendship I bear towards you, I have dedicated this book to you. Not that I think thereby to render you any great or important service, but

I hope to give some evidence and measure of my good will towards you. For though I cannot benefit you with my works, my heart has none the less been always ready to render back a return for your favours and the love you cherish towards me.

Euclid, that most acute man, put together the ground-work of Geometry. Whoever well understandeth the same hath no need of this here following writing. It has been written only for lads and such as have none to instruct them aright.

Seeing that it is useful for stone-masons, painters, and joiners to know how to set up a common sun-dial on towers, houses and walls, I will here write somewhat thereof. . . . Builders, painters, and others sometimes have to show writing on high walls, and so it is needful for them to know how to form their letters correctly.

From The Mathematics of Great Amateurs
by Julian L. Coolidge

This space curve does not lie on a cylinder, nor yet on a cone, but makes one turn around a cylinder and one around a cone, which stands on top of it. The azimuth θ shall run from 0° to 120°, the height shall be given by

$$z = a \, tan\frac{\theta}{24}.$$

For the first turn, the horizontal projection on the (X, Y) plane is $r = b$; for the second turn we have the much more complicated form

$$r = b \, \frac{\tan 30° - \tan\frac{\theta}{24}}{\tan 30° - \tan 15°}$$

The horizontal projection on the (Y, Z) plane is

$$y = r \sin \theta.$$

Figure 36.2 shows the projection of the ellipse. The projection on the (Y, Z) plane is a line segment bounded by two sloping lines. This is divided into twelve equal parts, and through the points of division vertical lines are drawn and numbered. Horizontal lines also are drawn through the points of division; the segments determined on them by the sloping lines will be the

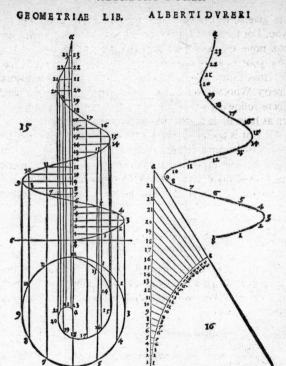

Fig. 36.1. A Dürer space curve.

diameters of the circular sections which horizontal planes cut
from the cone. We draw in the (X, Y) plane a series of circles with
these diameters, and when that is folded down to lie on the $(Y,
Z)$ plane we have a series of concentric circles just below the
figure in the (Y, Z) plane. Where each of these circles meets the
vertical line with the same number will be the folded-down projection of a point of the ellipse. We have in this way a rather un-
shapely projection of the original curve. However, Dürer does
not draw it in, but rather constructs the curve itself. The original
line segment has the length of the major axis. This we set upright
to the right of the figure, and divide into twelve equal parts.

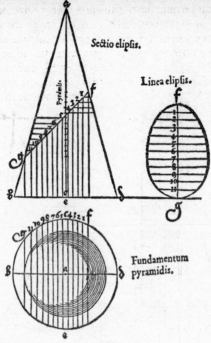

Fig. 36.2.

Through each point of division we draw a double ordinate equal to the diameter of the corresponding horizontal circle.

Figure 36.3 illustrates an even more complicated sort of twisting space-curve. This lies on three cylinders of revolution, each tangent to the next along a vertical element. The intersections with the (X, Y) plane, the vertical projection of the curve, are three tangent circles, which are treated as two spirals. . . .

The middle circle is tangent to the two others at points where $Y = 0$.

We start on the smallest circle at a point, not a point of contact, where $Y = 0$, and divide into six equal parts numbered 1 to 6; we pass then to the middle circle, continuing around in the same

110

sense of rotation and in a half-turn take twelve equal parts
numbered 7 to 18. Continuing always to turn in the same sense
we pass to the largest circle and make a complete circuit with
equal parts numbered 19 to 42, then around the other half of the
middle circle with parts numbered 43 to 54, and lastly on the
smallest circle with parts numbered 55 to 60. The space curve is
wound in this order, so that it goes around each imaginary

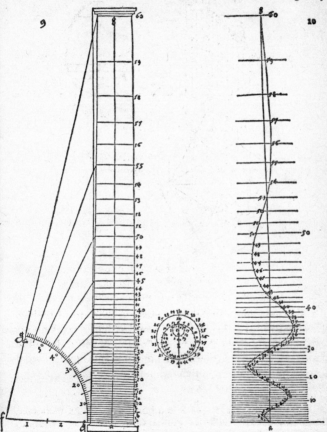

Fig. 36.3.

cylinder once. To write its equations we suppose r_1, r_2, r_3, the radii of the three circles, the azimuth θ. We also choose an arbitrary angle α which looks like 60° in the figure, and the height of the column h. If, then, the point numbered n have the azimuth θ, which will depend on whether the semicircle is divided into six parts or twelve, we have

$$y = r_i \sin \theta; \qquad z = \frac{h \tan(n\alpha/60)}{\tan \alpha}.$$

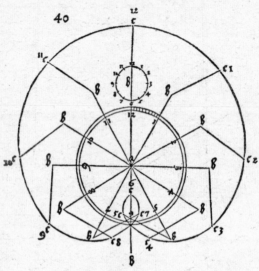

Fig. 36.4.

Dürer paid some attention to classical geometrical problems as well as to plane curves of his own devising. Here is his construction for a heart-shaped curve of the fourth order, drawn with a double ruler so constructed that one part makes twice the angle with the horizontal that the other does

$$ab = r; \qquad bc = p;$$
$$x = r \cos \theta + p \cos 2\theta; \qquad y = r \sin \theta + p \sin 2\theta;$$
$$x^2 + y^2 = r^2 + p^2 + 2rp \cos \theta;$$
$$2p(x+p) = [x^2 + y^2 - (r^2 + p^2)]\left[1 + \frac{1}{r^2}\{x^2 + y^2 - (r^2 + p^2)\} \right].$$

112

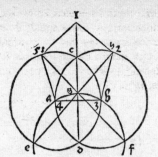

Fig. 36.5.

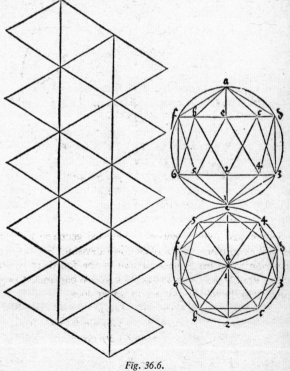

Fig. 36.6.

For the pentagon, given the side, find the circumscribed circle. α, β, δ are the centres of three equal circles. Find 2 and 5, the points where the lines from e and f to the middle point of the arc $\alpha\beta$ meet the circles about α and β. He takes 5α, $\alpha\beta$, $\beta2$ as three sides of the pentagon. This leads to the approximation

$$\sin 27° \doteq 2 \sin 60° \sin 15°,$$
$$0{\cdot}454 \doteq 0{\cdot}448.$$

An amusing construction appears at the beginning of his Book IV to construct regular solids by paper folding. This is the usual procedure taught today in our schools; so far as I can make out it is original with Dürer.[1]

I reproduce his picture for the icosahedron in Figure 36.6.

FOUR BOOKS ON HUMAN PROPORTION

From The Writings of Albrecht Dürer
translated by William M. Conway

To the honourable, highly esteemed Herr Wilibald Pirkheimer, his Imperial Majesty's Councillor, etc., my dear and gracious Master and most generous friend, I, Albrecht Dürer, present my service.

Many, I doubt not, gracious Master and friend, will find this undertaking of mine worthy of censure; that I, an untaught man of little learning and of little art, should write and teach something, which I myself have not been taught and wherein no one hath instructed me. Yet, after the repeated and, in a manner, urgent representations which you have addressed to me, that I should give this book of mine to the light, I have chosen rather to run the risk of vile calumny than to refuse your request. I hope therefore that no one, gifted with virtue and insight, will interpret it ill that I give the following to the light, freely, openly, and for the common use of all artists; for the work has been done with great pains, persistent care and labour, and with no small loss of temporal gain. I trust, on the contrary, that each and all will praise my good intention and kindly purpose, and will interpret the same in the best sense.

1. Cantor, vol. ii, p. 466.

As I feel sure that I am hereby doing a kindness to all who love Art and are eager to learn, I shall let Jealousy, which leaves nothing unreviled, take its accustomed way, returning it only this for answer, that it is very much easier to criticize a thing than to discover it for oneself. I certainly do not deny that, if the books of the Ancients, who wrote about the Art of Painting, still lay before our eyes, my design might be open to the false interpretation that I thought to find out something better than what was known unto them. These books, however, have been totally lost in the lapse of time; so I cannot be justly blamed for publishing my opinions and discoveries in writing, for that is exactly what the Ancients did. If other competent men are thereby induced to do the like, our descendants will have something which they may add to and improve upon, and thus the Art of Painting may in time advance and reach its perfection.

No one need blindly follow this theory of mine, as though it were quite perfect, for human nature has not yet so far degenerated, that another man can not discover something better. So each may use my teaching as long as it seems good to him or until he finds something better. Where he is not willing to accept it, he may well hold that this doctrine was not written for him but for others who are willing.

That must be a strangely dull head which never trusts itself to find out anything fresh, but only travels along the old path, simply following others and not daring to reflect for itself. For it beseems each Understanding, in following another, not to despair of, itself also, discovering something better. If that is done, there remaineth no doubt but that, in time, this Art will again reach the perfection it attained amongst the Ancients. For it is evident that, though the German painters are not a little skilful with the hand and in the use of colours, they have as yet been wanting in the arts of measurement, perspective and other like matters. It is therefore to be hoped that, if they learn these also and gain skill by knowledge and knowledge by skill, they will in time allow no other nation to take the prize before them.

Without proportion no figure can ever be perfect, even though it be made with all possible diligence. There is of course no need for all figures, especially quite little ones, to be constructed according to the canon, for that would involve too much labour. If, however, a man has a thorough knowledge of the canon and is well practised in the use of it, he will afterwards be able to make every figure so much the more easily, and without reference to the canon.

In order that this teaching of mine might be better understood I have already published a book about Measurements, that is to say of lines, planes, bodies and the like, without knowledge of which this my theory cannot be completely understood. It is therefore necessary for all who would try this art that they be well instructed in Measurements and know how to draw a plan and elevation of anything, in the manner in daily use amongst skilful stonemasons; otherwise they will not be able thoroughly to grasp my teaching.

No one should allow himself to be deterred from this study because he does not at once understand the whole, for what is quite easy can be no very high art, but what is full of art calls for diligence, pains and labour, before it can be understood and fixed in the memory.

If a work be incorrectly designed, however great the care and diligence spent on it, the labour is still in vain. If on the contrary it is rightly drawn, it cannot be condemned by anyone however simply it be finished.

I intend in these lessons to write only about the bounding outlines of forms and figures, and how to draw them from point to point. I shall say nothing at all about the parts within. Neither is this place to write about such matters as the antiquity of this art, its first discoverer, the respect and honour in which it was held by the Greeks and Romans, or how it should be used in the education of a good Painter or Workman. Whoever desireth to know about these matters should read Pliny and Vitruvius, where he will find enough information.

Considering, dear Sir, that I cannot make the work itself a token of my esteem and affection towards you, I have dedicated it to you, intending thereby to acknowledge the many proofs of love, friendship and goodwill, which you have given me so often and so long. I hope that hereby this book will acquire in you, noble Sir, a protector against ill reports. Confidently beseeching

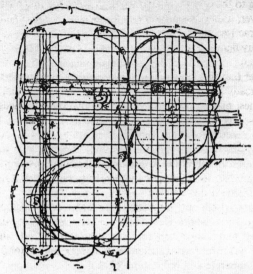

Fig. 36.7. The round-faced woman's head (B.M. MSS., vol. II, 21).

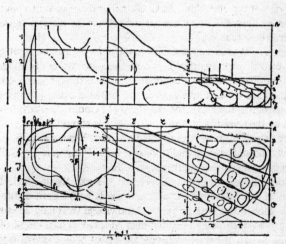

Fig. 36.8. The strong man's foot (B.M. MSS., vol. I, 187).

you to take it under your protection and ever to remain my gracious Master and helper, I continue eager to serve you whenever I may be able.

If we were to ask how we are to make a beautiful figure, some would give answer: According to human judgement (i.e. common taste). Others would not agree thereto, neither should I without a good reason. Who then will give us certainty in this matter? I believe that no man liveth who can grasp the whole beauty of the meanest living creature; I say not of a man, for he is an extraordinary creation of God, and other creatures are subject unto him. I grant, indeed, that one man will conceive and make a more beautiful figure and will explain the natural cause of its beauty more reasonably than another, but not to such an extent that there could not be anything more beautiful. For so fair a conception ariseth not in the mind of man; God alone knoweth such, and he to whom He revealeth it, he knoweth it likewise. That only, and nought else, containeth the perfect truth which is the most beautiful form and stature of a man that can be.

Men deliberate and hold numberless differing opinions about Beauty, and they seek after it in many different ways, although ugliness is thereby rather attained. Being then, as we are, in such a state of error, I know not certainly what the ultimate measure of true beauty is, and cannot describe it aright. But glad should I be to render such help as I can, to the end that the gross deformities of our work might be and remain pruned away and avoided, unless indeed anyone prefers to bestow great labour upon the production of deformities. We are brought back therefore, as aforesaid, to the judgement (or taste) of men, which considereth one figure beautiful at one time and another at another. When men demand a work of a master he is to be praised in so far as he succeeds in satisfying their likings. . . .

But it seemeth to me impossible for a man to say that he can point out the best proportions for the human figure; for the lie is in our perception, and darkness abideth so heavily about us that even our gropings fail. Howbeit if a man can prove his theory by Geometry and manifest forth its fundamental truth, him must all the world believe, for so is one compelled. And it were easy to hold such an one for endowed of God to be a master in such matters; and the demonstrations of his reasons are to be listened to with eagerness and still more gladly are his works to be beheld.

Because now we cannot altogether attain unto perfection, shall we therefore wholly cease from our learning? By no means. Let us not take unto ourselves thoughts fit for cattle. For evil and good lie before men, wherefore it behoveth a rational man to choose the good. In order therefore that we may approach unto the knowledge of how a good figure should be made, we must first order the whole figure well and nobly with all its limbs, and we must see next that every limb, regarded in itself, be made aright in all smallest as in greatest things, if so be that thus we may draw forth a part of the beauty given unto us and come so much the nearer to the perfect end. So then, as aforesaid, a man is one whole, made up of many parts, and as each of these parts hath its own proper form, so much equal care be given to all. Anything whereby they might be marred, that same must be shunned, and the true, natural character of each part must be very carefully maintained, neither must we swerve therefrom if we can help it.

. . . He therefore who, by a right understanding, hath attained a good style, hath it ever in his power to make something good, as far as that is possible to us; yet he will do so still better if he study from the life. But to make a good thing is impossible for the unpractised hand, for these things come not by chance. . . .

A man who hath not learnt anything about this art before, and who desireth to make a beginning from this book must read it with great diligence and learn to understand what he readeth; and, taking a little at a time, he must practise himself well in the same, until he can do it, and only then must he go on to do something else. For the understanding must begin to grow side by side with skill, so that the hand have power to do what the will in the understanding commands. By such means certainty of art and skill waxeth with the time; and these two must advance together, for the one is nought without the other. Further, it must be noted that though a common man knows the better from the worse, yet no man can perfectly judge a picture (in point of execution) except an understanding artist who hath often accomplished the like in his work.

Now a man might say: Who will devote continual labour and trouble, with consuming of much time, thus in tedious wise to measure out a single figure, seeing moreover that it often happeneth that he must make, it may be, twenty or thirty different figures in a short time? In answer to which, I do not mean that a man should at all times construct everything by measurements; but if thou hast well learnt the theory of measurements and attained understanding and skill in it, so that thou canst make a thing with free certainty of hand, and knowest how to do each thing aright, then it is not always needful always to measure everything, for the art which thou hast acquired giveth thee a good eye-measure, and the practised hand is obedient. And thus the power of art driveth away error from thy work and restraineth thee from making falsehood; for her thou knowest, and by thy knowledge thou art preserved from despair and become skilful in thy work so that thou makest no false touch or stroke. And this skill bringeth it to pass that thou hast no need to think, if thy head is full stored with Art. And thus thy work appeareth artistic, charming, powerful, free and good, and will receive manifold praise because rightness is infused into it.

But if thou lackest a true foundation it is impossible for thee to make aught aright and well. And although thou hadst the most skilful freedom of hand in the world, that is rather a slavery when it leads thee astray. Wherefore there must be no freedom without art, and art is lost without skill; so that, as aforesaid, the two must go together. It is therefore needful that a man learn to measure right artfully. He that can do so well maketh wonderful things. The human figure cannot be outlined with rule or compass, but must be drawn from point to point, as above explained; and without a true canon this cannot in any wise be done aright.

It might come to pass that a man, who wanted to use this above written 'canon of a figure' in some large work, might go wrong through his own want of skill, and then might lay the blame upon me, saying that my rules were right for small things but they were misleading for large works. Such, however, cannot be the case, for the small cannot be right and the large wrong, or the small bad and the large good, neither can the matter be divided in this wise. For a circle, whether small or large, abideth round, and so it is with a square. Each proportion therefore remaineth unaltered whether the scale be large or small, even as in music, a note answereth to its octave, the one high the other low, yet both are the same note.

A good figure cannot be made without industry and care; it should therefore be well considered before it is begun, so that it be correctly made. For the lines of its form cannot be traced by compass or rule, but must be drawn by the hand from point to point, so that it is easy to go wrong in them. And for such figures great attention should be paid to human proportions, and all their kinds should be investigated. I hold that the more nearly and accurately a figure is made to resemble a man, so much the better will the work be. If the best parts, chosen from many well-formed men, are united in one figure, it will be worthy of praise. But some are of another opinion and discuss how men *ought* to be made. I will not argue with them about that. I hold Nature for master in such matters and the fancy of man for delusion. The Creator fashioned men once for all as they *must* be, and I hold that the perfection of form and beauty is contained in the sum of all men. That man will I rather follow, who can extract this perfection aright, than one who invents some new body of proportions, not to be found amongst men. For the human figure must, once for all, remain different from those of other creatures, let them do with it what else they please. If I, however, were here to be attacked upon this point, namely, that I myself have set up strange proportions for figures, about that I will not argue with anyone. Nevertheless they are not really inhuman, I only exaggerate them so that all may see my meaning in them. Let anyone, who thinks that I alter the human form too much or too little, take care to avoid my error and to follow nature. There are many different kinds of men in various lands; whoso travels far will find this to be so and see it before his eyes. We are considering about the most beautiful human figure conceivable, but the Maker of the world knows how that should be. Even if we succeed well we do but approach towards it somewhat from afar. For we ourselves have differences of perception, and the vulgar who follow only their own taste usually err. Therefore I will not advise anyone to follow me, for I only do what I can, and that is not enough even to satisfy myself.

*

He now who trieth and followeth this my teaching, first giving figures a right proportion according to the canon, then arranging them orderly, laying out the outlines, giving the effect of depth by perspective, and so artistically drawing his picture or whatever it may be, will soon find out of how great service it will be unto him, and doubtless will discover much more than is here shown or handled. Notwithstanding that this my doctrine may be considered in some points difficult, it is nevertheless true; for what is hard to understand cannot be learnt without diligence and toil.

And herewith, gracious Master, I shall for this time bring my writing to an end. If God, in his own time, granteth me to write something further about matters connected with Painting I will do so in the hope that this Art may not rest upon use and wont alone, but that in time it may be taught on true and orderly principles and may be understood to the praise of God and the use and pleasure of all lovers of Art.

37

ROBERT RECORDE
(c. 1510–58)

ROBERT RECORDE, eminent physician and mathematician, founder of the English school of mathematics, lived at a time of social change, of economic expansion and of religious strife in England. An active participant in the turbulent life of his times, Recorde rose to a position of great trust and responsibility. The legacies he left behind in money and goods were very small. His scientific legacy to the English people is easily enumerated. It is less easily assessed in terms of its vast and ramified effects on the spread of education in England. A courageous man with the rare gift of loyalty and compassion, the span of his life was not without its tragedy.

Robert Recorde was born of a good family in Tenby, Pembrokeshire, Wales. The exact date of his birth is not known, but it is recorded that he entered the University of Oxford in 1525 and that he was made fellow of All Souls College in 1531. After teaching at Oxford for some time, he went to Cambridge to study medicine and there, in 1545, he received his M.D. degree. Recorde was at the height of his career in the reigns of Henry VIII, Edward VI and Mary I. It is said that he was physician to Edward VI and Queen Mary. As early as 1549, in King Edward's reign, Recorde was Comptroller of His Majesty's Mint at Bristol. In 1551 a record was made of the action of the Privy Council giving instructions to him for service as Surveyor of the King's Mines and Monies in Ireland. Recent investigations point to some irregularities or inefficiencies connected with his duties in the latter office, and we know that there were charges brought against him arising from the controversies and complaints concerning his administration of the office. It is known, too, that at the risk of incurring the displeasure of Queen Mary, should his action be found out, Recorde attended his friend Edward Underhill imprisoned in Newgate by the Queen's Council, during Underhill's illness. Recorde's work as a teacher, writer and physician was ended in 1557 by his imprisonment in the King's Bench Prison. His will, in which he described himself as 'Robert

Recorde, doctor of physicke, though sicke in body, yet whole of minde', was probated in June 1558. Such valuables as remained to him were bequeathed in part to Arthur Hilton, Under-Marshal of the King's Bench, 'Where,' Recorde stated, 'I now remaine prisoner.' Nothing has been found to indicate the reason for Recorde's imprisonment.

Robert Recorde published four mathematical text-books. Each of them was an example of his unparalleled genius for filling the needs of his times. The long sustained popularity of his texts rested on three principal bases. First, his books contained sound mathematics. Second, they were written in English which was capable of being readily understood. Third, they reflected the author's unusual capacity for presenting lucid, logical, interesting theoretical explanations and appropriate applications. His *The Grounde of Artes*, on the elements of arithmetic, gained an immediate popularity which it enjoyed undiminished through twenty-eight editions from 1542 to 1699. His *The Castle of Knowledge* (printed 1556) was the first astronomy text in English to present the Copernican theory. His *The Pathwaie to Knowledge* (printed 1551) was an excellent shortened version of Euclid's *Elements*. His *The Whetstone of Witte* (printed 1557) was the first text in English to be devoted to the principles of algebra, the 'Cossike Arte'. In *The Whetstone of Witte*, Recorde introduced the modern sign = for equality, and his method for extracting roots of algebraic expressions. The last pages of *The Whetstone* achieve, in addition, a totally unexpected dramatic value and emotional effect beginning with the knock at the door which brings the algebra lesson and the author's career to an end.

To help in understanding the extensive influence of Robert Recorde's texts, parts of his *The Whetstone of Witte* are given as printed in 1557. This original edition, exceedingly rare now, is accessible in modern libraries through the medium of microfilm. In perusing its beautiful pages of old 'black-letter' type the reader will need to be aware, for the most part, of only a few differences between Recorde's spelling and our own. Note, for example, that the letter *i* at the beginning of a word stands for the modern *j*, that although a *v* is used at the beginning of a word, it is replaced by a *u* in the interior parts of a word, that the letters *ie* are frequently used where we would use *y*, and that the long *s* of the Middle-English period is used in accordance with common practice in the sixteenth century printing.

THE PREFACE
to the gentle Reader.

Lthough nomber be in-
finite in increasyng : so that
there is not in all the worlde,
any thing that can excede the
quantitie of it : Nother the
grasse on the ground, nother
the droppes of water in the
sea, no not the small graines
of Sande through the whole
masse of the yearth: yet maie it seme by good reason,
that noe man is so experte in *Arithmetike*, that can no-
ber the commodities of it. Wherefore I maie truely
saie, that if any imperfection bee in nomber, it is bicause that nomber, can scarsely nomber, the commodities of it self. For the moare that any experte man,
doeth weigh in his mynde the benifites of it, the more
of them shall he see to remain behinde. And so shall he
well perceiue, that as nomber is infinite, so are the
commodities of it as infinite. And if any thyng doe or
maie excede the whole worlde, it is nomber, whiche
so farre surmounteth the measure of the worlde, that
if there were infinite worldes, it would at the full cō-
prehend them all. This nomber also hath other pre-
rogatiues, aboue all naturalle thynges, for neither is
there certaintie in any thyng without it, nother good
agremente where it wanteth. Whereof no man can
doubte, that hath been accustomed in the Bookes of
Plato, Aristotell, and other aunciente Philosophers,
where he shall see, how thei searche all secrete knowledge and hid misteries, by the aide of nomber.

*The excel-
lencie of
nomber.*

¶The seconde parte of Arithmetike,
containyng the extraction of Rootes in di=
verse kindes, with the Arte of Cossike
nombers, and of Surde nombes
also, in soudrie sortes.

¶The interlocutors, Master. Scholar.

The Master.

I See your desire can not
bee satisfied, neither your request staied, vntill I maie iu-
stly aunswere you, that I can
teache you no more : whiche
aunswere maie staie your request, althougi, it content not
your desire.

Scholar. I beseche God of
his mercie, to withstande all suche occasion: except it
maie be more to your owne contentation and profite,
then it would be pleasaunt to the louers of learning.

Master. Yet a iuste excuse maie stande for my de=
claration : As if ignorraunce doe inforce me to staie
my trauell.

Scholar. Your owne ignorraunce, I trust, you will
not allege: and as for the ignorraunce of other, it ought
to bee no staie : sith the ignorraunte multitude doeth,
but as it was euer wonte, enuie that knowledge,
whiche thei can not attaine, and wishe all men igno=
raunt, like vnto themself, but all gentle natures, con
temneth suche malice : and despiseth theim as blinde
wormes, whom nature doeth plague, to staie the poi=
sone of their venemous stynge.

Master. We shall not nede to stande on this talke,
but trauell with knowledge to vanquishe ignorraunce:
And beleue that the pricke of knowledge, is more of
force then the stynge of ignorraunce:

A.i.

Ombers *Coßike*, are soche
as bee contracte vnto a denomination of some *Coßike* signe
as 1. nomber. 1. roote. 1. square
1. Cube. ꝛc.

But as for compendiousnesse
in the vse of theim , there bee
certain figures set for to signifie them: so I thinke it good to
expresse vnto you those figures, before wee enter any
farther, to thintente we male procede alwaies in certentie , and knowe the thynges that wee intermedle
withall: for thei are the signes of all the arte, that foloweth here to be taught.

And although there be many kindes of irrationall
nombers, yet those figures that serue in *Coßike* nöbers,
bee the figures also of all irrrtionalle nombers, and
therfore being ones well knowen, thei serue in bothe
placcs commodiously.

These therfore be their signes , and significations
briefly touched: for their nature is partly declared before.

ꝯ.	Betokeneth nomber absolute: as if it had no signe.
℞.	Signifieth the roote of any nomber.
℥.	Represemteth a square nomber,
℥℥.	Expresseth a Cubike nomber.
℥℥ᵢ	Is the signe of a square of squares, or Zenzizenzike.
ſℬ.	Standeth for a Sursolide.
℥℞.	Doeth signifie a Zenzicubike, or a square of Cubes.
b/ℬ.	Doeth betoken a seconde Sursolide.
℥℥℥.	Doeth represent a square of squares squared

W,

127

ly, oz a zenzizenzizenzike.

æ̇ æ̇.	Significth a *Cube of Cubes.*
ȝ·ſȝ·.	Expzesseth a *Square of Surſolides.*
eſȝ·.	Betokeneth a *thirde Surſolide.*
ȝ·ȝ·æ̇.	Repzesenteth a *Square of Squared Cubes* : oz a *Zenzizenzicubike.*
Dſȝ·.	Standeth foz a *fourthe Surſolide.*
ȝ· ſbȝ·.	Is the ſigne of a *ſquare of ſeconde Surſolides*
æ̇ſȝ·.	Significth a *Cube of Surſolides.*
ȝ·ȝ·ȝ·ȝ·.	Betokeneth a *Square of ſquares, ſquaredly ſquared.*
Eſȝ·.	Is the firſte *Surſolide.*
ȝ·ȝ̇·æ̇.	Expzesseth a *ſquare of Cubike Cubes.*
Fſȝ·.	Is the ſixte *Surſolide.*
ȝ·ȝ·ſȝ·.	Doeth repzesente a *ſquare of ſquared ſur-ſolides.*
æ̇ bſȝ·.	Standeth foz a *Cube of ſeconde Surſolides.*
ȝ·eſȝ·.	Is a *ſquare of thirde Surſolides.*
gſȝ·.	Doeth betoken the *ſeuenthe Surſolide.*
ȝ·ȝ·ȝ·æ̇.	Significth a *ſquare of ſquares, of ſqua-red Cubes.*

And though I maie pzoceade infinitely in this
ſozte, yet I thinke it ſhall be a rare chaunce, that you
ſhall nede this moche: and therfoze this maie ſuffice.
Notwithſtandynge, I will anon tell you, how you
maie cotinue theſe nombers, by pzogreſſion, as farre
as you liſte.

And farther you ſhal vnderſtande, that many men
doe euer moze call ſquare nombers *zenzikes*, as a ſhoz-
ter and apter name, other men call thoſe ſquares the
firſte quantities, and the *cubes* thei call *ſeconde quantities*
ſquares of ſquares thei call *thirde quantities*, and ſurſo-
lides *fourthe quantities*. And ſo namīg thenrall quantities (excepte nombers and rootes) thei dooe adde to
them foz a difference, an ozdinall name of nomber, as
thei doe goe in ozder ſucceſſiuely.

Ʃ.ij.   Is

128

As here folowith in example.

<table>
<tr><td>℥.</td><td></td><td>First.</td></tr>
<tr><td>℥.℥.</td><td></td><td>Seconde.</td></tr>
<tr><td>℥.℥.℥.</td><td></td><td>Thirde.</td></tr>
<tr><td>ƒℨ℥.</td><td>ℨ.</td><td>Fourthe.</td></tr>
<tr><td>ℨℨ.</td><td>℥.</td><td>Fifte.</td></tr>
<tr><td>ƀℨℨ.</td><td></td><td>Sirte.</td></tr>
<tr><td>ℨℨℨ.</td><td>℥.</td><td>Seuenthe.</td></tr>
<tr><td>℥℥.</td><td>℥℥.</td><td>Eighte.</td></tr>
<tr><td>ℨℨℨ.</td><td>℥.</td><td>Nineth.</td></tr>
<tr><td>ℨℨ.</td><td></td><td>Tenthe.</td></tr>
<tr><td>ℨℨℨ.</td><td>℥.</td><td>Eleuenthe</td></tr>
<tr><td>ƀℨℨ.</td><td></td><td>Twelfthe.</td></tr>
</table>

} Quantities.

And so forthe, of as many as maie bee reckened.

But althoughe somemen accompte this the moze easie waie: bicause the other names be combcrouse, yet those other names befoze, do exprlse the qualitie of the nomber, better then these later names doe.

Scholar. I thanke you double, sith you are contente to teache me double names: foz so shall I be acquainted with bothe formes, as I shall chaunce on them in other mennes bookes.

Therfoze now you maie pzoceade to numeration: whiche I thinke it nexte.

Master. There be other .2. signes in often vse, of whiche the firste is made thus ——†—— and betokeneth moze: the other is thus made ———— and betokeneth lesse.

Ⱥ S in nombers *Abstracte*, euery nomber is not a rooted nomber , but some certaine onely emongest theim , so in nombers *Coßike*, all nombers haue not rootes: but soche onely emongest simple *Coßike* nombers are rooted , whose nomber hath a roote, agreable to the figure of his denomination.

So that. 16. ℂ. is not a Square nomber, nother hath any roote. For although. 16. bee a square nomber, and hath. 4. for his roote , yet the denomination (whiche is. ℂ.) hath noe square roote: but . 16. ℨ. is a square nomber: and hath. 4. ℨℯ ,for his roote.

Likewaies. 8. ℂ. is a *Cubike* nomber, and his roote is. 2. ℨℯ : but. 8. ℨ. hath noe roote. For bicause. 8. hath no square roote, agreable to the signe. ℨ. nother is it a *Cubike* nomber, although it haue a *Cubike* roote, bicause the roote is disagreable from the signe. ℨ.

Scholar. I perceiue that in these nombers, as wel as in all other, the roote beeyng multiplied by it self, will make the nomber, whose roote it is. And therfore can no nomber be called square, or *Cubike*, or any waies els a rooted nomber , excepte the roote of the nomber agree with his signe: Wherby I perceiue well, that. 32. ſℨ. is a rooted nomber, for bicause that 32. hath a *Surſolide* roote, agreable to the signe. So likewaies. 125. ℂ. is a rooted nomber, seyng 5. is the *Cubike* roote of. 125. But. 27. ℨ. is no rooted nöber.

Master. Thus you vnderstande sufficiently, the iudgemente of rooted nombers, and their knowlege, in simple *Coßike* nöbers, that be vtterly vncöpoude.

Wherfore, for extraction of their rootes, take this briefe order.

Extracte

Extracte the roote of your nomber, as if it were
absolute, and put to it. ℥ . for the denomination.

So.27. *Cubes* hath for his roote.3. ℥ .

And.49.℥ . hath.7.℥ . for his roote.

Again, the roote of.216. ℭℭ . is.6. ℥ .

Scholar. This I perceiue. And by like reason,
the roote of. 243./℥ . is. 3. ℥ . But why dooe you
name nobers *Cossike* vtterly vncompounde? For as I
vnderstande, that there bee nombers compounde, in
their signes, so I see that thei maie haue rootes also.

As. 16.℥℥ . hath for his roote.2.℥ . And like-
waies.64.℥ℭℭ. hath.2. ℥ . for his roote.

Master. And dooe you not see, that those compounde nombers, maie haue moare rootes then one?
Sith.16.℥℥.hath for his square roote.4.℥. as wel
as it hath.2.℥. for his *zenzizenzike* roote.

So.4.℥℥. hath for his Square roote.2. ℥. And
hath no *zenzizenzike* ℥ agreable to his whole signe.

Likewaies.9.℥ ℭℭ. hath no *zenzicubike* roote, according to his whole signe: but it hath a square roote
agreable to parte of the signe, and that is.3 ℭℭ.

Scholar. I see that also. And so hath.8.℥ ℭℭ. noe
zenzicubike roote, but a *Cubike* roote: whiche is.2.℥.

Master. Therfore in compounde signes, if the signe
maie haue soche a roote, as the nomber will yelde, it
is a rooted nomber, els not.

Whereby you maie perceiue, that if any nomber
compounde in signe, haue a roote agreable to his whole
signe, then maie it haue also, as many rootes, as ther
be partes in that compounde signe.

Cxaple xf. 529℥. ℀ —┼— 184℥.℥. —┼— 16℥. whiche is a *Square* nomber, made by multiplication of. 23. ℀ —┼— 4. ℀. by it self. This nomber maie haue his Roote oxderly extracted thus.

$$529.℥. ℀ —┼— 184℥.℥. —┼— 16℥. (23℀+4℀.$$
$$23 \qquad 46.℀.$$

In the firste nomber, J finde the *Square* roote to bee 23. And fox his denomination, J take halfe the *Coßike* signe ℥. ℀. and that is. ℀. Fox as. ℀. multiplied by ℀. doeth make. ℥. ℀. So in diuision by. 2. and in extraction of *Square* rootes, J shall take the . ℀. fox the halfe of ℥. ℀ and the denomination of his roote: and so set it doune in the *quotiente*.

Then J shall double the nomber *Abstracte* of that *quotiente* (kepyng his *Coßike* signe vnaltered) and that double shall J set euermoxe vnder the nexte nomber, toward the righte hande. As here, you see, J haue set 46(whiche is the double of 23) with his signe ℀. vnder the seconde nomber. And there J perceiue, J maie haue it. 4. tymes, if J doe diuide (as J oughte)184. by 46. And that. 4. J sette in the *quotiente*, with the signe —┼—, and the denomination.℀: seyng. ℥.℥. diuided by. ℀. doeth yelde. ℀.

Laste of all, J muste multiplie that parte of the *quotiente*. 4.℀. by it self, and it will yelde. 16.℥. whiche beyng subtracted also (as it should) leaueth nothyng remainyng of the square nomber.

This oxder must you kepe in all square nombers, how greate so euer thei be. As in this seconde exaple.

$$—90℥.℥.$$
$$253℀ —┼— 80℥℥ 26℥.℥. —— 144℀ —┼— 81℥ (5℀+8℥ —— 9℀$$
$$5.℀. \qquad 10℀ —┼— 64℥.℥.$$
$$—┼—10℀ —┼— 16℥. —— 9.℀.$$

The

The roote of the firſt nomber is.5 ℞.whiche I ſet
in a *quotiente*.

Then doe I double that.5,and it maketh.1 0,to be
ſette vnder.8.with his denomination,whiche is.℞.
like to the roote.

That.1 0.℞.maie be founde in.8 0.℥.8.times,t
therfoze I ſet.8.in the *quotiente*,with the ſigne—+—
and the denomination.℥.And then doe I multiplie
that.8 ℥.ſquaredly,whiche giueth.—+— 6 4.℥.℥.
to be ſubtracted out of———2 6.℥.℥.and ſo remaine-
neth———9 0.℥.℥.

After this I double all the *quotiente* again,where-
of commeth —+— 1 0.℞.—+— 1 6.℥. And bicauſe
there is a remainer,ouer the nomber that I wzought
laſte,I muſt ſet.1 0 ℞. vnder the remainer,and the
other nomber in ozder,as you ſee it ſet.

Then ſeke I how often tymes maie.1 0.℞.diuide
9 0.℥.℥,and I finde the *quotiente* to be———9.⅄.
And likewaies—+—1 6 ℥.multiplied by———9 ⅄.
doeth make———1 4 4.℞. equalle to the ſomme d
uer it:And ſo ſubtracteth it cleane.Wherfoze to ende
that woozke , I multiplie the laſte *quotiente* , by it ſelf
ſquare,and it yeldeth . —+— 8 1.℥. whiche is to bee
ſubtracted out of the like ſomme , in the ſquare nomber : and ſo reſteth nothyng . Wherefoze I iuſtly af-
firme,that the firſte nomber is a ſquare nomber,and
hath foz his roote.5.℞.—+— 8.℥.———9.⅄.

Scholar. That maie I ſone pzoue , if I multiplie

	5 ℞.	—+— 8 ℥.	———9.⅄.
	5 ℞.	—+— 8 ℥.	———9.⅄.
2 5 ℥ ℞.	—+— 4 0 ℟	———4 5 ℥ ℥.
	—+— 4 0 ℟.	—+— 6 4 ℥ ℥.
8 1. ℥.	———7 2 ℞.	———4 5 ℥ ℥.
	———7 2.℞.

2 5 ℥ ℞.—+— 8 0 ℟.———2 6 ℥ ℥.———1 4 4 ℞.—+— 8 1 ℟.

that roote by it self, as here I haue doen it. Wherby
I haue not onely confirmed it to be a square nomber:
but also I haue espied, that you vsed the nomber not
so plainly set doune, as the particulare multiplication did make it: but rather as a reasonable reduction
would expresse it. I meane in the $3 \cdot 3 \cdot$. where the
particulare multiplication hath —+—$64 \cdot 3 \cdot 3 \cdot$. and
———$90 3 \cdot 3 \cdot$. For whiche twoo nombers you sette
one, that resulteth of the bothe, that is ———$26 3 \cdot 3 \cdot$.

Master. But if you would take the nomber in that
sorte, the woorke would be not onely all one: but also
somewhat plainer to bee perceiued of a learner. And
therefore for your pleasure, I will set forthe here, the
example of that woorke. And loe, here it is.

$$3 \cdot \mathcal{C} + 80 \cdot \sqrt{3} + 64 3 \cdot 3 \cdot \quad 90 3 \cdot 3 \cdot \quad 144 3 \cdot + 81 3 \cdot \left(3 \mathcal{C} + 55 \cdot - 9\right)$$
$$\cdot \mathcal{C} \quad 10 \cdot \mathcal{C} + 64 3 \cdot 3 \cdot \quad 10 \cdot \mathcal{C} \cdot - + - 16.$$

Scholar. By comparynge these bothe formes of
woorke together, I doe better vnderstande, the reason of the firste woorke.

The rule of equation.

Hetherto haue I taughte you, the common formes of worke, in nombers Denominate. Which rules are vsed also in nōbers Abstracte, & likewaies in Surde nombers. Although the formes of these workes be seuerralle, in eche kinde of nomber. But now will I teache you that rule, that is the principall in Cossike woorkes: and for which all the other dooe serue.

This Rule is called the Rule of *Algeber*, after the name of the inuentoure, as some men thinke: or by a name of singular excellencie, as other iudge. But of his vse it is rightly called, the rule of *equation*: bicause that by *equation* of nombers, it doeth dissolue doubtefull questions: And vnfolde intricate ridles. And this is the order of it.

The somme of the rule of equation:

When any question is propoūded, appertemyng to this rule, you shall imagin a name for the nomber, that is to bee soughte, as you remember, that you learned in the rule of false position. And with that nomber shall you procede, accordyng to the question, vntil you finde a Cossike nomber, equalle to that nomber, that the question expresseth, whiche you shal reduce

reduce euer more to the leaste nombers. And then diuide the nomber of the lesser denomination, by the nomber of the greateste denomination, and the quotient doeth aunswere to the question Except the greater denominatió, doe beare the signe of some rooted nóber. For thenmust you extract the roote of that quotiente, accordyng to that signe of denomination.

Scholar. It semeth that this rule is all one, with the rule of false position: and therefore mighte so bee called: seyng it taketh a false nóber, to worke with al.

Maister. This rule doeth farre excell that other. And dooeth not take a false nomber, but a true nomber for his position, as it shall bee declared anon. Wherby it maie bee thoughte, to bee a rule of wonderfull inuention, that teacheth a manne at the first worde, to name a true nomber, before he knoweth resolutely, what he hath named.

But bicause that name is common to many nombers (although not in one question) and therefore the name is obscure, till the worke doe detect it, I thinke this rule might well bee called, the rule of darke position, or of straunge position: but not of false position:

And for the moze easie and apte worke in this arte wee dooe commonly name that darke position. !. —. And with it doe we worke, as the question intendeth, till we come to the equation.

Alwaies willyng you to remember, that you reduce your nombers, to their laste denominations, and smalleste formes, before you procede any farther.

And again, if your *equation* be soche, that the greateste denomination *Coßike*, be ioined to any parte of a compounde nomber, you shall tourne it so, that the nomber of the greateste signe alone, maie stande as equalle to the reste.

And this is all that neadeth to be taughte, concernyng this woorke.

Howbeit, for easie alteratiõ of *equations*. I will propounde a fewe exãples, bicause the extraction of their rootes, maie the more aptly bee wroughte. And to auoide the tediouse repetition of these woordes : is equalle to : I will sette as I doe often in woorke bse, a paire of paralleles, or Gemowe lines of one lengthe, thus: ━━━━ , bicause noe. 2. thynges, can be moare equalle. And now marke these nombers.

1. $14.\underline{z}. + .15.\%. ━━━ 71.\%.$

2. $20.\underline{z}. ━━━ .18.\%. ━━━ .102.\%.$

3. $26.\zeta. + 10\underline{z} ━━━ 9.\zeta. ━━━ 10\underline{z} + 213.\%.$

4. $19.\underline{z} + 192.\%. ━━━ 10\zeta. + 108\%. ━━━ 19\underline{z}.$

5. $18.\underline{z} + 24.\%. ━━━ 8.\zeta. + 2.\underline{z}.$

6. $34\zeta. ━━━ 12\underline{z} ━━━ 40\underline{z} + 480\%. ━━━ 9.\zeta.$

Scholar. Now I perceiue that in Addition , and Subtraction of *Surdes*, the last nombers that did result of that woozke, were *vniuersalle rootes*.

Master. You saie truthe. But harke what mea neth that hastie knockyng at the dooze?

Scholar. It is a messenger.

Master. what is the message? tel me in mine eare

Yea sir is that the mater? Then is there noe reme die, but that I must neglect all studies, and teaching, foz to withstande those daungers. My foztune is not so good, to haue quiete tyme to teache.

Scholar. But my foztune and my fellowes , is moche woze, that your vnquietnes, so hindereth our knowledge. I pzaie God amende it.

Master. I am inforzed to make an eande of this mater: But yet will I pzomise you , that whiche you shall chalenge of me, whē you see me at better laiser: That I will teache you the whole arte of *vniuersalle rootes*. And the extraction of rootes in all *Square Surdes*: with the demonstration of theim , and all the formēr woozkes.

If I mighte haue been quietly permitted , to reste but a litle whileloger, I had determined not to haue ceased, till I had ended all these thinges at large. But now

now farewell. And applie your studie diligently in this that you haue learned. And if I maie gette any quietnesse reasonable, I will not forget to performe my promise with an augementation.

Scholar. My harte is so oppressed with pēsīenes, by this sodaine vnquietnesse, that I can not expresse my grief. But I will praie, with all theim that loue honeste knowledge, that God of his mercie, will sone ende your troubles, and graunte you soche reste, as your trauell doeth merite. And al that loue lear- nyng: saie ther- to. Amen. Master. Amen, and Amen.

℥Imprinted at London, by Jhon Kyngston.

Anno domini. 1557.

38

SIMON STEVIN
(1548–1620)

THE extremely versatile mathematician, scientist and engineer, Simon Stevin, who, more than any other single individual, was responsible for the widespread adoption of the decimal system, was born in Bruges, Flanders, in 1548. A leading figure in the Dutch school of mathematics and science, and an outstanding representative of the great scholars of the closing years of the Late Renaissance, he combined a genius for theoretical investigation with extraordinary practical skill and inventiveness. He gave himself not only to writing theoretical treatises but also to the construction of mills, the improvement of sluices and locks, and to the organization of commercial enterprises whose business was based on his inventions. Numerous patents were issued to him. The invention which brought him his greatest fame during his lifetime was a 'sailing chariot' capable of carrying twenty-five people. Set on wheels, it had a steering mechanism and was equipped with sails under which it was propelled like a sailing vessel at sea. An account of a trip made by Stevin's sailing chariot made him a popular figure all over Europe.

Stevin served as the mathematical and scientific tutor of the Stadtholder, Prince Maurice of Orange. During the long and close relationship between Prince Maurice and Stevin, the prince exhibited the utmost confidence in his tutor, and Stevin's influence grew with the rising tide of the prince's fortunes. Stevin wrote text-books on all the mathematical and scientific subjects in which the prince was interested. Here, too, his inventiveness was evident, for each of his texts contained some innovation or improvement introduced by him. In a period of intermittent warfare, and of expanding economy, Stevin was sensitive to the growing need for technical instruction on the part of increasing numbers of merchants, surveyors, navigators and the like, to whom Latin was not available. Convinced that the Dutch language was excellently suited to scientific purposes, and that moreover the use of the Dutch language in science would contribute to the greatness of the nation, he wrote many of his works in Dutch, enriching the language with words of his own invention. In 1600, at the request of Prince Maurice, he directed the organization of a school of engineering at the University of Leyden, where Dutch, rather than Latin, was the language of instruction. Drawing upon his knowledge of commercial mathematics, Stevin urged that a separation be effected between governmental accounts and Prince Maurice's personal accounts, and on his advice, a system of double entry book-keeping was instituted.

Stevin held rather modest official posts throughout his life. In 1592 he was in charge of the waterways of the Delft. He served in the States Army for some years under the title of 'engineer'. In 1603, Maurice recommended him for the post of Quartermaster of the States Army, and under this appointment he was in charge of laying out military camps and of their internal organization. This was the post which he held to the end of his life.

During the years 1582–6 Stevin prepared a number of works for publication. These included his *Tables of Interest*, his *Problemata Geometria*, his *L'Arithmétique*, his *De Thiende*, his *Pratique d'Arithmetique* and four books on mechanics slanted towards mathematics. Of all these publications, *De Thiende* was by far the work of greatest importance. It contained a complete decimal system consistently applied to integers and fractions. In it Stevin demonstrated the simplicity, feasibility and advantage of the system, addressing himself to astronomers, surveyors, bankers, merchants, to any and all who dealt with measure. At the close of the treatise he added a plea for the application of the decimal system to all weights and measures and to coinage. *De Thiende*, written originally in Dutch, was published in 1585. In that same year, it was also published in French. An English translation by Robert Norton was printed in 1608 and soon thereafter numerous reprints and versions appeared, with the accompanying wide dissemination and implementation of Stevin's ideas.

DE THIENDE

From The Principal Works of Simon Stevin
Vol. II, edited by Dirk J. Struik

THE ART OF TENTHS
or
Decimal Arithmetic,
Teaching how to perform all computations
whatsoever by whole numbers without
fractions, by the four principles of
common arithmetic, namely: addition,
subtraction, multiplication, and
division.
Invented by the excellent mathematician,
SIMON STEVIN
Published in English with some additions
by
Robert Norton, Gentleman.
Imprinted at London by S.S. for Hugh
Astley, and are to be sold at his
shop at St Magnus' Corner. 1608.[1]

1. This English translation of *De Thiende* was prepared by Richard
Norton and published in 1608. The booklet contains a literal translation,
almost certainly from the French version, with some additions: (a) a short
preface 'to the courteous reader', (b) a table for the conversion of sexa-
gesimal fractions into decimal ones, and (c) a short exposition on integers,
how to write them, to perform the main species and to work with the rule
of three. This exposition is taken from Stevin's *L'Arithmétique*. In using
Norton's translation we have modernized the spelling and corrected some
misprints.

The translator, Richard Norton, was the son of the British lawyer and
poet Thomas Norton (1532–84) and a nephew of Archbishop Cranmer. The
father is remembered as the co-author of what is said to be the first English
tragedy in blank verse, *Gorboduc* (acted in 1561) and as a translator of
psalms and of Calvin's *Institutes*. The son, according to the *Dictionary of
National Biography* 41, (1895), was an engineer and gunner in the Royal
service, became engineer of the Tower of London in 1627 and died in 1635.
He wrote several texts on mathematics and artillery, supplied tables of
interest to the 1628 edition of Robert Recorde's *Grounde of Arts* and seems
to have been the author of the verses signed Ro: Norton, printed at the
beginning of Captain John Smith's *Generall historie of Virginia, New England
and the Summer Isles*, London, 1624.

THE PREFACE OF SIMON STEVIN

To Astronomers, Land-meters, Measurers of Tapestry,
Gaugers, Stereometers in general, Money-Masters,
and to all Merchants, SIMON STEVIN wishes health.

Many, seeing the smallness of this book and considering your
worthiness, to whom it is dedicated, may perchance esteem
this our conceit absurd. But if the *proportion* be considered, the
small quantity hereof compared to human imbecility, and the
great utility unto high and ingenious intendments, it will be
found to have made comparison of the extreme *terms*, which
permit not any conversion of proportion. But what of that? Is
this an admirable invention? No certainly: for it is so mean as
that it scant deserves the name of an invention, for as the country-
man by chance sometime finds a great treasure, without any use
of skill or cunning, so hath it happened herein. Therefore, if any
will think that I vaunt myself of my knowledge, because of the
explicitation of these utilities, out of doubt he shows himself to
have neither judgement, understanding nor knowledge, to discern simple things from ingenious inventions, but he (rather)
seems envious of the common benefit; yet howsoever, it were
not fit to omit the benefit hereof for the inconvenience of such
calumny. But as the mariner, having by hap found a certain
unknown island, spares not to declare to his Prince the riches and
profits thereof, as the fair fruits, precious minerals, pleasant
champion,[1] etc., and that without imputation of self-glorification,
even so shall we speak freely of the great use of this invention; I
call it great, being greater than any of you expect to come from
me. Seeing then that the *matter* of this Dime (the cause of the

---

On Norton see also E. J. R. Taylor, *The Mathematical Practitioners of
Tudor and Stuart England*. C. U. P. 1954, xi + 442 pp.

Norton calls Stevin's method both *Dime* and *The Art of Tenths* in the
title, but in the text only uses the term *Dime*.

We reproduce this translation of *De Thiende* through the courtesy of the
Houghton Library of Harvard University, Cambridge, Mass.

1. Champion, comp. French '*champagne*', field, landscape. Comp. e.g.
Deut. xi, 30, author. transl. of 1611: 'the Canaanites which dwell in the
campions'.

name whereof shall be declared by the first *definition* following)
is number, the use and effects of which yourselves shall sufficiently
witness by your continual experiences, therefore it were not
necessary to use many words thereof, for the *astrologer* knows
that the world is become by *computation astronomical* (seeing it
teaches the pilot the elevation of the *equator* and of the *pole*, by
means of the declination of the sun, to describe the true longitudes, latitudes, situations and distance of places, etc.) a paradise,
abounding in some places with such things as the earth cannot
bring forth in other. But as *the sweet is never without the sour*, so
the travail in such computations cannot be unto him hidden,
namely in the busy multiplications and divisions which proceed
of the 60th *progression* of *degrees, minutes, seconds, thirds, etc.*
And the surveyor or land-meter knows what great benefit the
world receives from his science, by which many dissensions and
difficulties are avoided which otherwise would arise by reason of
the unknown capacity of land; besides, he is not ignorant (especially whose business and employment is great) of the troublesome
multiplications of rods, feet, and oftentimes of inches, the one by
the other, which not only molests, but also (though he be very
well experienced) causes error, tending to the damage of both
parties, as also to the discredit of land-meter or surveyor, and so
for the money-masters, merchants, and each one in his business.
Therefore how much they are more worthy, and the means to
attain them the more laborious, so much the greater and better
is this *Dime*, taking away those difficulties. But how? It teaches
(to speak in a word) the easy performance of all reckonings,
computations and accounts, without broken numbers, which can
happen in man's business, in such sort as that the four principles
of arithmetic, namely addition, subtraction, multiplication and
division, by whole numbers may satisfy these effects, affording
the like facility unto those that use counters. Now if by those
means we gain the time which is precious, if hereby that be saved
which otherwise should be lost, if so the pains, controversy, error,
damage, and other inconveniences commonly happening therein
be eased, or taken away, then I leave it willingly unto your judgement to be censured; and for that, that some may say that
certain inventions at the first seem good, which when they come

to be practised effect nothing of worth, as it often happens to the searchers of strong moving,[1] which seem good in small proofs and models, when in great, or coming to the effect, they are not worth a button: whereto we answer that herein is no such doubt, the same, namely by the practice of divers expert land-meters of Holland, unto whom we have shown it, who (laying aside that which each of them had, according to his own manner, invented to lessen their pains in their computations) do use the same to their great contentment and by such fruit as the nature of it witnesses the due effect necessarily follows. The like shall also happen to each of yourselves using the same as they do. Meanwhile live in all felicity.

THE ARGUMENT

THE DIME has two parts, that is Definitions and Operations. By the first definition is declared what *Dime* is, by the second, third and fourth what *commencement, prime, second*, etc. and *dime numbers* are. The operation is declared by four propositions: the addition, subtraction, multiplication and division of dime numbers. The order whereof may be successively represented by this Table.

THE DIME *has two parts*	*Definitions, as what is*	*Dime* *Commencement,* *Prime, Second, etc.* *Dime number.*
	Operations or Practice of the	*Addition* *Subtraction,* *Multiplication,* *Division.*

And to the end the premises may the better be explained, there shall be hereunto an APPENDIX adjoined, declaring the use of

1. This is a translation of the Dutch '*roersouckers*', after Stevin's French version: '*chercheurs de fort mouvements*'. It probably stands for people who start moving things, take initiative, comp. the archaic Dutch expressions *roermaker, roerstichter* (information from Prof. Dr C. G. N. De Vooys). The Dutch has *vonden der roersouckers*, where *vonden* stands for 'findings, inventions', and the whole expression for something like 'widely proclaimed innovations'.

the Dime in many things by certain examples, and also definitions and operations, to teach such as do not already know the use and practice of numeration, and the four principles of common arithmetic in whole numbers, namely addition, subtraction, multiplication and division, together with the Golden Rule, sufficient to instruct the most ignorant in the usual practice of this art of Dime or decimal arithmetic.

THE FIRST PART
Of the Definitions of the Dimes

THE FIRST DEFINITION

Dime is a kind of arithmetic, invented by the tenth progression, consisting in character of ciphers, whereby a certain number is described and by which also all accounts which happen in human affairs are dispatched by whole numbers, without fractions or broken numbers.

Explication

Let the certain number be one thousand one hundred and eleven, described by the characters of ciphers thus 1111, in which it appears that each 1 is the 10th part of his precedent character 1; likewise in 2378 each unity of 8 is the tenth of each unity of 7, and so of all the others. But because it is convenient that the things whereof we would speak have names, and that this manner of computation is found by the consideration of such tenth or dime progression, that is that it consists therein entirely as shall hereafter appear, we call this treatise fitly by the name of *Dime*, whereby all accounts happening in the affairs of man may be wrought and effected without fractions or broken numbers, as hereafter appears.

THE SECOND DEFINITION

Every number propounded is called COMMENCEMENT, *whose sign is thus* ⓪.

146

Explication

By example, a certain number is propounded of three hundred sixty-four: we call them the 364 *commencements*, described thus 364 ⓪, and so of all other like.

THE THIRD DEFINITION

And each tenth of the unity of the COMMENCEMENT *we call the* PRIME, *whose sign is thus* ①, *and each tenth part of the unity of the prime we call the* SECOND, *whose sign is* ② *and so of the other; each tenth part of the unity of the precedent sign, always in order one further.*

Explication

As 3 ① 7 ② 5 ③ 9 ④, that is to say: 3 *primes*, 7 *seconds*, 5 *thirds*, 9 *fourths*, and so proceeding infinitely, but to speak of their value, you may note that according to this definition the said numbers are 3/10, 7/100, 5/1000, 9/10000, together 3759/10000, and likewise 8 ⓪ 9 ① 3 ② 7 ③ are worth 8, 9/10, 3/100, 7/1000, together 8-937/1000, and so of other like. Also you may understand that in this *dime* we use no fractions, and that the multitude of signs, except ⓪, never exceed 9, as for example not 7 ① 12 ②, but in their place 8 ① 2 ②, for they value as much.

THE FOURTH DEFINITION

The numbers of the second and third definitions before-going are generally called DIME NUMBERS.

The End of the Definitions

THE SECOND PART OF THE DIME
Of the Operation or Practice

THE FIRST PROPOSITION: OF ADDITION

Dime numbers being given, how to add them to find their sum.
THE EXPLICATION PROPOUNDED: There are 3 orders of dime

numbers given, of which the first 27 ⓪, 8 ①, 4 ②, 7 ③, the second 37 ⓪, 6 ①, 7 ②, 5 ③, the third 875 ⓪, 7 ①, 8 ②, 2 ③.

THE EXPLICATION REQUIRED: We must find their total sum.

CONSTRUCTION: The numbers given must be placed in order as here adjoining, adding them in the vulgar manner of adding of whole numbers in this manner. The sum (by the first problem of our French Arithmetic[1]) is 941304, which are (that which the signs above the numbers do show) 941 ⓪ 3 ① 0 ② 4 ③. I say they are the sum required.

DEMONSTRATION: The 27 ⓪ 8 ① 4 ② 7 ③ given make

⓪	①	②	③	
2	7	8	4	7
3	7	6	7	5
8	7	5	7	8	2
9	4	1	3	0	4

by the 3rd definition before 27, 8/10, 4/100, 7/1000, together 27 847/1000 and by the same reason the 37 ⓪ 6 ① 7 ② 5 ③ shall make 37 675/1000 and the 875 ⓪ 7 ① 8 ② 2 ③ will make 875 782/1000, which three numbers make by common addition of vulgar arithmetic 941 304/1000. But so much is the sum 941 ⓪ 3 ① 0 ② 4 ③; therefore it is the true sum to be demonstrated. Conclusion: The dime numbers being given to be added, we have found their sum, which is the thing required.

Note that if in the number given there want some signs of their natural order, the place of the defectant shall be filled. As, for example, let the numbers given be 8 ⓪ 5 ① 6 ② and 5 ⓪ 7 ②, in which the latter wanted the sign of ①; in the place thereof shall 0 ① be put. Take then for that latter number given 5 ⓪ 0 ① 7 ②, adding them in this sort.

⓪	①	②
8	5	6
5	0	7
1	3	6	3

This advertisement shall also serve in the three following

1. *L'Arithmétique* (1585), Work V, p. 81.

propositions, wherein the order of the defailing figures must be supplied, as was done in the former example.

THE SECOND PROPOSITION: OF SUBTRACTION

A dime number being given to subtract, another less dime number given: out of the same to find their rest.

EXPLICATION PROPOUNDED: Be the numbers given 237 ⓪ 5 ① 7 ② 8 ③ and 59 ⓪ 7 ① 3 ② 9 ③.

THE EXPLICATION REQUIRED: To find their rest.

CONSTRUCTION: The numbers given shall be placed in this sort, subtracting according to vulgar manner of subtraction of whole numbers, thus.

			⓪	①	②	③
2	3	7	5	7	8	
	5	9	7	3	9	
1	7	7	8	3	9	

The rest is 177839, which values as the signs over them do denote 177 ⓪ 8 ① 3 ② 9 ③. I affirm the same to be the rest required.

Demonstration: the 237 ⓪ 5 ① 7 ② 8 ③ make (by the third definition of this Dime) 237 5/10, 7/100, 8/1000, together 237 578/1000, and by the same reason the 59 ⓪ 7 ① 3 ② 9 ③ value 59 739/1000, which subtracted from 237 578/1000, there rests 177 839/1000, but so much doth 177 ⓪ 8 ① 3 ② 9 ③ value; that is then the true rest which should be made manifest. CONCLUSION: a dime being given, to subtract it out of another dime number, and to know the rest, which we have found.

THE THIRD PROPOSITION: OF MULTIPLICATION

A dime number being given to be multiplied, and a multiplicator given: to find their product.

THE EXPLICATION PROPOUNDED: Be the number to be multiplied 32 ⓪ 5 ① 7 ②, and the multiplicator 89 ⓪ 4 ① 6 ②.

THE EXPLICATION REQUIRED: To find the product.

CONSTRUCTION: The given numbers are to be placed as here is shown, multiplying according to the vulgar manner of multiplication by whole numbers, in this manner, giving the product 29137122. Now to know how much they value, join the two last signs together as the one ② and the other ② also, which together make ④, and say that the last sign of the product shall be ④, which being known, all the rest are also known by their continued order. So that the product required is 2913 ⓪ 7 ① 1 ② 2 ③ 2 ④.

$$\begin{array}{ccccccc} & & & & ⓪ & ① & ② \\ & & & 3 & 2 & 5 & 7 \\ & & & 8 & 9 & 4 & 6 \\ \hline & & & 1 & 9 & 5 & 4 & 2 \\ & & 1 & 3 & 0 & 2 & 8 \\ & 2 & 9 & 3 & 1 & 3 \\ 2 & 6 & 0 & 5 & 6 \\ \hline 2 & 9 & 1 & 3 & 7 & 1 & 2 & 2 \\ & & & ⓪ & ① & ② & ③ & ④ \end{array}$$

DEMONSTRATION: The number given to be multiplied, 32 ⓪ 5 ① 7 ② (as appears by the third definition of this Dime), 32, 5/10, 7/100, together 32 57/100; and by the same reason the multiplicator 89 ⓪ 4 ① 6 ② value 89 46/100 by the same, the said 32 57/100 multiplied gives the product 2913 7122/10000. But it also values 2913 ⓪ 7 ① 1 ② 2 ③ 2 ④.

It is then the true product, which we were to demonstrate. But to show why ② multiplied by ② gives the product ④, which is the sum of their numbers, also why ④ by ⑤ produces ⑨, and why ⓪ by ③ produces ③, etc., let us take 2/10 and 3/100, which (by the third definition of this Dime) are 2 ① and 3 ②, their product is 6/1000, which value by the said third definition 6 ③; multiplying then ① by ②, the product is ③, namely a sign compounded of the sum of the numbers of the signs given.

CONCLUSION: A dime number to multiply and to be multiplied being given, we have found the product, as we ought.

NOTE: If the latter sign of the number to be multiplied be unequal to the latter sign of the multiplicator, as, for example, the one 3 ④ 7 ⑤ 8 ⑥, the other 5 ① 4 ②, they shall be handled as aforesaid, and the disposition thereof shall be thus.

$$
\begin{array}{cccc}
 & ④ & ⑤ & ⑥ \\
 & 3 & 7 & 8 \\
 & & 5 & 4 \qquad ② \\
\hline
1 & 5 & 1 & 2 \\
1 & 8 & 9 & 0 \\
\hline
2 & 0 & 4 & 1 & 2 \\
④ & ⑤ & ⑥ & ⑦ & ⑧
\end{array}
$$

THE FOURTH PROPOSITION: OF DIVISION

A dime number for the dividend and divisor being given: to find the quotient.

EXPLICATION PROPOSED: Let the number for the dividend be 3 ⓪ 4 ① 4 ② 3 ③ 5 ④ 2 ⑤ and the divisor 9 ① 6 ②.

EXPLICATION REQUIRED: To find their quotient.

CONSTRUCTION: The numbers given divided (omitting the signs) according to the vulgar manner of dividing of whole numbers, gives the quotient 3587; now to know what they value, the latter sign of the divisor ② must be subtracted from the latter sign of the dividend, which is ⑤, rest ③ for the latter sign of the latter character of the quotient, which being so known, all the rest are also manifest by their continued order, thus 3 ⓪ 5 ① 8 ② 7 ③ are the quotient required.

DEMONSTRATION: The number dividend given 3 ⓪ 4 ① 4 ② 3 ③ 5 ④ 2 ⑤ makes (by the third definition of this Dime) 3, 4/10, 4/100, 3/1000, 5/10000, 2/100000, together 3 44352/100000, and by the same reason the divisor 9 ① 6 ② values 96/100, by which 3 44352/100000 being divided, gives the quotient 3 587/1000; but the said quotient values 3 ⓪ 5 ① 8 ② 7 ③, therefore it is the true quotient to be demonstrated.

CONCLUSION: A dime number being given for the dividend and divisor, we have found the quotient required.

NOTE: If the divisor's signs be higher than the signs of the dividend, there may be as many such ciphers 0 joined to the dividend as you will, or as many as shall be necessary: as for example 7 ② are to be divided by 4 ⑤, I place after the 7 certain 0, thus 7000, dividing them as aforesaid, and in this sort it gives for the quotient 1750 ⓪.

$$
\begin{array}{l}
3\ 2 \\
7\ 0\ 0\ 0 \qquad (1750\ ⓪ \\
4\ 4\ 4\ 4
\end{array}
$$

It happens also sometimes that the quotient cannot be expressed by whole numbers, as 4 ① divided by 3 ② in this sort, whereby appears that there will infinitely come

$$
\begin{array}{ll}
1\ 1\ 1\ (1 & ⓪\ ①\ ② \\
4\ 0\ 0\ 0\ 0\ 0\ 0 & (1\ \ 3\ \ 3\ \ 3 \\
3\ 3\ 3\ 3 &
\end{array}
$$

3's, and in such a case you may come so near as the thing requires, omitting the remainder. It is true, that 13 ⓪ 3 ① 3 1/3 ② or 13 ⓪ 3 ① 3 ② 3 1/3 ③ etc., shall be the perfect quotient required. But our invention in this Dime is to work all by whole numbers. For seeing that in any affairs men reckon not of the thousandth part of a mite, es, grain, etc., as the like is also used of the principal geometricians and astronomers in computations of great consequence, as Ptolemy and Johannes Montaregio,[1] have not described their tables of arcs, chords or sines in extreme perfection (as possibly they might have done by multinomial numbers), because that imperfection (considering the scope and end of those tables) is more convenient than such perfection.

NOTE 2 : The extraction of all kinds of roots may also be made by these dime numbers; as, for example, to extract the square

1. See the Introduction to *De Thiende*, esp. footnote 5, and to the *Driehouckhandel*. Johannes Montaregio, or Ian van Kuenincxberghe, Iehan de Montroial, is best known under his latinized name Iohannes Regiomontanus (1436–76). This craftsman, humanist, astronomer and mathematician of Nuremberg, born near Konigsberg in Franconia (hence his name), influenced the development of trigonometry as an independent science for more than a century by his tables and his *De Triangulis Omnimodis libri quinque* (first published in 1533). The sines, for Regiomontanus as well as for Stevin, were half chords, not ratios. On Regiomontanus see E. Zinner, *Leben und Wirken des Johannes Muller von Konigsberg genannt Regiomontanus*, Schriftenreihe zur buyr. Landesgesch. 31, Munchen, 1938, xiii + 294 pp.

root of 5 ② 2 ③ 9 ④, which is performed in the vulgar manner
of extraction in this sort, and the root shall be 2 ① 3 ②, for the
moiety or half of the latter sign of the root; wherefore, if the
latter sign given were of a number impair, the sign of the next

$$\begin{array}{c} 1 \\ 5\ 2\ 9 \\ \hline 2\quad 3 \\ \hline 4 \end{array}$$

following shall be added, and then it shall be a number pair; and
then extract the root as before. Likewise in the extraction of the
cubic root, the third part of the latter sign given shall be always
the sign of the root; and so of all other kinds of roots.

The end of the dime

THE APPENDIX

THE PREFACE

*Seeing that we have already described the Dime, we will now come
to the use thereof, showing by 6 articles how all computations which
can happen in any man's business may be easily performed thereby
beginning first to show how they are to be put in practice in the
casting up of the content or quantity of land, measured as follows.*

THE FIRST ARTICLE:
OF THE COMPUTATIONS OF LAND-METING

Call the perch or rod[1] also *commencement*, which is 1 ⓪, dividing
that into 10 equal parts, whereof each one shall be 1 ①; then
divide each prime again into 10 equal parts, each of which shall
be 1 ②; and again each of them into 10 equal parts, and each of
them shall be 1 ③, proceeding further so, if need be. But in land-
meting, divisions of seconds will be small enough. Yet for such
things as require more exactness, as roofs of lead, bodies, etc.,

---

1. The English 'perch' or 'rod' and the Dutch '*roede*' are both measures
of area and of length. For information on the precise meaning of the many
measures mentioned in Stevin's book one may consult the *O.E.D.*

there may be thirds used, and for as much as the greater number of land-meters use not the pole, but a chain line of three, four or five perch long, marking upon the yard of their cross staff[1] certain feet 5 or 6 with fingers, palms, etc. the like may be done here; for in the place of their fingers, they may put 5 or 6 *primes* with their *seconds*.

This being so prepared, these shall be used in measuring, without regarding the feet and fingers of the pole, according to the custom of the place; and that which must be added, subtracted, multiplied or divided according to this measure shall be performed according to the doctrine of the precedent examples.

As, for example, we are to add 4 triangles or surfaces of land, whereof the first 345 ⓪ 7 ① 2 ②, the second 872 ⓪ 5 ① 3 ②, the third 615 ⓪ 4 ① 8 ②, the fourth 956 ⓪ 8 ① 6 ②.

		⓪	①	②
3	4	5	7	2
8	7	2	5	3
6	1	5	4	8
9	5	6	8	6
2 7	9	0	5	9

1. The Dutch '*rechtcruys*', in Stevin's French version '*croix rectangulaire*', translated 'cross-staff', was an instrument used by surveyors for setting out perpendiculars by lines of sight, crossing each other at right angles. It was also known as 'surveyor's cross'. The cross was horizontal and supported by a pole, the yard of our text, on which Stevin wants to measure off a decimal scale. A variant of this cross was a graduated horizontal circle with a pointer (alidade) along which sighting could be performed, but even in the variations the basic rectangular cross remained.

Surveyors also used chains for measuring distances, or setting out perpendiculars, in which case they used the so-called 6, 8 and 10 rule, a popular application of Pythagoras' theorem.

The surveyor's cross is mentioned in many books on surveying. In N. Bion, *Traité de la construction et des principaux usages des instruments de mathématique*, Nouvelle edition, La Haye 1723, p. 133, we find it referred to as '*equerre d'arpenteur*', with a picture (information from Dr P. H. van Cittert).

THE SECOND ARTICLE:
OF THE COMPUTATIONS OF THE
MEASURES OF TAPESTRY OR CLOTH

The ell of the measurer of tapestry or cloth shall be to him 1 ⓪, the which he shall divide (upon the side whereon the partitions which are according to the ordinance of the town is not set out) as is done above on the pole of the land-meter, namely into 10 equal parts, whereof each shall be 1 ⓪, then each 1 ① into 10 equal parts, of which each shall be 1 ②, etc. And for the practice, seeing that these examples do altogether accord with those of the first article of land-meting, it is thereby sufficiently manifest, so as we need not here make any mention again of them.

THE THIRD ARTICLE: OF THE COMPUTATIONS
SERVING TO GAUGING, AND THE MEASURES OF
ALL LIQUOR VESSELS

One ame (which makes 100 pots Antwerp) shall be 1 ⓪, the same shall be divided in length and deepness into 10 equal parts (namely equal in respect of the wine, not of the rod; of which the parts of the depth shall be unequal), and each part shall be 1 ① containing 10 pots; then again each 1 ① into 10 parts equal as afore, and each will make 1 ② worth 1 pot; then each 1 ② into 10 equal parts, making each 1 ③.

THE FOURTH ARTICLE: OF COMPUTATIONS OF
STEREOMETRY IN GENERAL

True it is that gaugery, which we have before declared, is stereometry (that is to say, the art of measuring of bodies), but considering the divers divisions of the rod, yard or measure of the one and other, and that and this do so much differ as the genus and the species: they ought by good reason to be distinguished. For all stereometry is not gaugery. To come to the point, the stereometrian shall use the measure of the town or place, as the yard, ell, etc. with his ten partitions, as is described in the first and second articles; the use and practice thereof (as is before shown) is thus: Put case we have a quadrangular rectangular column to be measured, the length whereof is 3 ① 1 ②, the breadth 2 ① 4 ②, the height 2 ⓪ 3 ① 5 ②. The question is how much the substance or matter of that pillar is. Multiply (according to the doctrine of the 4th proposition of this Dime) the length by the breadth, and the product again by the height in this manner.

And the product appears to be 1 ① 8 ② 4 ④ 8 ⑤.

		①	②	
		3	2	
		2	4	
		1	2	8
		6	4	
		7	6	8	④
		2	3	5	②
	3	8	4	0
2	3	0	4	
1	5	3	6	
1	8	0	4	8	0
①	②	③	④	⑤	⑥

THE SIXTH ARTICLE: OF THE COMPUTATIONS OF MONEY-MASTERS, MERCHANTS, AND OF ALL ESTATES IN GENERAL

. . . The examples hereof are vulgar computations, which do almost continually happen to every man, to whom it were necessary that the solution so found were of each accepted for good and lawful. Therefore, considering the so great use, it would be a commendable thing, if some of those who expect the greatest commodity would solicit to put the same in execution to effect, namely that joining the vulgar partitions that are now in weight, measures, and moneys (continuing still each capital measure, weight, and coin in all places unaltered) that the same tenth progression might be lawfully ordained by the superiors for everyone that would use the same; it might also do well, if the values of moneys, principally the new coins, might be valued and reckoned upon certain *primes*, *seconds*, *thirds*, etc.

ALGEBRAIC GEOMETRY AND CALCULUS

39

RENÉ DESCARTES
(1596–1650)

RENÉ DESCARTES, celebrated philosopher and mathematician of the seventeenth century, was a descendant of old line lesser nobility or landed gentry on both sides of his family. The title of Seigneur du Perron which came to him through a small estate inherited from his mother was sometimes used to distinguish him from his older brother, but René appears not to have cared to be addressed by this title. From 1604 to 1612, René Descartes was a student at the newly established Jesuit College of La Flèche. A frail child, he was permitted some special privileges (such as being excused from early rising), and thus gently cared for, he proved to be an excellent student in all the usual scholastic studies, though he showed a marked preference for mathematics. A year spent at home after leaving La Flèche was followed by a short, wild year in Paris. Then, suddenly and almost in concealment, he returned to his studies. During this period, he was encouraged in his philosophical and mathematical investigations by the mathematician Mydorge and by Marin Mersenne, who had formerly been a pupil at La Flèche. He also studied at the university of Poitier where he received a degree in law in 1616. In 1618, Descartes entered the service of Prince Maurice of Orange, whose army encampment at Breda served as an officers' training ground for young men of the nobility of all the countries of Europe. Later, Descartes fought under Maximilian I, elector and duke of Bavaria. For about ten or twelve years, Descartes criss-crossed the whole of Europe, having, as he said in his *Discourse on Method*, '. . . abandoned the study of letters and resolved no longer to seek any other science than the knowledge of myself and the book of the world'. He settled at last in Holland where, in a rising tide of economic, artistic and intellectual development, civil and religious freedom and personal security were found to a degree unequalled anywhere else in the world. 'What other place could you choose in all the world,' he wrote to Balzac in 1631, 'where all the comforts of life and all the curiosities that can be desired are so easy to find as here?

What other country where you can enjoy such perfect liberty, where you can sleep with more security, where there are always armies afoot for our protection, where poisonings, treacheries, calumnies are less known, and where there has survived more of the innocence of our forefathers?'

When Descartes arrived in Holland he had already achieved an established reputation as a philosopher and scientist. The twenty-one years of fruitful meditation and writing in Holland raised him to the position of world leadership in both fields. Most of his published works were written in Holland. By means of voluminous correspondence carried on either directly, or through Mersenne and Clerselier of Paris as intermediaries, he maintained contact with all the great intellectual figures of his time. However, life in Holland was by no means a sedentary retreat for him uninterrupted by travel. Excellent Dutch roads made all parts of the country accessible to him and, attracted by the beauty of the land, he made frequent use of this mobility, writing enthusiastically of his travels. In 1631, Descartes made a swift trip to England, and in 1634, another to Denmark. He revisited France three times, in 1644, 1647 and 1648, in response to invitations from Princess Elizabeth, daughter of the Winter King, Frederick V. In the fall of 1649, he sailed to Sweden to gratify Queen Christina's ambition for the intellectual development of her country as well as her personal desire to learn his philosophy at first hand. Descartes arrived at Stockholm in October, 1649. His health had never been robust, and there he succumbed to the rigours of an exceptionally cold northern winter. He died on 11 February, 1650.

Descartes' *Discourse on Method* contained his creative theory of perception. Its importance for mathematics lies, first, in his use of mathematical method for the construction of his general methodology, and, second, in his creation of new mathematical procedures in the course of his illustration of the application of his general methodology to the field of mathematics. Three essays were appended to Descartes' *Discourse on Method*: *La Dioptrique*, *Les Météores* and *La Géométrie*. Each was intended as an illustration of the practical usefulness of Descartes' methodology in advancing scientific knowledge. In the third of the essays, *La Géométrie*, Descartes proposed to show how easily his *Method* yielded the solution of a problem considered difficult by ancient mathematicians as well as by moderns, selecting a problem for which no complete solution had heretofore been offered by anyone. In carrying out this purpose, Descartes introduced an improved mathematical notation, made advances in the theory of equations, used for the first time, the highly significant method of restating a geometrical problem in algebraic form and then solving it algebraically, and thereby earned for himself the title of founder of analytic geometry.

DISCOURSE ON METHOD

translated by John Veitch

PART I

Good sense is, of all things among men, the most equally distributed; for every one thinks himself so abundantly provided with it, that those even who are the most difficult to satisfy in everything else, do not usually desire a larger measure of this quality than they already possess. And in this it is not likely that all are mistaken: the conviction is rather to be held as testifying that the power of judging aright and of distinguishing Truth from Error, which is properly what is called Good Sense or Reason, is by nature equal in all men; and that the diversity of our opinions, consequently, does not arise from some being endowed with a larger share of Reason than others, but solely from this, that we conduct our thoughts along different ways, and do not fix our attention on the same objects. For to be possessed of a vigorous mind is not enough; the prime requisite is rightly to apply it.

From my childhood, I have been familiar with letters; and as I was given to believe that by their help a clear and certain knowledge of all that is useful in life might be acquired, I was ardently desirous of instruction. But as soon as I had finished the entire course of study, at the close of which it is customary to be admitted into the order of the learned, I completely changed my opinion. For I found myself involved in so many doubts and errors, that I was convinced I had advanced no farther in all my attempts at learning, than the discovery at every turn of my own ignorance. And yet I was studying in one of the most celebrated Schools in Europe, in which I thought there must be learned men, if such were anywhere to be found. I had been taught all that others learned there; and not contented with the sciences actually taught us, I had, in addition, read all the books that had fallen into my hands, treating of such branches as are esteemed the most curious and rare. I knew the judgement which others had formed of me; and I did not find that I was considered inferior to my fellows, although there were among them some who were already marked out to fill the places of our instructors. And, in fine, our age appeared to me as flourishing, and as fertile in powerful minds as any preceding one. I was thus led to take the liberty of judging of all other men by myself, and of concluding that there was no science in existence that was of such a nature as I had previously been given to believe.

I still continued, however, to hold in esteem the studies of the Schools. I was aware that the Languages taught in them are necessary to the understanding of the writings of the ancients; that the grace of Fable stirs the mind; that the memorable deeds of History elevate it; and, if read with discretion, aid in forming the judgement; that the perusal of all excellent books is, as it were, to interview with the noblest men of past ages, who have written them, and even a studied interview, in which are discovered to us only their choicest thoughts; that Eloquence has incomparable force and beauty; that Poesy has its ravishing graces and delights; that in the Mathematics there are many refined discoveries eminently suited to gratify the inquisitive, as well as further all the arts and lessen the labour of man; that numerous highly useful precepts and exhortations to virtue are contained in treatises on Morals; that Theology points out the path to heaven; that Philosophy affords the means of discoursing with an appearance of truth on all matters, and commands the admiration of the more simple; that Jurisprudence, Medicine, and the other Sciences, secure for their cultivators honours and riches; and, in fine, that it is useful to bestow some attention upon all, even upon those abounding the most in superstition and error, that we may be in a position to determine their real value, and guard against being deceived.

... I was especially delighted with the Mathematics, on account of the certitude and evidence of their reasonings: but I had not as yet a precise knowledge of their true use; and thinking that they but contributed to the advancement of the mechanical arts, I was astonished that foundations, so strong and solid, should have had no loftier superstructure reared on them. ...

Among the branches of Philosophy, I had, at an earlier period, given some attention to Logic, and among those of the Mathematics to Geometrical Analysis and Algebra, – three Arts or Sciences which ought, as I conceived, to contribute something to my design. But, on examination, I found that, as for Logic, its syllogisms and the majority of its other precepts are of avail rather in the communication of what we already know, or even as the Art of Lully, in speaking without judgement of things of which we are ignorant, than in the investigation of the unknown; and although this Science contains indeed a number of correct and very excellent precepts, there are, nevertheless, so many others, and these either injurious or superfluous, mingled with the former, that it is almost quite as difficult to effect a severance of the true from the false as it is| to extract a Diana or a Minerva from a rough block of marble. Then as to the Analysis of the ancients and the Algebra of the moderns, beside that they embrace only matters highly abstract, and, to appearance, of no use, the former is so exclusively restricted to the consideration of figures that it can exercise the Understanding only on condition of greatly fatiguing the Imagination; and, in the latter, there is so complete a subjection to certain rules and formulas, that there results an art full of confusion and obscurity calculated to embarrass, instead of a science fitted to cultivate the mind. By these considerations I was induced to seek some other Method which would comprise the advantages of the three and be exempt from their defects. And as a multitude of laws often only hampers justice, so that a state is best governed when, with few laws, these are rigidly administered; in like manner, instead of the great number of precepts of which Logic is composed, I believed that the four following would prove perfectly sufficient for me, provided I took the firm and unwavering resolution never in a single instance to fail in observing them.

The *first* was never to accept anything for true which I did not clearly know to be such; that is to say, carefully to avoid precipitancy and prejudice, and to comprise nothing more in my judgement than what was presented to my mind so clearly and distinctly as to exclude all ground of doubt.

The *second*, to divide each of the difficulties under examination into as many parts as possible, and as might be necessary for its adequate solution.

The *third*, to conduct my thoughts in such order that, by commencing with objects the simplest and easiest to know, I might ascend by little and little, and, as it were, step by step, to the knowledge of the more complex; assigning in thought a certain order even to those objects which in their own nature do not stand in a relation of antecedence and sequence.

At the *last*, in every case to make enumerations so complete, and reviews so general, that I might be assured that nothing was omitted.

The long chains of simple and easy reasonings by means of which geometers are accustomed to reach the conclusions of their most difficult demonstrations had led me to imagine that all things, to the knowledge of which man is competent, are mutually connected in the same way, and that there is nothing so far removed from us as to be beyond our reach, or so hidden that we cannot discover it, provided only we abstain from accepting the false for the true, and always preserve in our thoughts the order necessary for the deduction of one truth from another. And I had little difficulty in determining the objects with which it was necessary to commence, for I was already persuaded that it must be with the simplest and easiest to know, and considering that of all those who have hitherto sought truth in the Sciences, the mathematics alone have been able to find any demonstrations, that is, any certain and evident reasons, I did not doubt but that such must have been the rule of their investigations. I resolved to commence, therefore, with the examination of the simplest objects, not anticipating, however, from this any other advantage than that to be found in accustoming my mind to the love and nourishment of truth, and to a distaste for all such reasonings as were unsound. But I had no intention on that account of attempting to master all the particular Sciences commonly denominated Mathematics: but observing that, however different their objects, they all agree in considering only the various relations or proportions subsisting among those objects, I thought it best for my purpose to consider these proportions in the most general form possible, without referring them to any objects in particular, except such as would most facilitate the knowledge of them, and without by any means restricting them to these, that afterward I might thus be the better able to apply them to every other class of objects to which they are legitimately applicable. Perceiving further, that in order to understand these relations I should sometimes have to consider them one by one, and sometimes only to bear them in mind, or embrace them in the aggregate, I thought that, in order the better to consider them individually, I should view them as subsisting between straight lines, than which I could find no objects more simple, or capable of being more distinctly represented to my imagination and senses; and on the other hand, that in order to retain them in the memory, or embrace an aggregate of many, I should express them by certain characters the briefest possible. In this way I believed that I could borrow all that was best both in Geometrical Analysis and in Algebra, and correct all the defects of the one by help of the other.

And, in point of fact, the accurate observance of these few precepts gave me, I take the liberty of saying, such ease in unravelling all the questions embraced in these two sciences, that in the two or three months I devoted to their examination, not only did I reach solutions of questions I had formerly deemed exceedingly difficult, but even as regards questions of the solution of which I continued ignorant, I was enabled, as it appeared to me, to determine the means whereby, and the extent to which, a solution was possible; results attributable to the circumstance that I commenced with the simplest and most general truths, and that thus each truth discovered was a rule available in the discovery of subsequent ones. Nor in this perhaps shall I appear too vain, if it be considered that, as the truth on any particular point is one, whoever apprehends the truth, knows all that on that point can be known. The child, for example, who had been instructed in the elements of Arithmetic, and has made a particular addition, according to rule, may be assured that he had found, with respect to the sum of the numbers before him, all that in this instance is within the reach of human genius. Now, in conclusion, the Method which teaches adherence to the true order, and an exact enumeration of all the conditions of the thing sought includes all that gives certitude to the rules of Arithmetic.

But the chief ground of my satisfaction with this Method was the assurance I had of thereby exercising my reason in all matters if not with absolute perfection, at least with the greatest attainable by me: besides, I was conscious that by its use my mind was becoming gradually habituated to clearer and more distinct conceptions of its objects; and I hoped also, from not having restricted this Method to any particular matter, to apply it to the difficulties of the other Sciences, with not less success than to those of Algebra.

THE GEOMETRY OF RENÉ DESCARTES

translated by David Eugene Smith and
Marcia L. Latham

BOOK I: PROBLEMS THE CONSTRUCTION OF WHICH REQUIRES ONLY STRAIGHT LINES AND CIRCLES

Any problem in geometry can easily be reduced to such terms that a knowledge of the lengths of certain straight lines is sufficient for its construction.[1] Just as arithmetic consists of only four or five operations, namely, addition, subtraction, multiplication, division and the extraction of roots, which may be considered a kind of division, so in geometry, to find lines it is merely necessary to add or subtract other lines; or else, taking one line which I shall call unity in order to relate it as closely as possible to numbers, and which can in general be chosen arbitrarily, and having given two other lines, to find a fourth line which shall be to one of the given lines as the other is to unity (which is the same as multiplication); or, again, to find a fourth line which is to one of the given lines as unity is to the other (which is equivalent to division); or, finally, to find one, two, or several mean proportionals between unity and some other line (which is the same as extracting the

1. Large collections of problems of this nature are contained in the following works: Vincenzo Riccati and Girolamo Saladino, *Institutiones Analyticae*, Bologna, 1765; Maria Gaetana Agnesi, *Istituzioni Analitiche*, Milan, 1748; Claude Rabuel, *Commentaires sur la Géométrie de M. Descartes*, Lyons, 1730 (hereafter referred to as Rabuel); and other books of the same period or earlier.

LA
GEOMETRIE.
LIVRE PREMIER.

Des problesmes qu'on peut construire sans
y employer que des cercles & des
lignes droittes.

Ou s les Problesmes de Geometrie se
peuuent facilement reduire a tels termes,
qu'il n'est besoin par aprés que de connoi-
stre la longeur de quelques lignes droites,
pour les construire.

Et comme toute l'Arithmetique n'est composée, que
de quatre ou cinq operations, qui sont l'Addition, la
Soustraction, la Multiplication, la Diuision, & l'Extraction des racines, qu'on peut prendre pour vne espece
de Diuision : Ainsi n'at'on autre chose a faire en Geometrie touchant les lignes qu'on cherche, pour les pre-
parer a estre connuës, que leur en adiouster d'autres, ou
en oster, Oubien en ayant vne, que ie nommeray l'vnité
pour la rapporter d'autant mieux aux nombres , & qui
peut ordinairement estre prise a discretion, puis en ayant
encore deux autres, en trouuer vne quatriesme, qui soit
à l'vne de ces deux, comme l'autre est a l'vnité, ce qui est
le mesme que la Multiplication ; oubien en trouuer vne
quatriesme, qui soit a l'vne de ces deux, comme l'vnité

Commēc
le calcul
d'Ari-
thmeti-
que se
rapporte
aux ope-
rations de
Geome-
trie.

P p   est

Fig. 39.1.

167

square root, etc., of the given line).[1] And I shall not hesitate to introduce these arithmetical terms into geometry, for the sake of greater clearness.

For example, let *AB* be taken as unity, and let it be required to multiply *BD* by *BC*. I have only to join the points *A* and *C*, and draw *DE* parallel to *CA*; then *BE* is the product of *BD* and *BC*.

If it be required to divide *BE* by *BD*, I join *E* and *D*, and draw *AC* parallel to *DE*; then *BC* is the result of the division.

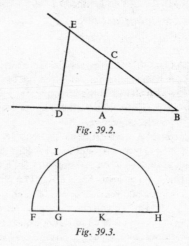

Fig. 39.2.

Fig. 39.3.

If the square root of *GH* is desired, I add, along the same straight line, *FG* equal to unity; then, bisecting *FH* at *K*, I describe the circle *FIH* about *K* as a centre, and draw from *G* a perpendicular and extend it to *I*, and *GI* is the required root. I do not speak here of cube root, or other roots, since I shall speak more conveniently of them later.

Often it is not necessary thus to draw the lines on paper, but it is sufficient to designate each by a single letter. Thus, to add the

1. While in arithmetic the only exact roots obtainable are those of perfect powers, in geometry a length can be found which will represent exactly the square root of a given line, even though this line be not commensurable with unity. Of other roots, Descartes speaks later.

lines BD and GH, I call one a and the other b, and write $a+b$. Then $a-b$ will indicate that b is subtracted from a; ab that a is multiplied by b; $\frac{a}{b}$ that a is divided by b; aa or a^2 that a is multiplied by itself; a^3 that this result is multiplied by a, and so on, indefinitely.[1] Again, if I wish to extract the square root of a^2+b^2, I write $\sqrt{a^2+b^2}$; if I wish to extract the cube root of $a^3-b^3+ab^2$, I write $\sqrt[3]{a^3-b^3+ab^2}$, and similarly for other roots.[2] Here it must be observed that by a^2, b^3, and similar expressions, I ordinarily mean only simple lines, which, however, I name squares, cubes, etc., so that I may make use of the terms employed in algebra.[3]

It should also be noted that all parts of a single line should always be expressed by the same number of dimensions, provided unity is not determined by the conditions of the problem. Thus, a^3 contains as many dimensions as ab^2 or b^3, these being the component parts of the line which I have called $\sqrt[3]{a^3-b^3+ab^2}$. It is not, however, the same thing when unity is determined, because unity can always be understood, even where there are too few dimensions; thus, if it be required to extract the cube root of a^2b^2-b, we must consider the quantity a^2b^2 divided once by unity, and the quantity b multiplied twice by unity.[4]

Finally, so that we may be sure to remember the names of these lines, a separate list should always be made as often as names are assigned or changed. For example, we may write,

---

1. Descartes uses a^3, a^4, a^5, a^6, and so on, to represent the respective powers of a, but he uses both aa and a^2 without distinction. For example, he often has $aabb$, but he also uses $\frac{3a^2}{4b^2}$.

2. Descartes writes: $\sqrt{C.a^3-b^3+abb}$.

3. At the time this was written, a^2 was commonly considered to mean the surface of a square whose side is a, and b^3 to mean the volume of a cube whose side is b; while b^4, b^5, . . . were unintelligible as geometric forms. Descartes here says that a^2 does not have this meaning, but means the line obtained by constructing a third proportional to 1 and a, and so on.

4. Descartes seems to say that each term must be of the third degree, and that therefore we must conceive of both a^2b^2 and b as reduced to the proper dimension.

$AB=1$, that is AB is equal to 1;[1] $GH=a$, $BD=b$, and so on.

If, then, we wish to solve any problem, we first suppose the solution already effected,[2] and give names to all the lines that seem needful for its construction – to those that are unknown as well as to those that are known. Then, making no distinction between known and unknown lines, we must unravel the difficulty in any way that shows most naturally the relations between these lines, until we find it possible to express a single quantity in two ways.[3] This will constitute an equation, since the terms of one of these two expressions are together equal to the terms of the other.

We must find as many such equations as there are supposed to be unknown lines; but if, after considering everything involved, so many cannot be found, it is evident that the question is not entirely determined. In such a case we may choose arbitrarily lines of known length for each unknown line to which there corresponds no equation.

If there are several equations, we must use each in order, either considering it alone or comparing it with the others, so as to obtain a value for each of the unknown lines; and so we must combine them until there remains a single unknown line[4] which is equal to some known line, or whose square, cube, fourth power, fifth power, sixth power, etc., is equal to the sum or difference of two or more quantities, one of which is known, while the others consist of mean proportionals between unity and this square, or

1. Descartes writes, $AB \propto 1$. He seems to have been the first to use this symbol. Among the few writers who followed him, was Hudde (1633–1704). It is very commonly supposed that $\propto$ is a ligature representing the first two letters (or diphthong) of 'aequare'. See, for example, M. Aubry's note in W. W. R. Ball's *Récréations mathématiques et problèmes des temps anciens et modernes*, French edition, Paris, 1909, part III, p. 164.

2. This plan, as is well known, goes back to Plato. It appears in the work of Pappus as follows: 'In analysis we suppose that which is required to be already obtained, and consider its connexions and antecedents, going back until we reach either something already known (given in the hypothesis), or else some fundamental principle (axiom or postulate) of mathematics.' *Pappi Alexandrini Collectiones quae supersunt e libris manu scriptis edidit Latina interpellatione et commentariis instruxit Fredericus Hultsch*, Berlin, 1876–8; vol. II, p. 635 (hereafter referred to as Pappus).

3. That is, we must solve the resulting simultaneous equations.

4. That is, a line represented by x, x^2, x^3, x^4, . . .

cube, or fourth power, etc., multiplied by other known lines. I may express this as follows:

$$z = b,$$
$$\text{or} \quad z^2 = -az + b^2,$$
$$\text{or} \quad z^3 = az^2 + b^2z - c^3,$$
$$\text{or} \quad z^4 = az^3 - c^3z + d^4, \text{ etc.}$$

That is, z, which I take for the unknown quantity, is equal to b; or, the square of z is equal to the square of b diminished by a multiplied by z; or, the cube of z is equal to a multiplied by the square of z, plus the square of b multiplied by z, diminished by the cube of c; and similarly for the others.

Thus, all the unknown quantities can be expressed in terms of a single quantity, whenever the problem can be constructed by means of circles and straight lines, or by conic sections, or even by some other curve of degree not greater than the third or fourth.

But I shall not stop to explain this in more detail, because I should deprive you of the pleasure of mastering it yourself, as well as of the advantage of training your mind by working over it, which is in my opinion the principal benefit to be derived from this science. Because, I find nothing here so difficult that it cannot be worked out by any one at all familiar with ordinary geometry and with algebra, who will consider carefully all that is set forth in this treatise.[1]

1. In the Introduction to the 1637 edition of *La Géométrie*, Descartes made the following remark: 'In my previous writings I have tried to make my meaning clear to everybody; but I doubt if this treatise will be read by anyone not familiar with the books on geometry, and so I have thought it superfluous to repeat demonstrations contained in them.' See *Oeuvres de Descartes*, edited by Charles Adam and Paul Tannery, Paris, 1897–1910, vol. VI, p. 368. In a letter written to Mersenne in 1637 Descartes says: 'I do not enjoy speaking in praise of myself, but since few people can understand my geometry, and since you wish me to give me your opinion of it, I think it well to say that it is all I could hope for, and that in *La Dioptrique* and *Les Météores*, I have only tried to persuade people that my method is better than the ordinary one. I have proved this in my geometry, for in the beginning I have solved a question which, according to Pappus, could not be solved by any of the ancient geometers.

'Moreover, what I have given in the second book on the nature and properties of curved lines, and the method of examining them, is, it seems

BOOK II: ON THE NATURE OF CURVED LINES

The ancients were familiar with the fact that the problems of geometry may be divided into three classes, namely, plane, solid and linear problems. This is equivalent to saying that some problems require only circles and straight lines for their construction, while others require a conic section and still others require more complex curves. I am surprised, however, that they did not go further, and distinguish between different degrees of these more complex curves, nor do I see why they called the latter mechanical, rather than geometrical.

*

In their treatment of the conic sections they did not hesitate to introduce the assumption that any given cone can be cut by a given plane. Now to treat all the curves which I mean to introduce here, only one additional assumption is necessary, namely, two or more lines can be moved, one upon the other, determining by

---

to me, as far beyond the treatment in the ordinary geometry, as the rhetoric of Cicero is beyond the a, b, c of children. . . .

'As to the suggestion that what I have written could easily have been got from Vieta, the very fact that my treatise is hard to understand is due to my attempt to put nothing in it that I believed to be known either by him or by any one else. . . . I begin the rules of my algebra with what Vieta wrote at the very end of his book, *De Emendatione Aequationum*. . . . Thus, I begin where he left off.' *Oeuvres de Descartes, publiées par Victor Cousin*, Paris, 1824, vol. VI, p. 294 (hereafter referred to as Cousin).

In another letter to Mersenne, written 20 April, 1646, Descartes writes as follows: 'I have omitted a number of things that might have made it (the geometry) clearer, but I did this intentionally, and would not have it otherwise. The only suggestions that have been made concerning changes in it are in regard to rendering it clearer to readers, but most of these are so malicious that I am completely disgusted with them.' Cousin, vol. IX, p. 553.

In a letter to the Princess Elizabeth, Descartes says: 'In the solution of a geometrical problem I take care, as far as possible, to use as lines of reference parallel lines or lines at right angles; and I use no theorems except those which assert that the sides of similar triangles are proportional, and that in a right triangle the square of the hypotenuse is equal to the sum of the squares of the sides. I do not hesitate to introduce several unknown quantities, so as to reduce the question to such terms that it shall depend only on these two theorems.' Cousin, vol. IX, p. 143.

their intersection other curves. This seems to me in no way more difficult.

It is true that the conic sections were never freely received into ancient geometry, and I do not care to undertake to change names confirmed by usage; nevertheless, it seems very clear to me that if we make the usual assumption that geometry is precise and exact while mechanics is not; and if we think of geometry as the science which furnishes a general knowledge of the measurement of all bodies, then we have no more right to exclude the more complex curves than the simpler ones, provided they can be conceived of as described by a continuous motion or by several successive motions, each motion being completely determined by those which precede; for in this way an exact knowledge of the magnitude of each is always obtainable.

*

I think the best way to group together all such curves and then classify them in order, is by recognizing the fact that all points of those curves which we may call 'geometric', that is, those which admit of precise and exact measurement, must bear a definite

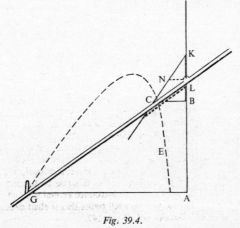

Fig. 39.4.

relation[1] to all points of a straight line, and that this relation must be expressed by means of a single equation.[2]

Suppose the curve *EC* to be described by the intersection of the ruler *GL* and the rectilinear plane figure *CNKL*, whose side *KN* is produced indefinitely in the direction of *C*, and which, being moved in the same plane in such a way that its side[3] *KL* always coincides with some part of the line *BA* (produced in both directions), imparts to the ruler *GL* a rotary motion about *G* (the ruler being hinged to the figure *CNKL* at *L*).[4] If I wish to find out to what class this curve belongs, I choose a straight line, as *AB*, to which to refer all its points, and in *AB* I choose a point *A* at which to begin the investigation.[5] I say 'choose this and that', because we are free to choose what we will, for while it is necessary to use care in the choice in order to make the equation as short and simple as possible, yet no matter what line I should take instead of *AB* the curve would always prove to be of the same class, a fact easily demonstrated.[6]

Then I take on the curve an arbitrary point, as *C*, at which we will suppose the instrument applied to describe the curve. Then I draw through *C* the line *CB* parallel to *GA*. Since *CB* and *BA* are unknown and indeterminate quantities, I shall call one of them *y* and the other *x*. To the relation between these quantities I must consider also the known quantities which determine the description of the curve, as *GA*, which I shall call *a*; *KL*, which

1. That is, a relation exactly known, as, for example, that between two straight lines in distinction to that between a straight line and a curve, unless the length of the curve is known.

2. It will be recognized at once that this statement contains the fundamental concept of analytic geometry.

3. '*Diamètre*'.

4. The instrument thus consists of three parts, (1) a ruler *AK* of indefinite length, fixed in a plane; (2) a ruler *GL*, also of indefinite length, fastened to a pivot, in the same plane, but not on *AK*; and (3) a rectilinear figure *BKC*, the side *KC* being indefinitely long, to which the ruler *GL* is hinged at *L*, and which is made to slide along the ruler *GL*.

5. That is, Descartes uses the point *A* as origin, and the line *AB* as axis of abscissas. He uses parallel ordinates, but does not draw the axis of ordinates.

6. That is, the nature of a curve is not affected by a transformation of coordinates.

I shall call b; and NL parallel to GA, which I shall call c. Then I say that as NL is to LK, or as c is to b, so CB or y, is to BK, which is therefore equal to $\frac{b}{c}y$. Then BL is equal to $\frac{b}{c}y-b$, and AL is equal to $x+\frac{b}{c}y-b$. Moreover, as CB is to LB, that is, as y is to $\frac{b}{c}y-b$, so AG or a is to LA or $x+\frac{b}{c}y-b$. Multiplying the second by the third, we get $\frac{ab}{c}y-ab$ equal to $xy+\frac{b}{c}y^2-by$, which is obtained by multiplying the first by the last. Therefore, the required equation is

$$y^2 = cy - \frac{cx}{b}y + ay - ac.$$

From this equation we see that the curve EC belongs to the first class, it being, in fact, a hyperbola.[1]

If in the instrument used to describe the curve we substitute for the rectilinear figure CNK this hyperbola or some other curve of the first class lying in the plane $CNKL$, the intersection of this curve with the ruler GL will describe, instead of the hyperbola EC, another curve, which will be of the second class.

Thus, if CNK be a circle having its centre at L, we shall describe the first conchoid of the ancients, while if we use a parabola having KB as axis we shall describe the curve which, as I have already said, is the first and simplest of the curves required in the problem of Pappus, that is, the one which furnishes the solution when five lines are given in position.

BOOK III: ON THE CONSTRUCTION OF SOLID AND SUPER SOLID PROBLEMS

Every equation can have[2] as many distinct roots (values of the unknown quantity) as the number of dimensions of the unknown

1. Cf. Briot and Bouquet, *Elements of Analytical Geometry of Two Dimensions*, trans. by J. H. Boyd, New York, 1896, p. 143.

The two branches of the curve are determined by the position of the triangle *CNKL* with respect to the directrix *AB*. See Rabuel, p. 119.

2. It is worthy of note that Descartes writes 'can have' ('*peut-il y avoir*'), not 'must have', since he is considering only real positive roots.

quantity in the equation.[1] Suppose, for example, $x=2$ or $x-2=0$, and again, $x=3$, or $x-3=0$. Multiplying together the two equations $x-2=0$ and $x-3=0$, we have $x^2-5x+6=0$, or $x^2=5x-6$. This is an equation in which x has the value 2 and at the same time[2] x has the value 3. If we next make $x-4=0$ and multiply this by $x^2-5x+6=0$, we have $x^3-9x^2+26x-24=0$ another equation, in which x, having three dimensions, has also three values, namely, 2, 3, and 4.

It often happens, however, that some of the roots are false[3] or less than nothing. Thus, if we suppose x to represent the defect[4] of a quantity 5, we have $x+5=0$ which multiplied by $x^3-9x^2+26x-24=0$, yields $x^4-4x^3-19x^2+106x-120=0$, an equation having four roots, namely three true roots, 2, 3, and 4, and one false root, 5.[5]

It is evident from the above that the sum[6] of an equation having several roots is always divisible by a binomial consisting of the unknown quantity diminished by the value of one of the true roots, or plus the value of one of the false roots. In this way,[7] the degree of an equation can be lowered.

On the other hand, if the sum of the terms of an equation[8] is not divisible by a binomial consisting of the unknown quantity plus or minus some other quantity, then this latter quantity is not a root of the equation. Thus the[9] above equation $x^4-4x^3-19x^2+$

1. That is, as the number denoting the degree of the equation.

2. '*Tout ensemble*' – not quite the modern idea.

3. '*Racines fausses*', a term formerly used for 'negative roots'. Fibonacci, for example, does not admit negative quantities as roots of an equation. *Scritti de Leonardo Pisano*, published by Boncompagni, Rome, 1857. Cardan recognizes them, but calls them '*aestimationes falsae*' or '*fictae*', and attaches no special significance to them. See Cardan, *Ars Magna*, Nurnberg, 1545, p. 2. Stifel called them '*Numeri absurdi*', as also in Rudolff's Coss, 1545.

4. '*Le défaut*'. If $x=-5$, -5 is the 'defect' of 5, that is, the remainder when 5 is subtracted from zero.

5. That is, three positive roots, 2, 3 and 4, and one negative root, -5.

6. '*Somme*', the left member when the right member is zero; that is, what we represent by $f(x)$ in the equation $f(x)=0$.

7. That is, by performing the division.

8. '*Si la somme d'un équation.*'

9. First member of the equation. Descartes always speaks of dividing the equation.

$106x - 120 = 0$ is divisible by $x - 2$, $x - 3$, $x - 4$ and $x + 5$,[1] but is not divisible by x plus or minus any other quantity. Therefore the equation can have only the four roots, 2, 3, 4, and 5.[2] We can determine also the number of true and false roots that any equation can have, as follows:[3] An equation can have as many true roots as it contains changes of sign, from + to − or from − to +; and as many false roots as the number of times two + signs or two − signs are found in succession.

Thus, in the last equation, since $+ x^4$ is followed by $-4x^3$, giving a change of sign from + to −, and $-19x^2$ is followed by $+106x$ and $+106x$ by -120, giving two more changes, we know there are three true roots; and since $-4x^3$ is followed by $-19x^2$ there is one false root.

It is also easy to transform an equation so that all the roots that were false shall become true roots, and all those that were true shall become false. This is done by changing the signs of the second, fourth, sixth, and all even terms, leaving unchanged the signs of the first, third, fifth, and other odd terms. Thus, if instead of

$$+x^4 - 4x^3 - 19x^2 + 106x - 120 = 0$$

we write

$$+x^4 + 4x^3 - 19x^2 - 106x - 720 = 0$$

we get an equation having one true root, 5, and three false roots, 2, 3, and 4.[4]

---

1. Incorrectly given as $x - 5$ in some editions.

2. Where 5 would now be written -5. Descartes neither states nor explicitly assumes the fundamental theorem of algebra, namely, that every equation has at least one root.

3. This is the well known 'Descartes's Rule of Signs'. It was known, however, before his time, for Harriot had given it in his *Artis Analyticae Praxis*, London, 1631. Cantor says Descartes may have learned it from Cardan's writings, but was the first to state it as a general rule. See Cantor, vol. II(1), pp. 496 and 725.

4. In absolute value.

40

ISAAC BARROW
(1630–77)

ISAAC BARROW, descended from an ancient Suffolk family, was born in London in October 1630. One of England's most brilliant and versatile scholars, he would have achieved lasting fame for his contribution to any single one of the fields of his interest, the classics, mathematics, science or theology. Reports of his behaviour at his first school, Charterhouse, in London, however, gave surprisingly little promise of the greatness to which he was destined. Here, it was said, he was given to fighting and even to promoting fighting among his fellow pupils. A transfer to Felstead, in Essex, provided a salutary change of environment and Barrow immediately proved to be an excellent student in all areas of work. He entered Trinity College, Cambridge, in 1644; he received the degree of B.A. in 1648, and was elected fellow of his College the following year.

In 1655, although he was eminently qualified for a professorship in Greek, his application for the position was denied because of a suspicion of sympathy with Arminianism. Deeply disappointed, he sold all his books and left England. The next four years were spent in travel, at times highly adventurous, over eastern Europe. His tour of the Continent was made feasible and successful to a large extent by the kind reception and assistance given him by the English ambassadors and others of his countrymen in the regions he visited. A small lean man of unusual strength and courage, Barrow was widely admired and respected for his learning and lively companionship.

Barrow returned to England in 1659. He was ordained by Bishop Brownrig the next year, and soon thereafter, in June 1660 (the year that Newton entered Trinity College), he was chosen for the Greek professorship at Cambridge. In 1662 he was elected professor of geometry in Gresham College. He also served as a substitute for Dr Pope, a professor of astronomy. In 1663, he became the first Lucasian professor of mathematics at Cambridge. In 1669, desiring to devote himself completely to the study of theology, Barrow resigned the Lucasian chair to his great pupil and friend Isaac Newton. Newton had already seen to the preparation of Barrow's *Lectiones Opticae* for publication (1669). In 1670, Newton also supervised the publication of Barrow's *Lectiones Geometricae*. During this year, too, Barrow became chaplain to Charles II. The king openly admired the outspoken Barrow as a scholar and a divine whose brilliant conversation enlivened and elevated the English court. Barrow was appointed Master of Trinity in 1672, and in 1675 he was chosen vice-chancellor of the university. He died suddenly of a fever on 4 May, 1677 and was interred in Westminster Abbey.

Barrow was in the first group of scientists to be elected to membership in the Royal Society. A prolific writer on theology, mathematics and poetry, his Latin and English translations of Euclid, Archimedes, Apollonius and Theodosius were widely read both in England and on the continent. Most of his contributions to mathematics appeared in the three published collections of his lectures, *Lectiones Mathematicae* (1683, posthumously), *Lectiones Opticae*, and *Lectiones Geometricae*. The last mentioned work contains the foundations of the calculus in geometrical form. It presents differentiation and integration as inverse processes, integration as a summation, and nomenclature and methods which were direct forerunners of the algorithmic procedures of the calculus. Lecture X, in particular, which contains Barrow's presentation of the differential triangle, clearly indicates the mutual influence of Barrow and Newton upon each other. Barrow's influence upon Leibniz, too, may be inferred from the fact that Leibniz is known to have purchased a copy of Barrow's *Lectiones Geometricae* in 1673. The *Lectiones Geometricae*, the greatest of Barrow's mathematical works, was translated into English from the Latin in 1916 by James Mark Child under the title, *The Geometrical Lectures of Isaac Barrow*.

THE GEOMETRICAL LECTURES OF ISAAC BARROW

translated, with notes and proofs, by James Mark Child

LECTURE X

Rigorous determination of ds/dx. Differentiation as the inverse of integration. Explanation of the 'Differential Triangle' method; with examples. Differentiation of a trigonometrical function.

1. Let *AEG* be any curve whatever, and *AFI* another curve so related to it that, if any straight line *EF* is drawn parallel to a straight line given in position (which cuts *AEG* in *E* and *AFI* in *F*), *EF* is always equal to the arc *AE* of the curve *AEG*, measured from *A*; also let the straight line *ET* touch the curve *AEG* at *E*, and let *ET* be equal to the arc *AE*; join *TF*; then *TF* touches the curve *AFI*.

2. Moreover, if the straight line *EF* always bears any the same ratio to the arc *AE*, in just the same way *FT* can be shown to touch the curve *AFI*.[1]

3. Let *AGE* be any curve, *D* a fixed point, and *AIF* be another curve such that, if any straight line *DEF* is drawn through *D*, the intercept *EF* is always equal to the arc *AE*; and let the straight line *ET* touch the curve *AGE*; make *TE* equal to the arc *AE*;[2] let

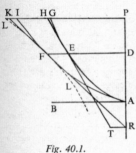

TKF be a curve such that, if any straight line *DHK* is drawn through *D*, cutting the curve *TKF* in *K* and the straight line *TE* in *H*, *HK* = *HT*; then let *FS* be drawn[3] to touch *TKF* at *F*; *FS* will touch the curve *AIF* also.

4. Moreover, if the straight line *EF* is given to bear any the same ratio to the arc *AE*, the tangent to it can easily be found from the above and Lect. VIII, § 8.

Fig. 40.1.

5. Let a straight line *AP* and two curves *AEG*, *AFI* be so related that, if any straight line *DEF* is drawn (parallel to *AB*, a straight line given in position), cutting *AP*, *AEG*, *AFI*, in the points *D*, *E*, *F* respectively *DF* is always equal to the arc *AE*; also let *ET* touch the curve *AEG* at *E*; take *TE* equal to the arc *AE*; and draw *TR* parallel to *AB* to cut *AP* in *R*; then, if *RF* is joined, *RF* touches the curve *AFI* (Figure 40.1).

1. Since the arc is a function of the ordinate, this is a special case of the differentiation of a sum, Lect. IX, §12; it is equivalent to $d(as+y)/dx = a \cdot ds/dx + dy/dx$; see note to §5.
2. *TE*, *AE* are drawn in the same sense.
3. By Lect. VIII, §19.

For, assume that *LFL* is a curve such that, if any straight line *PL* is drawn parallel to *AB*, cutting *AEG* in *G*, *TE* in *H*, and *LFL* in *L*, the straight line *PL* is always equal to *TH* and *HG* taken together. Then *PL* > arc *AEG* > *PI*; and therefore the curve *LFL* touches the curve *AFI*. Again, by Lect. VI, § 26, *PK* = *TH* (or *KL* = *GH*); hence the curve *LFL* touches the line *RFK* (by Lect. VII, § 3); therefore the line *RFK* touches the curve *AFI*.[1]

6. Also, if *DF* always bears any the same ratio to the arc *AE*, *RF* will still touch the curve *AFI*; as is easily shown from the above and Lect. VIII, § 6.

7. Let a point *D* and two curves *AGE*, *DFI* be so related that, if any straight line *DFE* is drawn through *D*, the straight line *DF* is always equal to the arc *AE*; also let the straight line *ET* touch the curve *AGE* at *E*; make *ET* equal to the arc *AE*; and assume that *DKK* is a curve such that, if any straight line *DH* is drawn through *D*, cutting *DKK* in *K* and *TE* in *H*, the straight line *DK* is always equal to *TH*. Then, if *FS* is drawn (by Lect. VIII, § 16) to touch the curve *DKK* at *F*, *FS* touches the curve *DIF* also.

8. Moreover, if *DF* always bears any the same ratio to the arc *AE*, the straight line touching the curve *DIF* can likewise be drawn; and in every case the tangent is parallel to *FS*.

9. By this method can be drawn not only the tangent to the Circular Spiral, but also the tangents to innumerable other curves produced in a similar manner.

10. Let *AEH* be a given curve, *AD* any given straight line in which there is a fixed point *D*, and *DH* a straight line given in position; also let *AGB* be a curve such that, if any point *G* is taken in it, and through *G* and *D* a straight line is drawn to cut the curve *AEH* in *E*, and *GF* is drawn parallel to *DH* to cut *AD* in *F*, the arc *AE* bears to *AF* a given ratio, *X* to *Y* say; also let *ET* touch the curve *AEH*; along *ET* take *EV* equal to the arc *AE*;

---

1. The proof of this theorem is given in full, since not only is it a fine example of Barrow's method, but also it is a *rigorous* demonstration of the principle of fluxions, that the motion along the path is the resultant of the two rectilinear motions producing it. Otherwise, for rectangular axes, $(ds/dx)^2 = 1 + (dy/dx)^2$; for $ds/dx = DF/DR = ET/DR = \text{cosec } DET$ and $dy/dx = \text{cot } DET$.

let *OGO* be a curve such that, if any straight line *DOL* is drawn, cutting the curve *OGO* in *O* and *ET* in *L*, and if *OQ* is drawn parallel to *GF*, meeting *AD* in *Q*, *LV*: *AQ* = *X*: *Y*. Then the curve *OGO* is a hyperbola (as has been shown).[1] Then, if *GS* touches this curve, *GS* will touch the curve *AGB* also.

If the curve *AEH* is a quadrant of a circle, whose centre is *D*, the curve *AGB* will be the ordinary Quadratrix. Hence the tangent to this curve (together with tangents to all curves produced in a similar way) can be drawn by this method.

I meant to insert here several instances of this kind; but really I think these are sufficient to indicate the method, by which, without the labour of calculation, one can find tangents to curves and at the same time prove the constructions. Nevertheless, I add one or two theorems, which it will be seen are of great generality, and not lightly to be passed over.

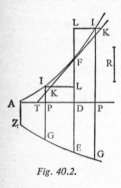

Fig. 40.2.

11. Let *ZGE* be any curve of which the axis is *AD*; and let ordinates applied to this axis, *AZ*, *PG*, *DE*, continually increase from the initial ordinate *AZ*; also let *AIF* be a line such that, if any straight line *EDF* is drawn perpendicular to *AD*, cutting the curves in the points *E*, *F*, and *AD* in *D*, the rectangle contained by *DF* and a given length *R* is equal to the intercepted space *ADEZ*; also let *DE*: *DF* = *R*: *DT*, and join *FT*. Then *TF* will touch the curve *AIF*.

For, if any point *I* is taken in the line *AIF* (first on the side of *F* towards *A*), and if through it *IG* is drawn parallel to *AZ*, and *KL* is parallel to *AD*, cutting the given lines as shown in the figure; then *LF*: *LK* = *DF*: *DT* = *DE*: *R*, or *R*.*LF* = *LK*.*DE*.

But, from the stated nature of the lines *DF*, *PK*, we have *R*.*LF* = area *PDEG*; therefore *LK*.*DE* = area *PDEG* < *DP*.*DE*; hence *LK* < *DP* < *LI*.

Again, if the point *I* is taken on the other side of *F*, and the

1. Only proved for a special case in Lect. VI, §17; but the method can be generalized without difficulty.

same construction is made as before, plainly it can be easily shown that $LK>DP<LI$.

From which it is quite clear that the whole of the line $TKFK$ lies within or below the curve $AIFI$.

Other things remaining the same, if the ordinates, AZ, PG, DE, continually decrease, the same conclusion is attained by similar argument; only one distinction occurs, namely, in this case, contrary to the other, the curve $AIFI$ is concave to the axis AD.

Cor. It should be noted that $DE.DT=R.DF=$area $ADEZ$.[1]

12. From the preceding we can deduce the following theorem.

Let ZGE, AKF be any two lines so related that, if any straight line EDF is applied to a common axis AD, the square on DF is always equal to twice the space $ADEZ$; also take DQ, along AD produced, equal to DE, and join FQ; then FQ is perpendicular to the curve AKF.

I will also add the following kindred theorems.

13. Let $AGEZ$ be any curve, and D a certain fixed point such that the radii, DA, DG, DE, drawn from D, decrease continually from the initial radius DA; then let DKE be another curve intersecting the first in E and such that, if any straight line DKG is drawn through D, cutting the curve AEZ in G and the curve DKE in K, the rectangle contained by DK and a given length R is equal to the area ADG; also let DT be drawn perpendicular to DE, so that $DT=2R$; join TE. Then TE touches the curve DKE.

Moreover, if any point, K say, is taken in the curve DKE, and through it DKG is drawn, and $DG:DK=R:P$; then if DT is taken equal to $2P$ and TG is joined, and also KS is drawn parallel to GT; KS will touch the curve DKE.

Observe that Sq. on DG: Sq. on $DK=2R:DS$.

Now, the above theorem is true, and can be proved in a similar way, even if the radii drawn from D, DA, DG, DE are equal (in which case the curve $AGEZ$ is a circle and the curve DKE is the Spiral of Archimedes), or if they continually increase from A.

14. From this we may easily deduce the following theorem.

Let AGE, DKE be two curves so related that, if straight lines DA, DG are drawn from some fixed point D in the curve DKE (of which the latter cuts the curve DKE in K), the square on DK

1. See note at end of this lecture.

is equal to four times the area *ADG*; draw *DH* perpendicular to *DG*, and make *DK* : *DG* = *DG* : *DH*; join *HK*; then *HK* is perpendicular to the curve *DKE*.

We have now finished in some fashion the first part, as we declared, of our subject. Supplementary to this we add, in the form of appendices, a method for finding tangents by calculation frequently used by us (*a nobis usitatum*). Although I hardly know, after so many well-known and well-worn methods of the kind above, whether there is any advantage in doing so. Yet I do so on the *advice of a friend*; and all the more willingly, because it seems to be more profitable and general than those which I have discussed.[1]

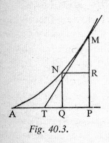

Fig. 40.3.

Let *AP*, *PM* be two straight lines given in position, of which *PM* cuts a given curve in *M*, and let *MT* be supposed to touch the curve at *M*, and to cut the straight line at *T*.

In order to find the quantity of the straight line *PT*, I set off an indefinitely small arc, *MN*, of the curve; then I draw *NQ*, *NR* parallel to *MP*, *AP*; I call *MP* = *m*, *PT* = *t*, *MR* = *a* *NR* = *e* and other straight lines, determined by the special nature of the curve, useful for the matter in hand, I also designate by name; also I compare *MR*, *NR* (and through them, *MP*, *PT*) with one another by means of an equation obtained by calculation; meantime observing the following rules.

Rule 1. In the calculation, I omit all terms containing a power of *a* or *e*, or products of these (for these terms have no value).

Rule 2. After the equation has been formed, I reject all terms consisting of letters denoting known or determined quantities, or terms which do not contain *a* or *e* (for these terms, brought over to one side of the equation, will always be equal to zero).

Rule 3. I substitute *m* (or *MP*) for *a*, and *t* (or *PT*) for *e*. Hence at length the quantity of *PT* is found.

Moreover, if any indefinitely small arc of the curve enters the calculation, an indefinitely small part of the tangent, or of any straight line equivalent to it (on account of the indefinitely small

1. See note at the end of this lecture.

size of the arc) is substituted for the arc. But these points will be made clearer by the following examples.

Note

Barrow gives five examples of this, the 'differential triangle' method. As might be expected, two of these are well-known curves, namely the Folium of Descartes, called by Barrow *La Galande*, and the *Quadratrix*; a third is the general case of the quasi-circular curves $x^n + y^n = a^n$; the fourth and fifth are the allied curves $r = a . \tan \theta$ and $y = a . \tan x$.

The fifth example, the case of the curve $y = a . \tan x$, I have selected for giving in full, for several reasons. It is the clearest and least tedious example of the method, it is illustrated by two diagrams, one being derived from the other, and therefore the demonstration is less confused, it suggests that Barrow was aware of the analogy of the differential form of the polar subtangent with the Cartesian subtangent, and that in this is to be found the reason why Barrow gives, as a rule, the polar forms of all his Cartesian theorems; and lastly, and more particularly, for its own intrinsic merits, as stated below. Barrow's enunciation and proof are as follows:

Example 5. Let *DEB* be a quadrant of a circle, to which *BX* is a tangent; then let the line *AMO* be such that, if in the straight line *AV* any part *AP* is taken equal to the arc *BE*, and *PM* is erected perpendicular to *AV*, then *PM* is equal to *BG* the tangent of the arc *BE*.

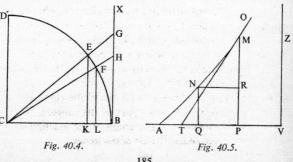

Fig. 40.4.   Fig. 40.5.

Take the arc BF equal to AQ and draw CFH; drop EK, FL perpendicular to CB. Let $CB=r$, $CK=f$, $KE=g$.

Then, since $CE:EK=$ arc $EF:LK=QP:LK$; therefore $r:g=e:LK$, or $LK=ge/r$, and $CL=f+ge/r$; hence also $LF=\sqrt{(r^2-f^2-2fge/r)}=\sqrt{(g^2-2fge/r)}$.

But $CL:LF=CB:BH$, or $f+ge/r:\sqrt{(g^2-2fge/r)}=r:m-a$; and squaring, we have

$$f^2+2fge/r:g^2-2fge/r=r^2:m^2-2ma.$$

Hence, omitting the proper terms, we obtain the equation

$$rfma=gr^2e+gm^2e;$$

and, on substituting m, t for a, e, we get

$$rfm^2=gr^2t+gm^2t, \quad \text{or} \quad rfm^2/(gr^2+gm^2)=t.$$

Hence since $m=rg/f$, we obtain

$$t=m.r^2/(r^2+m^2)=BG.CB^2/CG^2=BG.CK^2/CE^2.$$

In other words, this theorem states that, if $y=tan\,x$, where x is the circular measure of an 'angle' or an 'arc', then

$$dy/dx=m/t=CE^2/CK^2=\sec^2 x.$$

Moreover, although Barrow does not mention the fact, he must have known (for it is so self-evident) that the same two diagrams can be used for any of the trigonometrical ratios. Therefore *Barrow must be credited with the differentiation of the circular functions.* See Note to § 15 of App. 2 of Lect. XII.

As regards this lecture, it only remains to remark on the fact that the theorem of § 11 is a rigorous proof that differentiation and integration are inverse operations, where integration is defined as a summation. Barrow, as is well known, was the first to recognize this.

41

SIR ISAAC NEWTON
(1642–1727)

ISAAC NEWTON, one of the greatest mathematicians and physicists of all time, was born at Woolsthorpe, England, on Christmas Day in 1642. The inauspicious circumstances of his birth gave no hint of the brilliance of his career to come and of the eminence he was to achieve through his epoch-making scientific and mathematical discoveries. A few months before Newton was born, his father died. At birth Newton was so frail that small hope was entertained for his survival. His mother married a second time when he was only two years old, and he was left at Woolsthorpe at that tender age to be raised by his grandmother, Mrs Ayscough. He was fourteen when his mother, widowed again, withdrew him from school so that he might be of some help to her on her farm. It may be placed to the credit of the charm and good sense of both his mother and Mrs Ayscough, that a deep feeling of love and admiration between Newton and his mother persisted throughout his childhood and later, despite the long periods of separation. Quiet, introspective, and already inventive, Isaac showed little aptitude for the business of farming. In 1660, on the advice of his uncle, William Ayscough, he was returned to school to prepare for Cambridge.

In 1661 Newton entered Trinity College at Cambridge. Here his genius was soon recognized by the great Isaac Barrow, who accepted him first as a pupil, then as an assistant, and in 1669 as his successor in the Lucasian professorship of mathematics. Newton remained in continuous residence at Cambridge until 1696 except for a single period (1665–6) during the Black Plague in London, when he returned once more to Woolsthorpe. The months spent at Woolsthorpe in uninterrupted study saw the beginnings of Newton's magnificent creative achievements in mathematics and science. The concepts of the calculus were clear to him at that time. His optical researches, including his work in establishing the composition of white light were begun in 1665. The development of his laws of motion and his idea of universal gravitation, his theory of the moon and of tidal effects were well under way in 1666.

Newton's contributions to the development of mathematics and science were widely known for many years before their appearance in print. The substance of his early works, such as the *Arithmetica Universalis* (publ. 1707) and his *Opticks* (publ. 1704), was presented in the lectures given by him at Cambridge in his capacity as Lucasian professor (from 1669). During the period 1666–76, Newton wrote three important works presenting his method of fluxions, that is, his system of differential and integral calculus. Each of these treatises was in wide circulation among his friends long before publication. Newton's *De Analysi per Aequationes Numero Terminorem Infinitas* was written in 1669 and printed in 1711. His *Methodus Fluxionem et Serierum Infinitarum* was written about 1671 and printed in 1736, nine years after his death. His *Tractatus de Quadratura Curvarum*, written in 1676, was printed in 1704. The first of Newton's printed works to contain a presentation of his calculus was his *Philosophiae Naturalis Principia Mathematica* (1687, 'Mathematical Principles of Natural Philosophy'), one of the greatest systematic treatises ever written. In the *Principia*, Newton used the Euclidean method of definitions, axioms and theorems to construct a mathematical system of mechanics based upon his new mathematics. Newton's mechanics became the basis of the scientific developments and technical progress of the centuries which followed.

In 1668 Newton completed the invention of his reflecting telescope. He presented a model of his telescope to the Royal Society in 1671 and shortly thereafter he was elected Fellow of the Royal Society. In 1689 and again in 1701 he was elected Member of Parliament, representing Cambridge University. He became president of the Royal Society in 1703, a position to which he was annually re-elected to the end of his life. For about eighteen months in 1692–4, Newton suffered from a serious nervous disorder after which he seemed to have lost interest in scientific researches, preferring to devote himself to studies in theology. In 1696 he accepted an appointment as Warden of the Mint with the special assignment of reforming the coinage which had depreciated because of the adulteration of its precious metal content. An excellent administrator, he was appointed Master of the Mint in 1699. Newton remained in London for more than thirty years occupied with his duties at the Mint, his theological writings and his problems in alchemy and chemistry. Occasionally he would be aroused to interrupt the routine of these activities for the purpose of disposing swiftly and brilliantly of a challenge to solve difficult problems issued by the mathematicians of the Continent. The calculus, Newton's chief contribution to the development of mathematics, an invaluable aid in the forward-looking problems of satellites and space travel, continues to be applied to the analysis of the motion of all types of things, including rigid bodies, particles, fluids and gases.

TREATISE OF THE QUADRATURE OF CURVES

translated by John Stewart

INTRODUCTION TO THE QUADRATURE OF CURVES

1. I consider mathematical Quantities in this Place not as consisting of very small Parts; but as describ'd by a continued Motion. Lines are describ'd, and thereby generated not by the Apposition of Parts, but by the continued Motion of Points; Superficies's by the Motion of Lines; Solids by the Motion of Superficies's; Angles by the Rotation of the Sides; Portions of Time by a continual Flux: and so in other Quantities. These Geneses really take Place in the Nature of Things, and are daily seen in the Motion of Bodies. And after this Manner the Ancients, by drawing moveable right Lines along immoveable right Lines, taught the Genesis of Rectangles.

2. Therefore considering that Quantities, which increase in equal Times, and by increasing are generated, become greater or less according to the greater or less Velocity with which they increase and are generated; I sought a Method of determining Quantities from the Velocities of the Motions or Increments, with which they are generated; and calling these Velocities of the Motions or Increments *Fluxions*, and the generated Quantities *Fluents*, I fell by degrees upon the Method of Fluxions, which I have made use of here in the Quadrature of Curves, in the Years 1665 and 1666.

3. Fluxions are very nearly as the Augments of the Fluents generated in equal but very small Particles of Time, and, to speak accurately, they are in the *first Ratio* of the nascent Augments;

but they may be expounded by any Lines which are proportional to them.

4. Thus if the Area's *ABC*, *ABDG* be described by the Ordinates *BC*, *BD* moving along the Base *AB* with an uniform Motion, the Fluxions of these Area's shall be to one another as the describing Ordinates *BC* and *BD*, and may be expounded by these Ordinates, because that these Ordinates are as the nascent Augments of the Areas.

5. Let the Ordinate *BC* advance from its Place into any new Place *bc*. Complete the Parallelogram *BCEb*, and draw the right Line *VTH* touching the Curve in *C*, and meeting the two Lines *bc* and *BA* produc'd in *T* and *V*: and *Bb*, *Ec* and *Cc* will be *the* Augments now generated of the Absciss *AB*, the Ordinate *BC* and the Curve Line *ACc*; and the Side of the Triangle *CET* are in the *first Ratio* of these Augments considered as nascent, therefore the Fluxions of *AB*, *BC* and *AC* are as the Sides *CE*, *ET* and *CT* of that Triangle *CET*, and may be expounded by these same Sides, or, which is the same thing, by the Sides of the Triangle *VBC*, which is similar to the Triangle *CET*.

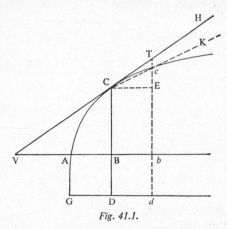

Fig. 41.1.

6. It comes to the same Purpose to take the Fluxions in the *ultimate Ratio* of the evanescent Parts. Draw the right Line *Cc*, and produce it to *K*. Let the Ordinate *bc* return into its former

Place *BC*, and when the Points *C* and *c* coalesce, the right Line *CK* will coincide with the Tangent *CH*, and the evanescent Triangle *CEc* in it's ultimate Form will become similar to the Triangle *CET*, and its evanescent Sides *CE*, *Ec* and *Cc* will be *ultimately* among themselves as the Sides *CE*, *ET* and *CT* of the other Triangle *CET*, are, and therefore the Fluxions of the Lines *AB*, *BC* and *AC* are in this same Ratio. If the Points *C* and *c* are distant from one another by any small Distance, the right Line *CK* will likewise be distant from the Tangent *CH* by a small Distance. That the right Line *CK* may coincide with the Tangent *CH*, and the ultimate Ratios of the Lines *CE*, *Ec* and *Cc* may be found, the Points *C* and *c* ought to coalesce and exactly coincide. The very smallest Errors in mathematical Matters are not to be neglected.

7. By the like way of reasoning, if a Circle describ'd with the Centre *B* and Radius *BC* be drawn at right Angles along the Absciss *AB*, with an uniform Motion, the Fluxion of the generated Solid *ABC* will be as that generating Circle, and the Fluxion of its Superficies will be as the Perimeter of that Circle and the Fluxion of the Curve Line *AC* jointly. For in whatever Time the Solid *ABC* is generated by drawing that Circle along the Length of the Absciss, in the same Time its Superficies is generated by drawing the Perimeter of that Circle along the Length of the Curve *AC*. You may likewise take the following Examples of this Method.

8. Let the right Line *PB*, revolving about the given Pole *P*, cut another right Line *AB* given in Position: it is required to find the Proportion of the Fluxions of these right Lines *AB* and *PB*.

Let the Line *PB* move forward from its Place *PB* into the new Place *Pb*. In *Pb* take *PC* equal to *PB*, and draw *PD* to *AB* in such manner that the Angle *bPD* may be equal to the Angle *bBC*; and because the Triangles *bBC*, *bPD* are similar, the Augment *Bb* will be to the Augment *Cb* as *Pb* to *Db*. Now let *Pb* return into its former Place *PB*, that these Augments may evanish, then the ultimate Ratio of these evanescent Augments, that is the ultimate Ratio being of *Pb* to *Db*, shall be the same with that of *PB* to *DB*, *PDB* then a right Angle, and therefore the Fluxion of *AB* is to the Fluxion of *PB* in that same Ratio.

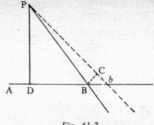

Fig. 41.2.

9. Let the right Line *PB*, revolving about the given Pole *P*, cut other two right Lines given in Position, *viz. AB* and *AE* in *B* and *E*: the Proportion of the Fluxions of these right Lines *AB* and *AE* is sought.

Let the revolving right Line *PB* move forward from its Place *PB* into the new Place *Pb*, so as to cut the Lines *AB, AE* in the Points *b* and *e*: and draw *BC* parallel to *AE* meeting *Pb* in *C*, and it will be *Bb : BC :: Ab : Ae*, and *BC : Ee :: PB : PE*, and

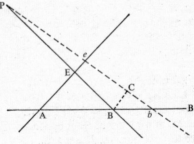

Fig. 41.3.

by joining the Ratios, *Bb : Ee :: Ab × PB : Ae × PE*. Now let *Pb* return into its former Place *PB*, and the evanescent Augment *Bb* will be to the evanescent Augment *Ee* as *AB × PB* to *AE × PE*; and therefore the Fluxion of the right Line *AB* is to the Fluxion of the right Line *AE* in the same Ratio.

10. Hence if the revolving right Line *PB* cut any curve Lines

given in Position in the Points B and E, and the right Lines AB, AE now becoming moveable, touch these Curves in the Points of Section B and E: the Fluxion of the Curve, which the right Line AB touches, shall be to the Fluxion of the Curve, which the right Line AE touches, as $AB \times PB$ to $AE \times PE$. The same thing would happen if the right Line PB perpetually touch'd any Curve given in Position in the moveable Point P.

11. Let the Quantity x flow uniformly, and let it be proposed to find the Fluxion of x^n.

In the same Time that the Quantity x, by flowing, becomes $x+o$, the Quantity x^n will become $(x+o)^n$, that is, by the Method of infinite Series's, $x^n + nox^{n-1} + \dfrac{n^2-n}{2} oox^{n-2} +$ etc. And the Augments o and $nox^{n-1} + \dfrac{n^2-n}{2}oox^{n-2} +$ etc., are to one another as 1 and $nx^{n-1} + \dfrac{n^2-n}{2}ox^{n-2} +$ etc.

Now let these Augments vanish, and their ultimate Ratio will be 1 to nx^{n-1}.

12. By like ways of reasoning, the Fluxions of Lines, whether right or curve in all Cases, as likewise the Fluxions of Superficies's, Angles and other Quantities, may be collected by the Method of *prime* and *ultimate* Ratios. Now to institute an Analysis after this manner in finite Quantities and investigate the *prime* or *ultimate* Ratios of these finite Quantities when in their nascent or evanescent State, is consonant to the Geometry of the Ancients: and I was willing to show that, in the Method of Fluxions, there is no necessity of introducing Figures infinitely small into Geometry. Yet the Analysis may be performed in any kind of Figures, whether finite or infinitely small, which are imagin'd similar to the evanescent Figures; as likewise in these Figures, which, by the Method of Indivisibles, use to be reckoned as infinitely small, provided you proceed with due Caution.

From the Fluxions to find the Fluents, is a much more difficult Problem, and the first Step of the Solution is equivalent to the Quadrature of Curves; concerning which I wrote what follows some considerable Time ago.

QUADRATURE OF CURVES

13. In what follows I consider indeterminate Quantities as increasing or decreasing by a continued Motion, that is, as flowing forwards, or backwards, and I design them by the Letters z, y, x, v, and their Fluxions or Celerities of increasing I denote by the same Letters pointed $\dot{z}$, $\dot{y}$, $\dot{x}$, $\dot{v}$. There are likewise Fluxions or Mutations more or less swift of these Fluxions, which may be call'd the second Fluxions of the same Quantities z, y, x, v, and may be thus design'd $\ddot{z}$, $\ddot{y}$, $\ddot{x}$, $\ddot{v}$: and the first Fluxions of these last, or the third Fluxions of z, y, x, v are thus denoted $\dddot{z}$, $\dddot{y}$, $\dddot{x}$, $\dddot{v}$: and the Fourth Fluxions thus $\ddddot{z}$, $\ddddot{y}$, $\ddddot{x}$, $\ddddot{v}$. And after the same manner that $\dot{z}$, $\dot{y}$, $\dot{x}$, $\dot{v}$ are the Fluxions of the Quantities $\ddot{z}$, $\ddot{y}$, $\ddot{x}$, $\ddot{v}$, and these the Fluxions of the Quantities $\dot{z}$, $\dot{y}$, $\dot{x}$, $\dot{v}$; and these last the Fluxions of the Quantities z, y, x, v; so the Quantities z, y, x, v may be considered as the Fluxions of others, which I shall design thus $\dot{z}$, $\dot{y}$, $\dot{x}$, $\dot{v}$; and these and the Fluxions of others $\ddot{z}$, $\ddot{y}$, $\ddot{x}$, $\ddot{v}$; and these last still as the Fluxions of others $\dddot{z}$, $\dddot{y}$, $\dddot{x}$, $\dddot{v}$. Therefore $\ddddot{z}$, $\dddot{z}$, z, $\dot{z}$, $\ddot{z}$, $\dddot{z}$, $\ddddot{z}$, $\dddddot{z}$, etc., design a Series of Quantities whereof every one that follows is the Fluxion of the one immediately preceding, and every one that goes before, is a flowing Quantity having that which immediately succeeds, for its Fluxion.

OF ANALYSIS BY EQUATIONS OF AN INFINITE NUMBER OF TERMS

translated by John Stewart

1. The General Method, which I had devised some considerable Time ago, for measuring the Quantity of Curves, by means of Series, infinite in the Number of Terms, is rather shortly explained, than accurately demonstrated in what follows.

2. Let the Base AB of any Curve AD have BD for its perpendicular Ordinate; and call $AB=x$, $BD=y$, and let a, b, c, etc., be given Quantities, and m and n whole Numbers. Then

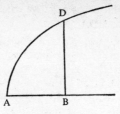

Fig. 41.4.

THE QUADRATURE OF SIMPLE CURVES

Rule I

3. If $ax^{m/n} = y$, it shall be $\dfrac{an}{m+n}x^{(m+n)/n} = $ Area ABD.

The thing will be evident by an Example.

1. If $x^2(=1x^{2/1})=y$, that is $a=1=n$, and $m=2$; it shall be $\frac{1}{3}x^3 = ABD$.

2. Suppose $4\sqrt{x}(=4x^{1/2})=y$; it will be $\frac{8}{3}x^{3/2}(=\frac{8}{3}\sqrt{x^3})=ABD$.

3. If $\sqrt[3]{x^5}(=x^{5/3})=y$; it will be $\frac{3}{8}x^{8/3}\ (=\frac{3}{8}\sqrt[3]{x^8})=ABD$.

4. If $\dfrac{1}{x^2}\ (=x^{-2})=y$, that is if $a=1=n$, and $m=-2$; it will be

$\dfrac{1}{-1}x^{-1/1}\ (=-x^{-1})=\dfrac{-1}{x}=\alpha BD$, infinitely extended towards α,

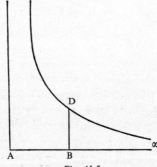

Fig. 41.5.

which the Calculation places negative, because it lies upon the other side of the Line BD.[1]

5. If $\dfrac{1}{\sqrt{x^3}}(=x^{-3/2})=y$; it will be $\dfrac{2}{-1}x^{-1/2}=\dfrac{2}{-\sqrt{x}}=BD\alpha$

6. If $\dfrac{1}{x}(=x^{-1})=y$; it will be $\dfrac{1}{0}x^{0/1}=\dfrac{1}{0}x^0=\dfrac{1}{0}\times1=\dfrac{1}{0}=$ an infinite

Quantity; such as is the Area of the Hyperbola upon both sides of the Line BD.

THE QUADRATURE OF CURVES COMPOUNDED OF SIMPLE ONES

Rule II

4. If the Value of y be made up of several such Terms, the Area

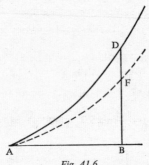

Fig. 41.6.

likewise shall be made up of the Areas which result from every one of the Terms.

The first Examples

5. If it be $x^2+x^{3/2}=y$; it will be $\frac{1}{3}x^3+\frac{2}{5}x^{5/2}=ABD$.

For if it be always $x^2=BF$ and $x^{3/2}=FD$, you will have by the preceding Rule $\frac{1}{3}x^3=$ Superficies AFB; described by the Line BF;

1. Whatever is laid down by our Author with respect to the Position of Areas of Curves in this and the following Rules is explained at full length in §5 of the preceding Treatise, which see.

and $\frac{2}{5}x^{5/2}=AFD$ described by DF; wherefore $\frac{1}{3}x^3+\frac{2}{5}x^{5/2}=$ the whole Area ABD.

Thus if it be $x^2-x^{3/2}=y$; it will be $\frac{1}{3}x^3-\frac{2}{5}x^{5/2}=ABD$. And if it be $3x-2x^2+x^3-5x^4=y$; it will be $\frac{3}{2}x^2-\frac{2}{3}x^3+\frac{1}{4}x^4-x^5=ABD$.

The second Examples

6. If $x^{-2}+x^{-3/2}=y$; it will be $x^{-1}-2x^{-1/2}=\alpha BD$. Or if it be $x^{-2}-x^{-3/2}=y$; it will be $-x^{-1}+2x^{-1/2}=\alpha BD$.

And if you change the Signs of the Quantities, you will have the affirmative Value $(x^{-1}+2x^{-1/2}$, or $x^{-1}-2x^{-1/2})$ of the Superficies αBD, provided the whole of it fall above the Base $AB\alpha$.

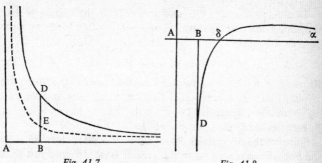

Fig. 41.7.   Fig. 41.8.

7. But if any Part fall below (which happens when the Curve decussates or crosses its Base betwixt B and α, as you see here in δ) you are to subtract that Part from the Part above the Base; and so you shall have the Value of the Difference: but if you would have their Sum, seek both the Superficies's separately, and add them. And the same thing I would have observed in the other Examples belonging to this Rule.

The third Examples

8. If $x^2+x^{-2}=y$; it will be $\frac{1}{3}x^3-x^{-1}=$ the Superficies described.

But here it must be remarked that the Parts of the said Superficies so found, lie upon opposite Sides of the Line BD.

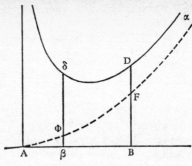

Fig. 41.9.

That is, putting $x^2 = BF$, and $x^{-2} = FD$; it shall be $\frac{1}{3}x^3 = ABF$ the Superficies described by BF, and $-x^{-1} = DF\alpha$ the Superficies described by DF.

9. And this always happens when the Indexes $\frac{m+n}{n}$ of the Ratios of the Base x in the Value of the Superficies sought, are affected with different Signs. In such Cases any middle part $BD\delta\beta$ of the Superficies (which only can be given, when the Superficies is infinite upon both Sides) is thus found.

Subtract the Superficies belonging to the lesser Base $A\beta$ from the Superficies belonging to the greater Base AB, and you shall have $\beta BD\delta$ the Superficies insisting upon the difference of the Bases. Thus in this Example (see the preceding diagram).

If $AB = 2$, and $A\beta = 1$; it will be $\beta BD\delta = 17/6$.

For the Superficies belonging to AB (viz. $ABF - DF\alpha$) will be $8/3 - 1/2$ or $13/6$; and the Superficies belonging to $A\beta$ (viz. $A\phi\beta - \delta\phi\alpha$) will be $1/3 - 1$, or $-2/3$: and their Difference (viz. $ABF - DF\alpha - A\phi\beta + \delta\phi\alpha = \beta BD\delta$) will be $13/6 + 2/3$ or $17/6$.

After the same manner, if $A\beta = 1$, and $AB = x$; it will be $\beta BD\delta = 2/3 + 1/3x^3 - x^{-1}$

Thus if $2x^3 - 3x^5 - \frac{2}{3}x^{-4} + x^{-3/5} = y$, and $A\beta = 1$; It will be $\beta BD\delta = \frac{1}{2}x^4 - \frac{1}{2}x^6 + \frac{2}{9}x^{-3} + \frac{5}{2}x^{2/5} - \frac{49}{18}$.

10. Finally it may be observed, that if the Quantity x^{-1} be found in the Value of y, that Term (since it generates an hyperbolical Surface) is to be considered apart from the rest.

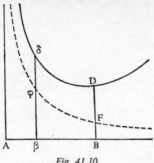

Fig. 41.10.

As if it were $x^2 + x^{-3} + x^{-1} = y$: let it be $x^{-1} = BF$, and $x^2 + x^{-3} = FD$; and $A\beta = 1$; and it will be $\delta\phi FD = \frac{1}{6} + \frac{1}{3}x^3 - \frac{1}{2}x^{-2}$, as being that which is generated by the Terms $x^2 + x^{-3}$.

Wherefore if the remaining Superficies $\beta\phi FB$, which is hyperbolical, be given by any Method of Computation, the whole βBDd will be given.

THE QUADRATURE OF ALL OTHER CURVES

Rule III

11. But if the Value of y, or any of its Terms be more compounded than the foregoing, it must be reduced into more simple Terms; by performing the Operation in Letters, after the same Manner as Arithmeticians divide in Decimal Numbers, extract the Square Root, or resolve affected Equations; and afterwards by the preceding Rules you will discover the Superficies of the Curve sought.

Examples, where you divide

12. Let $\dfrac{aa}{b+x} = y$; Viz. where the Curve is an Hyperbola.

Now that that Equation may be freed from its Denominator, I make the Division thus.

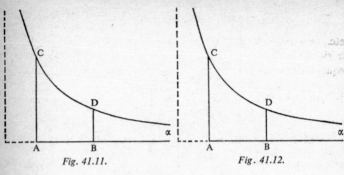

Fig. 41.11.   *Fig. 41.12.*

$$b+x)aa+0\left(\frac{aa}{b}-\frac{aax}{b^2}+\frac{aax^2}{b^3}-\frac{aax^3}{b^4}\text{ etc.}\right.$$

$$aa+\frac{aax}{b}$$

$$0-\frac{aax}{b}+0$$

$$-\frac{aax}{b}-\frac{aax^2}{b^2}$$

$$0+\frac{aax^2}{b^2}+0$$

$$+\frac{aax^2}{b^2}+\frac{aax^3}{b^3}$$

$$0-\frac{aax^3}{b^3}+0$$

$$-\frac{aax^3}{b^3}-\frac{aax^4}{b^4}$$

$$0+\frac{aax^4}{b^4}$$

etc.

And thus in Place of this $y=\dfrac{aa}{b+x}$, a new Equation arises, viz.

$$y=\frac{a^2}{b}-\frac{a^2x}{b^2}+\frac{a^2x^2}{b^3}-\frac{a^2x^3}{b^4}\text{ etc.}$$

this Series being continued infinitely; and therefore (by the second Rule).

The Area sought $ABDC$ will be equal to $\dfrac{a^2x}{b}-\dfrac{a^2x^2}{2b^2}+\dfrac{a^2x^3}{3b^3}-\dfrac{a^2x^4}{4b^4}$ etc., an infinite Series likewise but yet such, that a few of the initial Terms are exact enough for any Use, provided that b be equal to x repeated some few times.

13. After the same Manner if it be $\dfrac{1}{1+xx}=y$ by dividing there arises $y=1-xx+x^4-x^6+x^8$ etc. Whence (by the second Rule) You will have $ABDC=x-\frac{1}{3}x^3+\frac{1}{5}x^5-\frac{1}{7}x^7+\frac{1}{9}x^9$ etc.

Or if x^2 be made the first Term in the Divisor, viz. (thus: x^2+1) there will arise $x^{-2}-x^{-4}+x^{-6}-x^{-8}$ etc., for the Value of y; whence (by the second Rule).

It will be $BD\alpha=-x^{-1}+\frac{1}{3}x^{-3}-\frac{1}{5}x^{-5}+\frac{1}{7}x^{-7}$ etc. You must proceed in the first Way when x is small enough, but the second Way, when it is supposed great enough.

MATHEMATICAL PRINCIPLES OF NATURAL PHILOSOPHY

NEWTON'S PREFACE TO THE FIRST EDITION

Since the ancients (as we are told by Pappus) esteemed the science of mechanics of greatest importance in the investigation of natural things, and the moderns, rejecting substantial forms and occult qualities, have endeavoured to subject the phenomena of nature to the laws of mathematics, I have in this treatise cultivated mathematics as far as it relates to philosophy. The ancients considered mechanics in a twofold respect; as rational, which proceeds accurately by demonstration, and practical. To practical mechanics all the manual arts belong, from which mechanics took its name. But as artificers do not work with perfect accuracy, it comes to pass that mechanics is so distinguished from geometry that what is perfectly accurate is called geometrical; what is less so, is called mechanical. However, the errors are not in the art, but in the artificers. He that works with less accuracy is an imperfect mechanic; and if any could work with perfect accuracy, he would be the most perfect mechanic of all, for the description of right lines and circles, upon which geometry is founded, belongs to mechanics. Geometry does not teach us to draw these lines, but requires them to be drawn, for it requires that the learner should first be taught to describe these accurately before he enters upon geometry, then it shows how by these operations problems may be solved. To describe right lines and circles are problems, but not geometrical problems. The solution of these problems is required from mechanics, and by geometry the use of them, when so solved, is shown; and it is the glory of geometry that from those few principles, brought from without, it is able to produce so many things. Therefore geometry is founded in mechanical practice, and is nothing but that part of universal mechanics which accurately proposes and demonstrates the art of measuring. But since the manual arts are chiefly employed in the moving of bodies, it happens that geometry is commonly referred to their magnitude, and mechanics to their motion. In this sense rational mechanics will be the science of motions resulting from any forces whatsoever, and of the forces required to produce any motions, accurately proposed and demonstrated. This part of mechanics, as far as it extended to the five powers which relate to manual arts, was cultivated by the ancients, who considered gravity (it not being a manual power) no otherwise than in moving weights by those powers. But I consider philosophy rather than arts and write not concerning manual but natural powers, and consider chiefly those things which relate to gravity, levity, elastic force, the resistance of fluids, and the like forces, whether attractive or impulsive; and therefore I offer this work as the mathematical principles of philosophy, for the whole burden of philosphy seems to consist in this – from the phenomena of motions to investigate the forces of nature, and then from these forces to demonstrate the other phenomema; and to this end the general propositions in the first and second Books are directed. In the third Book I give an example of this in the explication of the System of the World; for by the propositions mathematically demonstrated in the former Books, in the third I derive from the celestial phenomena the forces of gravity with which bodies tend to the sun and the several planets. Then from these forces, by other propositions which are also mathematical,

I deduce the motions of the planets, the comets, the moon, and the sea. I wish we could derive the rest of the phenomena of Nature by the same kind of reasoning from mechanical principles, for I am induced by many reasons to suspect that they may all depend upon certain forces by which the particles of bodies, by some causes hitherto unknown, are either mutually impelled towards one another, and cohere in regular figures, or are repelled and recede from one another. These forces being unknown, philosophers have hitherto attempted the search of Nature in vain; but I hope the principles here laid down will afford some light either to this or some truer method of philosophy.

BOOK ONE: THE MOTIONS OF BODIES

Section I

*The method of first and last ratios of quantities,
by the help of which we demonstrate
the propositions that follow.*

LEMMA I

Quantities, and the ratios of quantities, which in any finite time converge continually to equality, and before the end of that time approach nearer to each other than by any given difference, become ultimately equal.

If you deny it, suppose them to be ultimately unequal, and let D be their ultimate difference. Therefore they cannot approach nearer to equality than by that given difference D; which is contrary to the supposition.

LEMMA II

If in any figure $AacE$, terminated by the right lines Aa, AE, and the curve acE, there be inscribed any number of parallelograms Ab, Bc, Cd, etc., comprehended under equal bases AB, BC, CD, etc. and the sides, Bb, Cc, Dd, etc., parallel to one side Aa of the figure; and the parallelograms $aKBl$, $bLcm$, $cMdn$, etc., are completed: then if the breadth of those parallelograms be supposed to be diminished, and their number to be augmented *in infinitum*, I say, that the ultimate ratios which the inscribed figure

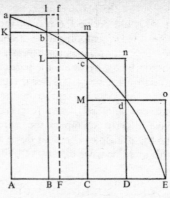

Fig. 41.13.

AKbLcMdD, the circumscribed figure *AalbmcndoE*, and curvilinear figure *AabcdE*, will have to one another, are ratios of equality.

For the difference of the inscribed and circumscribed figures is the sum of the parallelograms *Kl, Lm, Mn, Do*, that is (from the equality of all their bases), the rectangle under one of their bases *Kb* and the sum of their altitudes *Aa*, that is, the rectangle *ABla*. But this rectangle, because its breadth *AB* is supposed diminished *in infinitum*, becomes less than any given space. And therefore (by Lem. I) the figures inscribed and circumscribed become ultimately equal one to the other; and much more will the intermediate curvilinear figure be ultimately equal to either. Q.E.D.

LEMMA III

The same ultimate ratios are also ratios of equality, when the breadths *AB*, *BC*, *DC* etc., of the parallelograms are unequal, and are all diminished *in infinitum*.

For suppose *AF* equal to the greatest breadth, and complete the parallelogram *FAaf*. This parallelogram will be greater than the difference of the inscribed and circumscribed figures; but because its breadth *AF* is diminished *in infinitum*, it will become less than any given rectangle. Q.E.D.

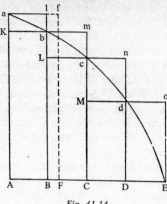

Fig. 41.14.

Cor. I. Hence the ultimate sum of those evanescent parallelograms will in all parts coincide with the curvilinear figure.

Cor. II. Much more will the rectilinear figure comprehended under the chords of the evanescent arcs *ab*, *bc*, *cd*, etc. ultimately coincide with the curvilinear figure.

Cor. III. And also the circumscribed rectilinear figure comprehended under the tangents of the same arcs.

Cor. IV. And therefore these ultimate figures (as to their perimeters *acE*) are not rectilinear, but curvilinear limits of rectilinear figures.

*

Those things which have been demonstrated of curved lines, and the surfaces which they comprehend, may be easily applied to the curved surfaces and contents of solids. These Lemmas are premised to avoid the tediousness of deducing involved demonstrations *ad absurdum*, according to the method of the ancient geometers. For demonstrations are shorter by the method of indivisibles; but because the hypothesis of indivisibles seems somewhat harsh, and therefore that method is reckoned less geometrical, I chose rather to reduce the demonstrations of the following Propositions to the first and last sums and ratios of nascent and evanescent quantities, that is, to the limits of those sums and ratios, and so to premise, as short as I could, the demonstrations of those limits. For hereby the same thing is performed as by the method of indivisibles; and now those principles being demonstrated, we may use them with greater safety. Therefore if hereafter I should happen to consider quantities as made up of particles, or should use little curved lines for right ones, I would not be understood to mean indivisibles, but evanescent divisible quantities; not the sums and ratios of determinate parts, but always the limits of sums and ratios; and that the force of such demonstrations always depends on the method laid down in the foregoing Lemmas.

Perhaps it may be objected, that there is no ultimate proportion of evanescent quantities; because the proportion, before the quantities have vanished, is not the ultimate, and when they are vanished, is none. But by the same argument it may be alleged that a body arriving at a certain place, and there stopping, has no ultimate velocity; because the velocity, before the body comes to the place, is not its ultimate velocity; when it has arrived, there is none. But the answer is easy; for by the ultimate velocity is meant that with which the body is moved, neither before it arrives at its last place and the motion ceases, nor after, but at the very instant it arrives; that is, that velocity with which the body arrives at its last place, and with which the motion ceases. And in like manner, by the ultimate ratio of evanescent quantities is to be understood the ratio of the quantities not before they vanish, nor afterwards, but with which they vanish. In like manner the first ratio of nascent quantities is that with which they begin to be. And the first or last sum is that with which they begin and cease to be (or to be augmented or diminished). There is a limit which the velocity at the end of the motion may attain, but not exceed. This is the ultimate velocity. And there is the like limit in all quantities and proportions that begin and cease to be. And since such limits are certain and definite, to determine the same is a problem strictly geometrical. But whatever is geometrical we may use in determining and demonstrating any other thing that is also geometrical.

BOOK TWO: SECTION II, LEMMA II

The moment of any *genitum* is equal to the moments of each of the generating sides multiplied by the indices of the powers of those sides, and by their coefficients continually.

I call any quantity a *genitum* which is not made by addition or subtraction of divers parts, but is generated or produced in arithmetic by the multiplication, division or extraction of the root of any terms whatsoever; in geometry by the finding of contents and sides, or of the extremes and means of proportionals. Quantities of this kind are products, quotients, roots, rectangles, squares, cubes, square and cubic sides, and the like. These quantities I here consider as variable and indetermined, and increasing or decreasing, as it were, by a continual motion or flux; and I understand their momentary increments or decrements by the name of moments; so that the increments may be esteemed as added or affirmative moments; and the decrements as subtracted or negative ones. But take care not to look upon finite particles as such. Finite particles are not moments, but the very quantities generated by the moments. We are to conceive them as the just nascent principles of finite magnitudes. Nor do we in this Lemma regard the magnitude of the moments, but their first proportion, as nascent. It will be the same thing, if, instead of moments, we use either the velocities of the increments and decrements (which may also be called the motions, mutations, and fluxions of quantities), or any finite quantities proportional to those velocities. The coefficient of any generating side is the quantity which arises by applying the genitum to that side.

Wherefore the sense of the Lemma is, that if the moments of any quantities A, B, C, etc., increasing or decreasing by a continual flux, or the velocities of the mutations which are proportional to them, be called a, b, c etc., the moment or mutation of the generated rectangle AB will be $aB + bA$; the moment of the generated content ABC will be $aBC + bAC + cAB$; and the moments of the generated powers A^2, A^3, A^4, $A^{1/2}$, $A^{3/2}$, $A^{1/3}$, $A^{2/3}$, A^{-1}, A^{-2}, $A^{-1/2}$ will be $2aA$, $3aA^2$, $4aA^3$, $\frac{1}{2}aA^{-1/2}$, $\frac{3}{2}aA^{1/2}$, $\frac{1}{3}aA^{-2/3}$, $\frac{2}{3}aA^{-1/3}$, $-aA^{-2}$, $-2aA^{-3}$, $-\frac{1}{2}aA^{-3/2}$ respectively; and, in general, that the moment of any power $A^{n/m}$ will be $\frac{n}{m}aA^{(n-m)/m}$. Also, that the moment of the generated quantity A^2B will be $2aAB+bA^2$; the moment of the generated quantity $A^3B^4C^2$ will be $3aA^2B^4C^2+4bA^3B^3C^2+2cA^3B^4C$; and the moment of the generated quantity $\frac{A^3}{B^2}$ or A^3B^{-2} will be $3aA^2B^{-2}-2bA^3B^{-3}$; and so on. The Lemma is thus demonstrated.[1]

Case 1. Any rectangle, as AB augmented by a continual flux, when, as yet, there wanted of the sides A and B half their moments $\frac{1}{2}a$ and $\frac{1}{2}b$, was $A-\frac{1}{2}a$ into $B-\frac{1}{2}b$, or $AB-\frac{1}{2}aB-\frac{1}{2}bA+\frac{1}{4}ab$; but as soon as the sides A and B are augmented by the other half-moments, the rectangle becomes $A+\frac{1}{2}a$ into $B+\frac{1}{2}b$, or $AB+\frac{1}{2}aB+\frac{1}{2}bA+\frac{1}{4}ab$. From this rectangle subtract the former rectangle, and there will remain the excess $aB+bA$. Therefore with the whole increments a and b of the sides, the increment $aB+bA$ of the rectangle is generated. Q.E.D.

Case 2. Suppose AB always equal to G, and then the moments of the content ABC or GC (by Case 1) will be $gC+cG$, that is (putting AB and $aB+bA$ for G and g), $aBC+bAC+cAB$. And the reasoning is the same for contents under ever so many sides. Q.E.D.

Case 3. Suppose the sides A, B, and C, to be always equal among themselves; and the moment $aB+bA$, of A^2, that is, of the rectangle AB, will be $2aA$; and the moment $aBC+bAC+cAB$ of A^3, that is, of the content ABC, will be $3aA^2$. And by the same reasoning the moment of any power A^n is naA^{n-1}. Q.E.D.

Case 4. Therefore since $\frac{1}{A}$ into A is 1, the moment of $\frac{1}{A}$ multiplied by A, together with $\frac{1}{A}$ multiplied by a, will be the moment of 1, that is, nothing. Therefore the moment of $\frac{1}{A}$, or of A^{-1}, is $\frac{-a}{A^2}$. And generally $\frac{1}{A^n}$ into A^n is 1, the moment of $\frac{1}{A^n}$ multiplied by A^n together with $\frac{1}{A^n}$ into naA^{n-1} will be nothing. And, therefore, the moment of $\frac{1}{A^n}$ or A^{-n} will be $-\frac{na}{A^{n+1}}$. Q.E.D.

1. Appendix. Note 31.

Case 5. And since $A^{1/2}$ into $A^{1/2}$ is A, the moment of $A^{1/2}$ multiplied by $2A^{1/2}$ will be a (by Case 3); and therefore, the moment of $A^{1/2}$ will be $\dfrac{a}{2A^{1/2}}$ or $\frac{1}{2}aA^{-1/2}$. And generally, putting $A^{m/n}$ equal to B, then A^m will be equal to B^n, and therefore maA^{m-1} equal to nbB^{n-1}, and maA^{-1} equal to nbB^{-1}, or $nbA^{-m/n}$; and therefore $\dfrac{m}{n}aA^{(m-n)/n}$ is equal to b, that is, equal to the moment of $A^{m/n}$. Q.E.D.

Case 6. Therefore the moment of any generated quantity A^mB^n is the moment of A^m multiplied by B^n, together with the moment of B^n multiplied by A^m, that is, $maA^{m-1}B^n+nbB^{n-1}A^m$; and that whether the indices m and n of the powers be whole numbers or fractions, affirmative or negative. And the reasoning is the same for higher powers. Q.E.D.

Cor. I. Hence in quantities continually proportional, if one term is given, the moments of the rest of the terms will be as the same terms multiplied by the number of intervals between them and the given term. Let A, B, C, D, E, F be continually proportional; then if the term C is given, the moments of the rest of the terms will be among themselves as $-2A$, $-B$, D, $2E$, $3F$.

Cor II. And if in four proportionals the two means are given, the moments of the extremes will be as those extremes. The same is to be understood of the sides of any given rectangle.

Cor. III. And if the sum or difference of two squares is given, the moments of the sides will be inversely as the sides.

LOGIC

42

GOTTFRIED WILHELM von LEIBNIZ
(1646–1716)

THE universal genius of Gottfried von Leibniz was early in evidence and his extraordinary power marked all of his varied activities in law, diplomacy, philosophy, theology, mathematics and science. Leibniz was born on 21 June, 1646, at Leipzig. He was the only son of his father's third wife. His father died when Gottfried was only six years old. Before he was twelve years of age, Leibniz was fluent in Latin and he had begun the study of Greek. Permission to use his father's law library was then granted to him and he entered upon preparatory studies for a career at law. At fifteen he was enrolled in the university at Leipzig, where he studied law, philosophy and mathematics. At the age of seventeen, he wrote his *De Principiis Individuis*. At twenty he experienced the sharp turns of fortune that shaped the course of his eventful and turbulent life. Refused a law degree at Leipzig in 1666 because of his youth, it was said, he went to Altdorf where, in the same year, he was not only granted a degree but also given an offer of a professorship at that university. At the age of twenty-one, Leibniz had already joined the ranks of serious writers on philosophy, mathematics and jurisprudence. Refusing the position at Altdorf, he accompanied the Baron von Boyneburg to Frankfort. There, as the Baron's protégé and friend, he entered upon a political career in the service of the Elector of Mainz. Two of his political writings, *Thoughts on Public Safety* and *Consilium Aegyptiacum*, intended to benefit the German States, brought him an appointment as an aide to a diplomatic mission to Paris on behalf of the Elector. Strongly attracted to the society of the leading scientists and mathematicians in Paris, Leibniz renewed his mathematical studies at this time, and attacked the current problems in mathematics and science with characteristic gusto. By 1673 he had published articles on logic, natural philosophy, mathematics, mechanics and optics, as well as on theology, law and politics, making some original contributions in each of these fields. Improving on Pascal's calculating machine he devised one which performed the four fundamental operations and also extracted roots. Nevertheless a professorship at Paris and the academic life which he now sought, were denied to him, and in 1676 Leibniz entered the service of the Duke of Hanover. He returned to Germany from Paris by a route which enabled him to visit London and Amsterdam and afforded him an opportunity of meeting with the great mathematicians and scientists of England and Holland.

Leibniz remained in the service of the Brunswick family for forty years to the day of his death. His duties as librarian of the great Hanover library did not preclude his continuing political and scientific activities. Under his aegis the *Akademie der Wissenschaften* was established in Berlin. His genealogical researches in Italy and elsewhere in Europe established the Hanoverian claim of succession to the throne of Great Britain. He planned to write an encyclopaedia. An enormous mass of his writings found after his death attested to his preparatory efforts. Unfortunately these writings were in a disorganized state and some were no more than fragments of projected articles. Leibniz's heirs released his entire literary remains to the Hanover National Library for a nominal sum. In addition to this large literary legacy, Leibniz also left a fairly large amount of money, most of which was found cached in a chest in his home.

It was Leibniz's lot to be involved almost continually in one controversy or another. The most widely publicized of these was his dispute with Newton over priority in the discovery of the calculus, but every field in which he wrote offered some feature of controversy. His death in 1716 interrupted his famous disputative correspondence with Samuel Clarke. During the period immediately after his death, Leibniz was known principally as a philosopher. In the modern view, his contributions to mathematics take on a larger significance. Broad applications of the science of symbolism to philosophy, to the art of invention, and to other fields of learning, continued to interest him throughout his life. His first thoughts on the concept of a symbolic logic appeared in his *De Arte Combinatoria* (1666), and he returned to his fundamental idea again and again, clarifying, emending and implementing it. Two fragments found among his papers contain Leibniz's introduction to symbolic logic. They clearly establish him as one of the founders of the science.

GOTTFRIED WILHELM V. LEIBNIZ

TWO FRAGMENTS FROM LEIBNIZ

translated from the Latin of Gehrhardt's text, Die
Philosophischen Schriften von G. W. Leibniz, Band VII,
'Scientia Generalis. Characteristica', XIX *and* XX,
from A Survey of Symbolic Logic, *by Clarence I. Lewis*

These two fragments represent the final form of Leibniz's
'universal calculus': their date is not definitely known, but almost
certainly they were written after 1685. Of the two, XX is in all
respects superior, as the reader will see, but XIX also is included
because it contains the operation of 'subtraction' which is
dropped in XX. Leibniz's comprehension of the fact that $+$ and $-$
(or, in the more usual notation, 'multiplication' and 'division')
are not simple inverses in this calculus, and his appreciation of
the complexity thus introduced, is the chief point of interest in
XIX. The distinction of 'subtraction' (in intension) and negation,
is also worthy of note. It will be observed that, in both these
fragments, $A+B$ (or $A \oplus B$) may be interpreted in two ways: (1)
As 'both A and B' in intension; (2) as 'either A or B', the class
made up of the two classes A and B, in extension. The 'logical'
illustrations mostly follow the first interpretation, but in XX
(see esp. *scholium to defs.* 3, 4, 5 *and* 6) there are examples of the
application to logical classes in extension. The illustration of the
propositions by the relations of line-segments also exhibits the
application to relations of extension. Attention is specifically
called to the parallelism between relations of intension and
relations of extension in the remark appended to prop. 15, in XX.
The *scholium to axioms* 1 *and* 2, in XX, is of particular interest as
an illustration of the way in which Leibniz anticipates later
logistic developments.

The Latin of the text is rather careless, and constructions are
sometimes obscure. Gehrhardt notes (p. 232) that the manuscript contains numerous interlineations and is difficult to read
in many places.

XIX: NON INELEGANS SPECIMEN DEMONSTRANDI IN ABSTRACTIS[1]

Def. 1. Two terms are the *same* (*eadem*) if one can be substituted for the other without altering the truth of any statement (*salva veritate*). If we have A and B, and A enters into some true proposition, and the substitution of B for A wherever it appears, results in a new proposition which is likewise true, and if this can be done for every such proposition, then A and B are said to be the *same*; and conversely, if A and B are the same, they can be substituted for one another as I have said. Terms which are the same are also called *coincident* (*coincidentia*); A and A are, of course, said to be the same, but if A and B are the same, they are called *coincident*.

Def. 2. Terms which are not the same, that is, terms which cannot always be substituted for one another, are *different* (*diversa*). *Corollary*. Whence also, whatever terms are not different are the same.

Charact. 1.[2] $A = B$ signifies that A and B are the *same*, or *coincident*.

Charact. 2.[3] $A \neq B$, or $B \neq A$, signifies that A and B are *different*.

Def. 3. If a plurality of terms taken together coincide with one, then any one of the plurality is said to *be in* (*inesse*) or to *be contained in* (*contineri*) that one with which they coincide, and that one is called the *container*. And conversely, if any term be contained in another, then it will be one of a plurality which taken together coincide with that other. For example, if A and B taken together coincide with L, then A, or B, will be called the *inexistent* (*inexistens*) or the *contained*; and L will be called the *container*. However, it can happen that the container and the contained coincide, as for example, if $(A$ and $B) = L$, and A and L coincide, for in that case B will contain nothing which is different from A. . . .[4]

---

1. This title appears in the manuscript, but Leibniz has afterward crossed it out. Although pretentious, it expresses admirably the intention of the fragment, as well as of the text.

2. We write $A = B$ where the text has $A \infty B$.

3. We write $A \neq B$ where the text has A non∞B.

4. Lacuna in the text, followed by '*significet A, significabit Nihil*'.

Scholium. Not every inexistent thing is a part, nor is every container a whole – e.g. an inscribed square and a diameter are both in a circle, and the square, to be sure, is a certain part of the circle, but the diameter is not a part of it. We must, then, add something for the accurate explanation of the concept of whole and part, but this is not the place for it. And not only can those things which are not parts be contained in, but also they can be subtracted (or 'abstracted', *detrahi*); e.g. the centre can be subtracted from a circle so that all points except the centre shall be in the remainder; for this remainder is the locus of all points within the circle whose distance from the circumference is less than the radius, and the difference of this locus from the circle is a point, namely the centre. Similarly the locus of all points which are moved, in a sphere in which two distinct points on a diameter remain unmoved, is as if you should subtract from the sphere the axis or diameter passing through the two unmoved points.

On the same supposition [that A and B together coincide with L], A and B taken together are called *constituents* (*constituentia*), and L is called *that which is constituted* (*constitutum*).

Charact. 3. $A+B=L$ signifies that A *is in* or *is contained in L*.

Scholium. Although A and B may have something in common, so that the two taken together are greater than L itself, nevertheless what we have here stated, or now state, will still hold. It will be well to make this clear by an example: Let L denote the straight line RX, and A denote a part of it, say the line RS, and B denote another part, say the line XY. Let either of these parts, RS or

$$R \qquad Y \qquad S \qquad X$$

XY, be greater than half the whole line, RX; then certainly it cannot be said that $A+B$ equals L, or $RS+XY$ equals RX. For inasmuch as YS is a common part of RS and XY, $RS+XY$ will be equal to $RX+SY$. And yet it can truly be said that the lines RS and XY together *coincide* with the line RS.[1]

Def. 4. If some term M is in A and also in B, it is said to be common to them, and they are said to be *communicating* (*com-*

1. Italics ours.

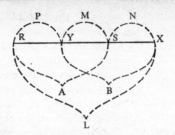

Fig. 42.1.

municantia).[1] But if they have nothing in common, as *A* and *N* (the lines *RS* and *XS*, for example), they are said to be *non-communicating* (*incommunicantia*).

Def. 5. If *A* is in *L* in such wise that there is another term, *N*, in which belongs everything in *L* except what is in *A*, and of this last nothing belongs in *N*, then *A* is said to be *subtracted* (*detrahi*) or taken away (*removeri*), and *N* is called the *remainder* (*residuum*).

Charact. 4. $L-A=N$ signifies that *L* is the container from which if *A* be *subtracted* the *remainder* is *N*.

Def. 6. If some one term is supposed to coincide with a plurality of terms which are added (*positis*) or subtracted (*remotis*), then the plurality of terms are called the *constituents*, and the one term is called the thing constituted.[2]

Scholium. Thus all terms which are in anything are constituents, but the reverse does not hold; for example, $L-A=N$, in which case *L* is not in *A*.

Def. 7. Constitution (that is, addition or subtraction) is either tacit or expressed, $-N$ or $-M$ the tacit constitution of *M* itself, as *A* or $-A$ in which *N* is. The expressed constitution of *N* is obvious.[3]

Def. 8. *Compensation* is the operation of adding and subtracting the same thing in the same expression, both the addition

1. The text here has '*communicatia*', clearly a misprint.
2. Leibniz's idea seems to be that if $A+N=L$ then *L* is 'constituted' by *A* and *N*, and also if $L-A=N$ then *L* and *A* 'constitute' *N*. But it may mean that if $L-A=N$, then *A* and *N* 'constitute' *L*.
3. This translation is literal: the meaning is obscure, but see the diagram above.

and the subtraction being expressed [as $A+M-M$]. *Destruction* is the operation of dropping something on account of compensation, so that it is no longer expressed, and for $M-M$ putting Nothing.

Axiom 1. If a term be added to itself, nothing new is constituted or $A+A=A$.

Scholium. With numbers, to be sure, $2+2$ makes 4, or two coins added to two coins make four coins, but in that case the two added are not identical with the former two; if they were, nothing new would arise, and it would be as if we should attempt in jest to make six eggs out of three by first counting 3 eggs, then taking away one and counting the remaining 2, and then taking away one more and counting the remaining 1.

Axiom 2. If the same thing be added and subtracted, then, however, it enter into the constitution of another term, the result coincides with Nothing. Or A (however many times it is added in constituting any expression) $-A$ (however many times it is subtracted from that same expression) $=$ Nothing.

Scholium. Hence $A-A$ or $(A+A-)-A$ or $A (A+A)$, etc. $=$ Nothing. For by axiom 1, the expression in each case reduces to $A-A$.

Postulate 1. Any plurality of terms whatever can be added to constitute a single term; as for example, if we have A and B, we can write $A+B$, and call this L.

Post. 2. Any term, A, can be subtracted from that in which it is, namely $A+B$ or L, if the remainder be given as B, which added to A constitutes the container L – that is, on this supposition [that $A+B=L$] the remainder $L-A$ can be found.

Scholium. In accordance with this postulate, we shall give, later on, a method for finding the difference between two terms, one of which, A, is contained in the other, L, even though the remainder, which together with A constitutes L, should not be given – that is, a method for finding $L-A$, or $A+B-A$, although A and L only are given, and B is not.

Theorem 1

Terms which are the same with a third, are the same with each other.

If $A=B$ and $B=C$, then $A=C$. For if in the proposition $A=B$ (true by hyp.) C be substituted for B (which can be done by def. 1, since, by hyp., $B=C$), the result is $A=C$. Q.E.D.

Theorem 2

If one of two terms which are the same be different from a third term, then the other of the two will be different from it also.

If $A=B$ and $B\neq C$, then $A\neq C$. For if in the proposition $B\neq C$ (true by hyp.) A be substituted for B (which can be done by def. 1, since, by hyp., $A=B$), the result is $A\neq C$. Q.E.D.

[Theorem in the margin of the manuscript.]

Here might be inserted the following theorem: *Whatever is in one of two coincident terms, is in the other also.*

If A is in B and $B=C$, then also A is in C. For in the proposition A is in B (true by hyp.) let C be substituted for B.

Theorem 3

If terms which coincide be added to the same term, the results will coincide.

If $A=B$, then $A+C=B+C$. For if in the proposition $A+C=A+C$ (true *per se*) you substitute B for A in one place (which can be done by def. 1, since $A=B$), it gives $A+C=B+C$. Q.E.D.

Corollary. If terms which coincide be added to terms which coincide, the results will coincide. If $A=B$ and $L=M$, then $A+L=B+M$. For (by the present theorem) since $L=M$, $A+L=A+M$, and in this assertion putting B for A in one place (since by hyp. $A=B$) gives $A+L=B+M$. Q.E.D.

Theorem 4

A container of the container is a container of the contained; or if that in which something is, be itself in a third thing, then that which is in it will be in that same third thing – that is, if A is in B and B is in C, then also A is in C.

For A is in B (by hyp.), hence (by def. 3 or charact. 3) there is some term, which we may call L, such that $A+L=B$. Similarly, since B is in C (by hyp.), $B+M=C$, and in this assertion putting $A+L$ for B (since we show that these coincide) we have $A+L+M=C$. But putting N for $L+M$ (by post. 1) we have $A+N=C$. Hence (by def. 3) A is in C. Q.E.D.

Theorem 5

Whatever contains terms individually contains also that which is constituted of them.

If A is in C and B is in C, then $A+B$ (constituted of A and B, def. 4) is in C. For since A is in C, there will be some term M such that $A+M=C$ (by def. 3). Similarly, since B is in C, $B+N=C$. Putting these together (by the corollary to th. 3), we have $A+M+B+N=C+C$. But $C+C=C$ (by ax. 1), hence $A+M+B+N=C$. And therefore (by def. 3) $A+B$ is in C. Q.E.D.[1]

Theorem 6

Whatever is constituted of terms which are contained, is in that which is constituted of the containers.

If A is in M and B is in N, then $A+B$ is in $M+N$. For A is in M (by hyp.) and M is in $M+N$ (by def. 3), hence A is in $M+N$ (by th. 4). Similarly, B is in N (by hyp.) and N is in $M+N$ (by def. 3), hence B is in $M+N$ (by th. 4). But if A is in $M+N$ and B is in $M+N$, then also (by th. 5) $A+B$ is in $M+N$. Q.E.D.

Theorem 7

If any term be added to that in which it is, then nothing new is constituted; or if B is in A, then $A+B=A$.

For if B is in A, then [for some C] $B+C=A$ (def. 3). Hence (by th. 3) $A+B=B+C+B=B+C$ (by ax. 1) $=A$ (by the above). Q.E.D.

1. In the margin of the manuscript at this point Leibniz has an untranslatable note, the sense of which is to remind him that he must insert illustrations of these propositions in common language.

Converse of the Preceding Theorem

If by the addition of any term to another nothing new is constituted, then the term added is in the other.

If $A+B=A$, then B is in A; for B is in $A+B$ (def. 3), and $A+B=A$ (by hyp.). Hence B is in A (by the principle which is inserted between ths. 2 and 3). Q.E.D.

Theorem 8

If terms which coincide be subtracted from terms which coincide, the remainders will coincide.

If $A=L$ and $B=M$, then $A-B=L-M$. For $A-B=A-B$ (true *per se*), and the substitution, on one or the other side, of L for A and M for B, gives $A-B=L-M$. Q.E.D.

[Note in the margin of the manuscript.] In dealing with concepts, *subtraction* (*detractio*) is one thing, negation another. For example, '½non-rational man' is absurd or impossible. But we may say; An ape is a man except that it is not rational. [They are men except in those respects in which man differs from the beasts, as in the case of Grotius's Jumbo[1] (*Homines nisi qua bestiis differt homo, ut in Jambo Grotii*). 'Man'—'rational' is something different from 'non-rational man'. For 'man'—'rational'='brute'. But 'non-rational man' is impossible. 'Man'—'animal'—'rational' is Nothing. Thus subtractions can give Nothing or simple non-existence – even less than nothing – but negations can give the impossible.[2]

1. Apparently an allusion to some description of an ape by Grotius.
2. This is not an unnecessary and hair-splitting distinction, but on the contrary, perhaps the best evidence of Leibniz's accurate comprehension of the logical calculus which appears in the manuscripts. It has been generally misjudged by the commentators, because the commentators have not understood the logic of intension. The distinction of the merely non-existent and the impossible (self-contradictory or absurd) is absolutely essential to any calculus of relations in intension. And this distinction of subtraction (or in the more usual notation, division) from negation, is equally necessary. It is by the confusion of these two that the calculuses of Lambert and Castillon break down.

XX

Def. 1. Terms which can be substituted for one another wherever we please without altering the truth of any statement (*salva veritate*), are the *same* (*eadem*) or *coincident* (*coincidentia*). For example, 'triangle' and 'trilateral', for in every proposition demonstrated by Euclid concerning 'triangle', 'trilateral' can be substituted without loss of truth.

$A = B$[1] signifies that A and B are the same, or as we say of the straight line XY and the straight line YX, $XY = YX$, or the shortest path of a [point] moving from X to Y coincides with that from Y to X.

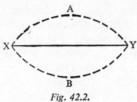

Fig. 42.2.

Def. 2. Terms which are not the same, that is, terms which cannot always be substituted for one another, are *different* (*diversa*). Such are 'circle' and 'triangle', or 'square' (supposed perfect, as it always is in Geometry) and 'equilateral quadrangle', for we can predicate this last of a rhombus, of which 'square' cannot be predicated.

$A \neq B$ signifies that A and B are different,[2] as for example, the straight lines XY and RS.

$$R \qquad Y \qquad S \qquad X$$

Prop. 1. *If $A = B$, then also $B = A$. If anything be the same with another, then that other will be the same with it.* For since $A = B$ (by hyp.), it follows (by def. 1) that in the statement $A = B$ (true by hyp.) B can be substituted for A and A for B; hence we have $B = A$.

1. $A = B$ for $A \infty B$, as before.
2. $A \neq B$ for A non∞B, as before.

Prop. 2. *If $A \neq B$, then also $B \neq A$. If any term be different from another, then that other will be different from it.* Otherwise we should have $B = A$, and in consequence (by the preceding prop.) $A = B$, which is contrary to hypothesis.

Prop. 3. *If $A = B$ and $B = C$, then $A = C$. Terms which coincide with a third term coincide with each other*. For if in the statement $A = B$ (true by hyp.) C be substituted for B (by def. 1, since $A = B$), the resulting proposition will be true.

Coroll. If $A = B$ and $B = C$ and $C = D$, then $A = D$; and so on. For $A = B = C$, hence $A = C$ (by the above prop.). Again, $A = C = D$; hence (by the above prop.) $A = D$.

Thus since equal things are the same in magnitude, the consequence is that things equal to a third are equal to each other. The Euclidean construction of an equilateral triangle makes each side equal to the base, whence it results that they are equal to each other. If anything be moved in a circle, it is sufficient to show that the paths of any two successive periods, or returns to the same point, coincide, from which it is concluded that the paths of any two periods whatever coincide.

Prop. 4. *If $A = B$ and $B \neq C$, then $A \neq C$. If of two things which are the same with each other, one differ from a third, then the other also will differ from that third.* For if in the proposition $B \neq C$ (true by hyp.) A be substituted for B, we have (by def. 1, since $A = B$) the true proposition $A \neq C$.

Def. 3. *A is in L, or L contains A,* is the same as to say that L can be made to coincide with a plurality of terms, taken together, of which A is one.

Def. 4. Moreover, all those terms such that whatever is in them is in L, are together called *components* (*componentia*) with respect to the L thus *composed* or constituted.

$B \oplus N = L$ signifies that B is in L; and that B and N together compose or constitute L.[1] The same thing holds for a larger number of terms.

Def. 5. I call terms one of which is in the other *sub-alternates* (*subalternantia*), as A and B if either A is in B or B is in A.

1. In this fragment, as distinguished from XIX, the logical or 'real' sum is represented by $\oplus$. Leibniz has carelessly omitted the circle in many places, but we write $\oplus$ wherever this relation is intended.

Def. 6. Terms neither of which is in the other [I call] *disparate* (*disparata*).

Axiom 1. $B \oplus N = N \oplus B$, or transposition here alters nothing.

Post. 2. Any plurality of terms, as A and B, can be added to compose a single term, $A \oplus B$ or L.

Axiom 2. $A \oplus A = A$. If nothing new be added, then nothing new results, or repetition here alters nothing. (For 4 coins and 4 other coins are 8 coins, but not 4 coins and the same 4 coins already counted).

Prop. 5. If A is in B and $A = C$, then C is in B. *That which coincides with the inexistent, is inexistent.* For in the proposition, A is in B (true by hyp.), the substitution of C for A (by def. 1 of coincident terms, since, by hyp., $A = C$) gives, C is in B.

Prop. 6. If C is in B and $A = B$, then C is in A. *Whatever is in one of two coincident terms, is in the other also.* For in the proposition, C is in B, the substitution of A for C (since $A = C$) gives A is in B. (This is the converse of the preceding.)

Prop. 7. A is in A. *Any term whatever is contained in itself.* For A is in $A \oplus A$ (by def. of 'inexistent', that is, by def. 3) and $A \oplus A$ (by ax. 2). Therefore (by prop. 6), A is in A.

Prop. 8. If $A = B$, then A is in B. *Of terms which coincide, the one is in the other.* This is obvious from the preceding. For (by the preceding) A is in A – that is (by hyp.), in B.

Prop. 9. If $A = B$, then $A \oplus C = B \oplus C$. *If terms which coincide*

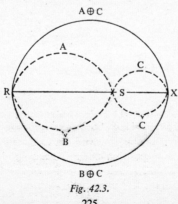

B ⊕ C

Fig. 42.3.

be added to the same term, the results will coincide. For if in the
proposition, $A \oplus C = A \oplus C$ (true *per se*), for A in one place
be substituted B which coincides with it (by def. 1), we have
$A \oplus C = B \oplus C$.

$$A \text{ 'triangle'} \atop B \text{ 'trilateral'} \Big\} \text{coincide}$$

$$A \oplus C \text{ 'equilateral triangle'} \atop B \oplus C \text{ 'equilateral trilateral'} \Big\} \text{coincide}$$

Scholium. This proposition cannot be converted, much less,
the two which follow.

Prop. 10. *If $A = L$ and $B = M$, then $A \oplus B = L \oplus M$. If terms
which coincide be added to terms which coincide, the results will
coincide.* For since $B = M$, $A \oplus B = A \oplus M$ (by the preceding),
and putting L for the second A (since, by hyp., $A = L$) we have
$A \oplus B = L \oplus M$.

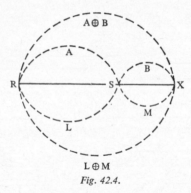

Fig. 42.4.

A 'triangle', and L 'trilateral' coincide. B 'regular' coincides
with M 'most capacious of equally-many-sided figures with equal
perimeters'. 'Regular triangle' coincides with 'most capacious of
trilaterals making equal peripheries out of three sides'.

Scholium. This proposition cannot be converted, for if $A \oplus B$
$= L \oplus M$ and $A = L$, still it does not follow that that $B = M$,
and much less can the following be converted.

Prop. 11. *If $A=L$ and $B=M$ and $C=N$, then $A \oplus B \oplus C=L \oplus M \oplus N$. And so on. If there be any number of terms under consideration, and an equal number of them coincide with an equal number of others, term for term, then that which is composed of the former coincides with that which is composed of the latter.* For (by the preceding, since $A=L$ and $B=M$) we have $A \oplus B = L \oplus M$. Hence, since $C=N$, we have (again by the preceding) $A \oplus B \oplus C=L \oplus M \oplus N$.

Prop. 12. *If B is in L, then $A \oplus B$ will be in $A \oplus L$. If the same term be added to what is contained and to what contains it, the former result is contained in the latter.* For $L=B \oplus N$ (by def. of 'inexistent'), and $A \oplus B$ is in $B \oplus N \oplus A$ (by the same), that is, $A \oplus B$ is in $L \oplus A$.

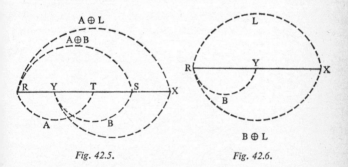

Fig. 42.5.   Fig. 42.6.

B 'equilateral', L 'regular', A 'quadrilateral'. 'Equilateral' is in or is attribute of 'regular'. Hence 'equilateral quadrilateral' is in 'regular quadrilateral' or 'perfect square'. YS is in RX. Hence $RT \oplus YS$, or RS, is in $RT \oplus RX$, or in RX.

Scholium. This proposition cannot be converted; for if $A \oplus B$ is in $A \oplus L$, it does not follow that B is in L.

Prop. 13. *If $L \oplus B=L$, then B is in L. If the addition of any term to another does not alter that other, then the term added is in the other.* For B is in $L \oplus B$ (by def. of 'inexistent') and $L \oplus B= L$ (by hyp.), hence (by prop. 6) B is in L.

$RY \oplus RX=RX$. Hence RY is in RX.

RY is in RX. Hence $RY \oplus RX=RX$.

Let L be 'parallelogram' (every side of which is parallel to some side),[1] B be 'quadrilateral'.

'Quadrilateral parallelogram' is in the same as 'parallelogram'. Therefore to be quadrilateral is in [the intension of] 'parallelogram'.

Reversing the reasoning, to be quadrilateral is in 'parallelogram'.

Therefore, 'quadrilateral parallelogram' is the same as 'parallelogram'.

Prop. 14. *If B is in L, then $L \oplus B = L$. Subalternates compose nothing new; or if any term which is in another be added to it, it will produce nothing different from that other.* (*Converse of the preceding.*) If B is in L, then (by def. of 'inexistent') $L = B \oplus P$. Hence (by prop. 9) $L \oplus B = B \oplus P \oplus B$, which (by ax. 2) is $= B \oplus P$, which (by hyp.) is $= L$.

Prop. 15. *If A is in B and B is in C, then also A is in C. What is contained in the contained, is contained in the container.* For A is in B (by hyp.), hence $A \oplus L = B$ (by def. of 'inexistent'). Similarly, since B is in C, $B \oplus M = C$, and putting $A \oplus L$ for B in this statement (since we have shown that these coincide), we have $A \oplus L \oplus M = C$. Therefore (by def. of 'inexistent') A is in C.

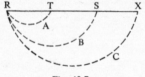

Fig. 42.7.

RT is in RS, and RS in RX.

Hence RT is in RX.

A 'quadrilateral', B 'parallelogram', C 'rectangle'.

To be quadrilateral is in [the intension of] 'parallelogram', and to be parallelogram is in 'rectangle' (that is, a figure every angle of which is a right angle). If instead of concepts *per se*, we consider individual things comprehended by the concept, and put

1. Leibniz uses 'parallelogram' in its current meaning, though his language may suggest a wider use.

A for 'rectangle', *B* for 'parallelogram', *C* for 'quadrilateral', the relations of these can be inverted. For all rectangles are comprehended in the number of the parallelograms, and all parallelograms in the number of the quadrilaterals. Hence also, all rectangles are contained amongst (*in*) the quadrilaterals. In the same way, all men are contained amongst (*in*) all the animals, and all animals amongst all the material substances, hence all men are contained amongst the material substances. And conversely, the concept of material substance is in the concept of animal, and the concept of animal is in the concept of man. For to be man contains [or implies] being animal.

Scholium. This proposition cannot be converted, and much less can the following.

Coroll. If $A \oplus N$ is in B, N also is in B. For N is in $A \oplus N$ (by def. of 'inexistent').

Prop. 16. If A is in B and B is in C and C is in D, then also A is in D. And so on. *That which is contained in what is contained by the contained, is in the container.* For if A is in B and B is in C, A also is in C (by the preceding). Whence if C is in D, then also (again by the preceding) A is in D.

Prop. 17. If A is in B and B is in A, then A = B. Terms which contain each other coincide. For if A is in B, then $A \oplus N = B$ (by def. of 'inexistent'). But B is in A (by hyp.), hence $A \oplus N$ is in A (by prop. 5). Hence (by coroll. prop. 15) N also is in A. Hence (by prop. 14) $A = A \oplus N$, that is, $A = B$.

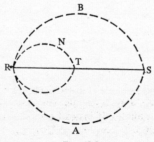

Fig. 42.8.

RT, N; RS, A; $SR \oplus RT$, B.

To be trilateral is in [the intension of] 'triangle', and to be triangle is in 'trilateral'. Hence 'triangle' and 'trilateral' coincide. Similarly, to be omniscient is to be omnipotent.

Prop. 18. *If A is in L and B is in L, then also A $\oplus$ B is in L. What is composed of two, each contained in a third, is itself contained in that third.* For since A is in L (by hyp.), it can be seen that $A \oplus M = L$ (by def. of 'inexistent'). Similarly, since B is in L, it can be seen that $B \oplus N = L$. Putting these together, we have (by prop. 10) $A \oplus M \oplus B \oplus N = L \oplus L$. Hence (by ax. 2)[1] $A \oplus M \oplus B \oplus N = L$. Hence (by def. of 'inexistent') $A \oplus B$ is in L.

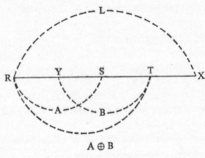

$A \oplus B$

Fig. 42.9.

RYS is in RX.

YST is in RX.

Hence RT is in RX.

A 'equiangular', B 'equilateral', $A \oplus B$ 'equiangular equilateral' or 'regular', L 'square'. 'Equiangular' is in [the intension of] 'square', and 'equilateral' is in 'square'. Hence 'regular' is in 'square'.

Scholium to defs. 3, 4, 5 *and* 6. We say that the concept of the genus *is in* the concept of the species; the individuals of the species amongst (*in*) the individuals of the genus; a part of the whole; and indeed the ultimate and indivisible in the continuous, as a point is in a line, although a point is not a part of the line. Likewise the concept of the attribute or predicate is in the concept of the

1. The number of the axiom is given in the text as 5, a misprint.

subject. And in general this conception is of the widest application. We also speak of that which is in something as contained in that in which it is. We are not here concerned with the notion of 'contained' in general – with the manner in which those things which are 'in' are related to one another and to that which contains them. Thus our demonstrations cover also those things which compose something in the distributive sense, as all the species together compose the genus. Hence all the inexistent things which suffice to constitute a container, or in which are all things which are in the container, are said to compose that container; as for example, $A \oplus B$ are said to *compose L*, if A, B, and L denote the straight lines RS, YX, and RX, for $RS \oplus YX = RX$. And such parts which complete the whole, I am accustomed to call 'cointegrants', especially if they have no common part; if they have a common part, they are called 'co-members', as RS and RX. Whence it is clear that the same thing can be composed in many different ways if the things of which it is composed are themselves composite. Indeed, if the resolution could finally be carried to infinity, the variations of composition would be infinite. Thus all synthesis and analysis depends upon the principles here laid down. And if those things which are contained are homogeneous with that in which they are contained, they are called parts and the container is called the whole. If two parts, however chosen, are such that a third can be found having a part of one and a part of the other in common, then that which is composed of them is continuous. Which illustrates by what small and simple additions one concept arises from another. And I call by the name 'subalternates' those things one of which is in the other, as the species in the genus, the straight line RS in the straight line RX; 'disparates' where the opposite is the case, as the straight lines RS and YX, two species of the same genus, perfect metal and imperfect metal – and particularly, members of the different divisions of the same whole, which (members) have something in common, as for example, if you divide 'metal' into 'perfect' and 'imperfect', and again into 'soluble in *aqua fortis*' and 'insoluble', it is clear that 'metal which is insoluble in *aqua fortis*' and 'perfect metal' are two disparate things, and there is metal which is perfect, or is always capable of being fulminated in a cupel,[1] and yet is soluble in *aqua fortis*, as silver, and on the other hand, there is imperfect metal which is insoluble in *aqua fortis*, as tin.

Scholium to axioms 1 *and* 2. Since the ideal form of the general [or ideal form in general, *speciosa generalis*] is nothing but the representation of combinations by means of symbols, and their manipulation, and the discoverable laws of combination are various,[2] it results from this that various modes of computation arise. In this place, however, we have nothing to do with the theory of the variations which consist simply in changes of order [i.e. the theory of permutations], and AB [more consistently, $A \oplus B$] is for us the same as BA [or $B \oplus A$]. And also we here take no account of repetition – that is AA [more consistently, $A \oplus A$] is for us the same as A. Thus wherever these laws just mentioned can be used, the present calculus can be applied. It is obvious that it can also be used in the composition of absolute concepts, where neither laws of order nor of repetition obtain; thus to say 'warm and light' is the same as to say 'light and warm', and to say 'warm fire' or 'white milk', after the fashion of the poets, is pleonasm; white milk is nothing different from milk, and rational man – that is, rational animal which is rational – is nothing different from rational animal. The same thing is true when certain given things are said to be contained in (*inexistere*) certain things. For the real addition of the same is a useless repetition. When 2 and 2 are said to make 4, the latter 2 must be different from the former. If they were the same, nothing new would arise, and it would be as if one should in jest attempt to make 6 eggs out of 3 by first counting 3 eggs, then taking away one and counting the remaining 2, and then taking away one more and counting the remaining 1. But in the calculus of numbers and magnitudes, A or B or any other symbol does not signify a certain object but anything you please with that number of

1. The text here has '. . . *fulminabile persistens in capella*': the correction is obvious.

2. '. . . *variaeque sint combinandi leges excogitabiles,* . . .' '*Excogitabiles*', 'discoverable by imagination or invention', is here significant of Leibniz's theory of the relation between the 'universal calculus' and the progress of science.

congruent parts, for any two feet whatever are denoted by 2; if foot is the unit or measure, then $2 \oplus 2$ makes the new thing 4 and 3 times 3 the new thing 9, for it is presupposed that the things added are always different (although of the same magnitude); but the opposite is the case with certain things, as with lines. Suppose we describe by a moving [point] the straight line, $RY \oplus YX = RYX$ or $P \oplus B = L$, going from R to X. If we suppose this same [point] then to return from X to Y and stop there, although it does indeed describe YX or B a second time, it produces nothing different than if it had described YX once. Thus $L \oplus B$ is the same as L – that is, $P \oplus B \oplus B$ or $RY \oplus YX \oplus XY$ is the same as $RY \oplus YX$. This caution is of much importance in making judgements, by means of the magnitude and motion of those things which generate[1] or describe, concerning the magnitude of those things which are generated or described. For care must be taken either that one [step in the process] shall not choose the track of another as its own – that is, one part of the describing operation follow in the path of another – or else [if this should happen] this [reduplication] must be subtracted so that the same thing shall not be taken too many times. It is clear also from this that 'components', according to the concept which we here use, can compose by their magnitudes a *magnitude* greater than the magnitude of the *thing* which they compose.[2] Whence the composition of things differs widely from the composition of magnitudes. For example, if there are two parts, A or RS and B or RX, of the whole line L or RX, and each of these is greater than half of RX itself – if, for example, RX is 5 feet and RS 4 feet and YX 3 feet – obviously the magnitudes of the parts compose a magnitude of 7 feet, which is greater than that of the whole; and yet the lines RS and YX themselves compose nothing different from RX – that is, $RS \oplus YX = RX$. Accordingly I here denote this real addition by $\oplus$, as the addition of magnitudes is denoted by $+$. And finally, although it is of much importance, when it is a question of the actual generation of things, what their order is

---

1. Reading 'generant' for '*generantur*' – a correction which is not absolutely necessary, since a motion which generates a line is also itself generated; but, as the context shows, '*generare*' and '*describere*' are here synonymous.

2. Italics ours.

(for the foundations are laid before the house is built), still in the mental construction of things the result is the same whichever ingredient we consider first (although one order may be more convenient than another), hence the order does not here alter the thing developed. This matter is to be considered in its own time and proper place. For the present, however, $RY \oplus YS \oplus SX$ is the same as $YS \oplus RY \oplus SX$.

43

AUGUSTUS DE MORGAN
(1806–71)

AUGUSTUS DE MORGAN, one of the greatest teachers of mathematics of all time, possessed a rare combination of profound insight, good humour and originality. Born in Madura, India, he was educated at private schools, entering Trinity College, Cambridge at the age of sixteen. An excellent student, he nevertheless failed to qualify for his M.A. degree because of his conscientious objection to the religious tests required of candidates for the degree at Cambridge at that time. In 1828 he received an appointment as professor of mathematics in the newly established London University. His highly informative lectures hold a unique position in university teaching for their lively extemporaneous delivery and sparkling interest. The lustre of his reputation was enhanced by his whimsy and love of the unexpected, clearly evidenced in his *Budget of Paradoxes*, by his gentle nature, and his genius for logical presentation and lucidity of expression. Extremely independent in mind and spirit, De Morgan encouraged genuine aptitude wherever it appeared. Increasingly liberal as he grew older, he became more sympathetic towards women's rights and he lectured to women's classes without charge.

De Morgan's texts and treatises on algebra, trigonometry, differential and integral calculus, on the calculus of variations, and on probability, charmingly written in a leisurely style, remain valuable sources to this day for students who wish to acquire mathematical techniques and basic understanding. Many of his writings appeared in periodicals, learned journals and encyclopaedias, where they were in great demand. A variety of his works, some of which are great treatises in mathematics and logic, may be found in the Cambridge Philosophical Society Transactions, in Lardner's *Cyclopaedia*, in the *Encyclopaedia Metropolitana*, in the *Quarterly Journal of Education*, in the *Penny Encyclopaedia* and others.

In 1837 De Morgan married Sophia Elizabeth Frend. Their home in Chelsea, enlivened by their five children, became the centre of a large circle of friends who shared their intellectual and artistic interests. Prominent among these was George Boole who side by side with De Morgan pioneered in the development of symbolic logic. De Morgan's *Formal Logic* appeared in the same year (1847) as Boole's work on the algebra of logic. Both works were concerned with an algebra or calculus of classes. In 1860 De Morgan published his *Syllabus of a Proposed System of Logic* in which he constructed a similar calculus of relations. The two dually related theorems of the algebra of classes (De Morgan's Laws), first enunciated by him in 1847, were clearly presented again in his article *On the Syllogism, No. III, and On Logic in General* (1864). De Morgan laid great stress on the need and value of the symbolism in logic. Through the power of his personal magnetism and the popularity of his numerous publications he was a stimulating force in the development of mathematics and logic.

ON THE SYLLOGISM NO. III AND ON LOGIC IN GENERAL

SECTION I. GENERAL CONSIDERATIONS

The syllogism itself is the web of an argument, on which the tapestry of thought is woven; the *primed* canvas on which the picture is painted. The logician presents it to the world as the tapestry or the picture: he does this in effect by the position he makes it occupy; for he sends the primed canvas to the exhibition. And the world does not see that, though the syllogism be a mere canvas, it stands to the thinker in a very different position from that in which the canvas stands to the painter. Call the historian or the moralist a practised artist at a thousand a year, and I am well content that his structure of the canvas shall be valued at ten shillings a week: it would not hurt my argument if it were valued at a half-penny. For the painter can and does delegate the preparation of the canvas; the historian cannot *put out* his logic. He must do it himself as he goes on; and he must do it well, or his whole work is spoiled.

I will take an example from one of the unusual forms of syllogism. Say ' The time is past in which the transmission of news can be measured by the speed of animals or even of steam; for the telegraph is not approached by either'. Is this a syllogism? Many would say it is not; but wrongly. Throw out the *charges*, the modal reference to past falsehood and present truth, the advantage of the telegraph, its *superior* speed, the reference to progress conveyed in *even* – and we rub off the whole design of the picture. But the ground which carried the design is a syllogism. In old form it is *Darapti*, awkwardly.

> All telegraph speed is (not steam speed)
> All telegraph speed is (not animal speed)
Therefore   Some (not animal speed) is (not steam speed).

In the system which admits contraries it is a syllogism with two negative premises, and a form of conclusion unknown to Aristotle: it is, in the symbols I use, the deduction of)(from)·()·(

> No animal speed is telegraph speed
> No steam speed is telegraph speed
Therefore   Some speed is neither animal nor steam speed.

When this is presented, a person would naturally ask, What then? The answer to this question is seen when the charges are restored, and the sentence takes its proper place in the whole argument.

v. A great objection has been raised to the employment of mathematical symbols: and it seems to be taken for granted that any symbols used by me must be mathematical. The truth is that I have not made much use of symbols actually employed in algebra; and the use which I have made is in one instance seriously objectionable, and must be discontinued. But it has been left to me to discover this mistake, into which I was led, as I shall show, by the ordinary school of logicians. If A and B be the premises of a syllogism, and C the conclusion, the representation $A+B=C$ is faulty in two points. The premises are compounded, not aggregated; and AB should have been written: the relation of joint premises to conclusion is that (speaking in extension) of contained and containing, and $AB<C$ should have been the symbol. Nevertheless, $A+B=C$, with all its imperfections, made a suggestion of remarkable character to an inventive friend of mine: while $AB<C$ was both a *suggestio veri* and a *suppressio*

falsi to myself. For these things see the second part of this paper.

As to symbols in general, it is not necessary to argue in their favour: mine or better ones will make their way, under all the usual difficulties of new language. There was a time when logic had more peculiar symbols than algebra. Every system of signs, before it has become familiar, as we all remember when we look back to *ABC*, is repulsive, difficult, unmeaning, full of signs of difference which are practical synonyms[1] by combination of want of comprehension with ignorance of the want. But it is too certain to need argument that the separation of form and matter requires as many symbols as there are separations. . . .

SECTION II. FIRST ELEMENTS OF A SYSTEM OF LOGIC

XXII. A *name* is a sign by which we distinguish one object, process or product of thought from another.

A name has four applications: two *objective*, signs of what the mind can (be it right or wrong in so doing) conceive to exist though thought were annihilated; two *subjective*, signs of what the mind cannot but conceive to be annihilated with thought. (§ VI.)

The objective applications are: first, to individual *objects* external to the mind; secondly, to individual notions attaching to or connected with objects, called *qualities*, which are said *and thought* to inhere in the objects. The object itself is but a compound of qualities. The *quality* is, in logic, any appurtenance whatever: thus to be called Caesar is a quality.

The subjective applications are: to designate *class*, collection of objects having similar qualities, or put into one notion by some similar *class-mark*; and a class may consist of one individual only, if only one have the mark: secondly, to designate *attribute*, the class-name of *quality*, the notion of quality in the mind.

1. A Cambridge tutor of high reputation was once trying to familiarize a beginner with the difference between na and a^n. After repeated illustration, he asked the pupil whether he saw the point. 'Thank you very much, Mr —' was the answer; 'I now see perfectly what you mean: but, Mr —, between ourselves, now, and speaking candidly, *don't* you think it's a *needless refinement*.'

Class and *attribute* are two modes of thinking many in the manner of one; two reductions of plurality to unity. The class *man* is one notion in the mind, the receptacle of many individuals: the attribute *human* is also one in thought, being the notion derived from similar qualities possessed by many individuals. *Class* is a noun of multitude, but not a multitude of nouns, nor even one noun selected from a multitude.

Quality, as a class name, may be distributed over any number of individuals. But it has a division peculiar to itself. As merely an appurtenant notion, it may be the compound of several notions. And similarly, attribute may be the compound of several attributes.

In grammar, class is often a substantive capable of designating the individual also, and used in both ways, as man, the man Plato. Quality is often an adjective, as *human*; attribute is often an abstract substantive, as *humanity*, which cannot designate an individual. Language would be more perfect if these distinctions were made by inflexion in all cases: logic would do well to think of introducing such[1] words as X-ic and X-ity.

The class is an aggregate of individuals; the individual is a compound of qualities; the attribute is a compound of attributes.

Objective names, representing objects and qualities, were once called names of *first intention* or *first notions*, as being used according to the mind's first *bent* towards names. Subjective names, representing classes and attributes, were names of *second intention*, or *second notions*. Thus, 'every crow is black' considered merely as a collation of cases, is of first intention: but 'the class crow has the attribute black' is of second intention. Nevertheless, the first sentence, spoken or written, may be thought under the second form. (§ XII.)

Logic is the science and art, the theory and practice, of the

1. The boldest symbolic word ever made was proposed – I think in print, but I cannot find the reference and I am not sure – by one of the young analysts (I speak of 1813, or thereabouts) who first cultivated the continental analysis in England. It was such as 'iso-*x*-ical', to signify that in which x has always one value. Thus every circle which has its centre at the origin is iso-$(x^2 + y^2)$-ical. When the crude form is not too complicated, the word might be useful.

form of thought, the law of its action, the working of its machinery; independently of the matter thought on. It considers different kinds of matter only when, if ever, and so far as, they necessitate different forms of thought. It must deal with names, and therefore should deal with all the forms of thought demanded in the four uses of names. It has no right to reject any use of a name: for every such use appertains to a form of thought. (§§ II, III.)

Logic considers both first and second intentions, because both are forms of thought; but the first chiefly as leading to the second: and in both it considers *quae non debentur rebus secundum se, sed secundum esse quod habent in anima*. That is, logic belongs to psychology, not to metaphysics.

It is not to be assumed that the practice of the logic of first intentions is the common property of mankind, and that second intentions form a science to which the student is to be led. The actual form of thought is an unanalysed mixture of first and second intention, with the latter in decided predominance.

A name may be formed from other names as follows. First, by *extension*, symbolized in $X = (A, B, C)$ where the *aggregate X* includes as much as can be spoken of under each and all of the *aggregants A, B, C*. Secondly, by *intension*, symbolized in $X = A\text{-}B\text{-}C$, or ABC, where the *compound X* includes *no more than* can be spoken of under *all* the *component* names *A, B, C*. Thirdly, by combinations of the two.

Increase of extension is generally diminution of intension, never increase: and diminution of extension is generally increase of intension, never diminution. And *vice versa*.

The disjunctive particle, *or*, expresses aggregation: 'either *A* or *B*' means 'in the class (A, B)'.

Class is most connected with extension, and attribute with intension. Extended attribute is merely class of qualities, and there is some effective use in the distinction between class and its subdivisions on the one hand, and the whole class-mark and its subdivision into qualities on the other hand. These two forms of thought, though closely related, must not be confounded with the relations arising out of comparison of extension and intension. (§ VIII.)

Aggregation of the impossible does not destroy the notion; composition of the impossible does. (§ x.)

The universe is the whole sphere of thought within which the matter in hand is contained: usually not the whole possible universe of thought, but a limited portion of it. In the last syllogism of § IV, the universe is *speed of transmission of news*. The universe being U, the class X introduces with itself the class not-X (or x). Let X and x be called *contraries* or *contradictories* (I make no distinction between these words). It is understood that every class is only *part* of the universe. The symbol (X, x) is equivalent to U: in extension, it contains everything; in intension, it belongs to everything. Thus A, if in the universe, is in (X, x); and A is A-(X, x) aggregate of A-X and A-x. The universe is the maximum of extension, and the minimum of intension. (§ IX.)

The contrary of an aggregate is the compound of the contraries of the aggregants: the contrary of a compound is the aggregate of the contraries of the components. Thus (A, B) and AB have ab and (a, b) for contraries.

XXIII. When two objects, qualities, classes, or attributes, viewed together by the mind, are seen under some connexion, that connexion is called a *relation*. To make very perfect parallelism, we should say that *relation* may be either of the four: that a boat towed by a ship, for instance, has the tow-rope for an *object* of relation. But relation, for all useful logical purposes, is a word of second intention, used only of class and attribute.

A *proposition* is the presentation of two names under a relation. A *judgement* is the sentence of the mind upon a proposition, true, false, more or less probable. The distinction of judgements, other than the simple *true* or *false*, is referred to the theory of probabilities, as a matter of practical convenience. The absolute exclusion of this distinction from logic is an error: the difference between certainty and uncertainty is of the form of thought; the amount of uncertainty is of the matter of this form. Full belief is a logical whole, which is divided into parts in the theory of probabilities: and the division of a logical whole into parts is of logic, whether it be convenient or not to treat it in the same book which treats the syllogism.

The distinction of *subject* and *predicate* is the distinction between the *notion in relation* and the *notion to which it is in relation*.

Every relation has its counter-relation, or *converse* relation: thus if *X* be in the relation *A* to *Y*, *Y* is therefore in some relation *B* to *X*: and *A* and *B* are converse relations, and the propositions are converse propositions. Every proposition has its converse, of meaning identical with itself.

When a relation is its own converse, the *proposition* is said to be *convertible*: meaning that the converse exhibits no change of relation. It is the *terms* (subject and predicate) which are convertible, strictly speaking.

When *X* has a relation (*A*) to that which has a relation (*B*) to *Y*, *X* has to *Y* a *combined* relation: the *combinants* are *A* and *B*. Relations have both extension and intension. Thus, to take one of those *relations* which have appropriated the word in common life, the relation of *first-cousin* is the *aggregate* of son of uncle, daughter of uncle, son of aunt, daughter of aunt. The relation of the minister to the crown is the *compound* of *subordinate* and *adviser*.

XXIV. Certain relations take precedence of all others, because they are presented by the notion of naming, and spring out of its purpose, if indeed they do not themselves constitute the purpose. They may be called *onomatic* or *onymatic* (*nominative* and *nominal* being engaged).

The excessive importance of these relations has enabled them to drive all others out of common logic, on the pretext of every other relation being expressible in terms of these: as must be the case, since these relations exist wherever names exist which apply to the same object.

The onymatic relations, to which in *this* paper I confine myself, are those of *whole and part* in the two aspects of *containing and contained* and *compounded and component*; and also the relations which the notion of contraries, and the notion of true and false, introduce in connexion with them.

Subordinate to, and necessarily compounded in, the notion of whole and part, is that of *more and less*, in the matter of which are the incidents of *quantity*. But *more and less* is only a component of *whole and part*, in the form of thought: and except as such component, is of no logical import. Thus *infusorium* is no doubt a larger class than *man*, and no doubt for reasons: but if we knew the extent of superiority, and the reasons, neither would be of any logical effect, for our present purpose; because the more and less is not that of containing and contained, and the relation is not onymatic.

The distinction of *aggregation* and *composition* is the most important distinction in the subdivisions of logic. Our knowledge does not suffice to define it by full description: we can only illustrate it. To the mathematician we may say that it has the distinctive character of $a+b$ and ab: to the chemist, of mechanical mixture and chemical combination: to the lawyer it appears in the distin ction between 'And be it further enacted' and 'provide always'.

44

GEORGE BOOLE
(1815–64)

GEORGE BOOLE, founder of the generalized class-calculus, sometimes called *Boolean algebra* in his honour, was animated throughout his life by a zeal for learning which was fulfilled only through his indefatigable efforts and resolution. Born in Lincoln, England, Boole received such early education as could be obtained partly from his father, a tradesman, who was ingenious in mechanics and exceptionally skilled in elementary mathematics, partly at the national school in Lincoln and partly under private tutelage in Greek and Latin. Some additional training was also afforded to him at a small commercial school in Lincoln. When Boole was sixteen years old his basically friendly nature, gentle manner, his integrity as well as his excellence in scholarship were already apparent and he was employed as a teacher, first in a school in Lincoln and later in another in nearby Waddington. During this period the study of modern languages, French, German and Italian, in addition to ancient Greek and Latin, occupied every moment of his spare time. At the age of twenty, after four years of teaching, Boole established his own school in Lincoln. Here he soon felt the need for further study in mathematics on his own part and from this time on he engaged in extensive research in mathematics, wholly self-directed and self-taught. The vigorous individuality of his way of thinking, nurtured by independent investigation, yielded not only a thorough knowledge of the subject but also some highly original results.

In 1849, on the recommendation of his friend, Augustus De Morgan, Boole was appointed professor of mathematics at the newly organized Queen's College in Cork, Ireland. In contrast to the privations and discouragements which marked his early years of struggle, the period of his activity at Queen's College was one of brilliant success. He was a popular lecturer. In later years he served as a public examiner for degrees in the Queen's university. His memoirs and treatises were well received. The eminence which he attained was recognized by such awards as the Keith Medal bestowed by the Royal Society of Edinburgh. The LL.D. degree was conferred upon him by the University of Dublin and the D.C.L. degree by Oxford. In 1855 Boole married Mary Everest, niece of the great surveyor, Sir George Everest, for whom Mount Everest was named. Their marriage was one of perfect happiness. When Boole died suddenly in 1864 a few days after his exposure in a rainstorm on his way to the college, he was survived by his wife and five daughters. His widow continued to be his admiring disciple after his death, expounding Boole's theories of education as she had learned them from him.

Many valuable memoirs in pure mathematics written by Boole appeared in various journals. His contributions to the theory of algebraic invariants, and to symbolical methods applicable to operative symbols separated from their operands, laid the groundwork on which future theories were built. Two systematic works, The *Treatise on Differential Equations* (1859) and the *Treatise on the Calculus of Finite Differences* (1860), contain original contributions which appeared in papers written and published by him much earlier. In his *The Mathematical Analysis of Logic being an Essay Towards a Calculus of Deductive Reasoning* (1847) and in his *An Investigation of the Laws of Thought on Which are Founded the Mathematical Theories of Logic and Probabilities* (1854), his theory of an algebra or a calculus of classes is constructed almost entirely on foundations originated by him, and these are the works that have brought Boole ever increasing fame.

———

THE MATHEMATICAL ANALYSIS OF LOGIC

FIRST PRINCIPLES

Let us employ the symbol 1, or unity, to represent the Universe, and let us understand it as comprehending every conceivable class of objects whether actually existing or not, it being premised that the same individual may be found in more than one class, inasmuch as it may possess more than one quality in common with other individuals. Let us employ the letters X, Y, Z, to represent the individual members of classes, X applying to every member of one class, as members of that particular class, and Y to every member of another class as members of such class, and so on, according to the received language of treatises on Logic.

Further let us conceive a class of symbols x, y z, possessed of the following character.

The symbol x operating upon any subject comprehending individuals or classes, shall be supposed to select from that subject all the Xs which it contains. In like manner the symbol y, operating upon any subject, shall be supposed to select from it all individuals of the class Y which are comprised in it, and so on.

When no subject is expressed, we shall suppose 1 (the Universe) to be the subject understood, so that we shall have

$$x = x \ (1),$$

the meaning of either term being the selection from the Universe of all the Xs which it contains, and the result of the operation being in common language, the class X, i.e. the class of which each member is an X.

From these premises it will follow, that the product xy will represent, in succession, the selection of the class Y, and the selection from the class Y of such individuals of the class X as are contained in it, the result being the class whose members are both Xs and Ys. And in like manner the product xyz will represent a compound operation of which the successive elements are the selection of the class Z, the selection from it of such individuals of the class Y as are contained in it, and the selection from the result thus obtained of all the individuals of the class X which it contains, the final result being the class common to X, Y, and Z.

From the nature of the operation which the symbols x, y, z are conceived to represent, we shall designate them as elective symbols. An expression in which they are involved will be called an elective function, and an equation of which the members are elective functions, will be termed an elective equation.

It will not be necessary that we should here enter into the analysis of that mental operation which we have represented by the electic symbol. It is not an act of Abstraction according to the common acceptation of that term, because we never lose sight of the concrete, but it may probably be referred to an exercise of the faculties of Comparison and Attention. Our present concern is rather with the laws of combination and of succession, by which its results are governed, and of these it will suffice to notice the following.

1st. The result of an act of election is independent of the grouping or classification of the subject.

Thus it is indifferent whether from a group of objects considered as a whole, we select the class X, or whether we divide the group into two parts, select the Xs from them separately, and then connect the results in one aggregate conception.

We may express this law mathematically by the equation

$$x(u+v)=xu+xv,$$

$u+v$ representing the undivided subject, and u and v the component parts of it.

2nd. It is indifferent in what order two successive acts of election are performed.

Whether from the class of animals we select sheep, and from the sheep those which are horned, or whether from the class of animals we select the horned, and from these such as are sheep, the result is unaffected. In either case we arrive at the class *horned sheep*.

The symbolical expression of this law is

$$xy=yx$$

3rd. The result of a given act of election performed twice, or any number of times in succession, is the result of the same act performed once.

If from a group of objects we select the Xs, we obtain a class of which all the members are Xs. If we repeat the operation on this class no further change will ensue: in selecting the Xs we take the whole. Thus we have

$$xx=x.$$

or

$$x^2=x;$$

and supposing the same operation to be n times performed, we have

$$x^n=x,$$

which is the mathematical expression of the law above stated.[1]

The laws we have established under the symbolical forms

$$x(u+v) = xu + xv \tag{1}$$

$$xy = yx \tag{2}$$

$$x^n = x \tag{3}$$

are sufficient for the basis of a Calculus. From the first of these, it appears that elective symbols are *distributive*, from the second that they are *commutative*; properties which they possess in common with symbols of *quantity*, and in virtue of which, all the processes of common algebra are applicable to the present system. The one and sufficient axiom involved in this application is that equivalent operations performed upon equivalent subjects produce equivalent results.

The third law (3) we shall denominate the index law. It is peculiar to elective symbols, and will be found of great importance in enabling us to reduce our results to forms meet for interpretation.

From the circumstance that the processes of algebra may be applied to the present system, it is not to be inferred that the interpretation of an elective equation will be unaffected by such processes. The expression of a truth cannot be negatived by a legitimate operation, but it may be limited. The equation $y = z$ implies that the classes Y and Z are equivalent, member for member. Multiply it by a factor x, and we have

$$xy = xz,$$

which expresses that the individuals which are common to the classes X and Y are also common to X and Z, and *vice versa*. This is a perfectly legitimate inference, but the fact which it

1. The office of the elective symbol x is to select individuals comprehended in the class X. Let the class X be supposed to embrace the universe; then, whatever the class Y may be, we have

$$xy = y.$$

The office which x performs is now equivalent to the symbol $+$, in one at least of its interpretations, and the index law (3) gives

$$+^n = +,$$

which is the known property of that symbol.

declares is a less general one than was asserted in the original proposition.

OF EXPRESSION AND INTERPRETATION

1. To express the class, not-X, that is, the class including all individuals that are not Xs.

The class X and the class not-X together make the Universe. But the Universe is 1, and the class X is determined by the symbol x, therefore the class not-X will be determined by the symbol $1-x$.

Hence the office of the symbol $1-x$ attached to a given subject will be, to select from it all the not-Xs which it contains.

And in like manner, as the product xy expresses the entire class whose members are both Xs and Ys, the symbol $y(1-x)$ will represent the class whose members are Ys but not Xs, and the symbol $(1-x)(1-y)$ the entire class whose members are neither Xs nor Ys.

2. To express the Proposition, All Xs are Ys.

As all the Xs which exist are found in the class Y, it is obvious that to select out of the Universe all Ys, and from these to select all Xs, is the same as to select at once from the Universe all Xs.

Hence $$xy = x,$$

or $$x(1-y) = 0 \qquad (4)$$

3. To express the Proposition, No Xs are Ys.

To assert that no Xs are Ys, is the same as to assert that there are no terms common to the classes X and Y. Now all individuals common to those classes are represented by xy. Hence the Proposition that No Xs are Ys, is represented by the equation

$$xy = 0 \qquad (5)$$

4. To express the Proposition, Some Xs are Ys.

If some Xs are Ys, there are some terms common to the classes X and Y. Let those terms constitute a separate class V, to which there shall correspond a separate elective symbol v, then

$$v = xy \qquad (6)$$

And as v includes all terms common to the classes X and Y we can indifferently interpret it, as Some Xs, or Some Ys.

5. To express the Proposition, Some Xs are not Ys. In the last equation write $1-y$ for y, and we have

$$v=(x1-y) \qquad (7)$$

the interpretation of v being indifferently Some Xs or Some not-Ys.

The above equations involve the complete theory of categorical Propositions, and so far as respects the employment of analysis for the deduction of logical inferences, nothing more can be desired. But it may be satisfactory to notice some particular forms deducible from the third and fourth equations, and susceptible of similar application.

If we multiply the equation (6) by x, we have

$$vx=x^2y=xy \text{ by (3).}$$

Comparing with (6), we find

$$v=vx,$$

or $$v(1-x)=0 \qquad (8)$$

And multiplying (6) by y, and reducing in a similar manner, we have

$$v=vy,$$

or $$v(1-y)=0, \qquad (9)$$

Comparing (8) and (9),

$$vx=vy=v \qquad (10)$$

And further comparing (8) and (9) with (4), we have as the equivalent of this system of equations the Propositions

All Vs are Xs $\Big\}$.
All Vs are Ys

The system (10) might be used to replace (6), or the single equation

$$vx=vy \qquad (11)$$

might be used, assigning to vx the interpretation, Some Xs, and to vy the interpretation, Some Ys. But it will be observed that this system does not express quite so much as the single equation

(6), from which it is derived. Both, indeed, express the Proposition, Some Xs are Ys, but the system (10) does not imply that the class V includes *all* the terms that are common to X and Y.

In like manner, from the equation (7) which expresses the Proposition Some Xs are not Ys, we may deduce the system

$$vx = v(1-y) = v \tag{12}$$

in which the interpretation of $v(1-y)$ is Some not-Ys. Since in this case $vy = 0$, we must of course be careful not to interpret vy as Some Ys.

If we multiply the first equation of the system (12), viz.

$$vx = v(1-y),$$

by y, we have

$$vxy = vy(1-y);$$
$$\therefore \quad vxy = 0 \tag{13}$$

which is a form that will occasionally present itself. It is not necessary to revert to the primitive equation in order to interpret this, for the condition that vx represents Some Xs, shows us by virtue of (5), that its import will be

Some Xs are not Ys,

the subject comprising *all* the Xs that are found in the class V.

Universally in these cases, difference of form implies a difference of interpretation with respect to the auxiliary symbol v, and each form is interpretable by itself.

Further, these differences do not introduce into the Calculus a needless perplexity. It will hereafter be seen that they give a precision and a definiteness to its conclusions, which could not otherwise be secured.

Finally, we may remark that all the equations by which particular truths are expressed, are deducible from any one general equation, expressing any one general Proposition, from which those particular Propositions are necessary deductions. This has been partially shown already, but it is much more fully exemplified in the following scheme.

The general equation $x = y$, implies that the classes X and Y are equivalent, member for member; that every individual belonging to the one, belongs to the other also. Multiply the equation by x, and we have

$$x^2 = xy;$$
$$\therefore \; x = xy,$$

which implies, by (4), that all Xs are Ys. Multiply the same equation by y, and we have in like manner

$$y = xy;$$

the import of which is, that all Ys are Xs. Take either of these equations, the latter for instance, and writing it under the form

$$(1 - x)y = 0,$$

we may regard it as an equation in which y, an unknown quantity, is sought to be expressed in terms of x. Now it will be shown when we come to treat of the Solution of Elective Equations (and the result may here be verified by substitution) that the most general solution of this equation is

$$y = vx,$$

which implies that all Ys are Xs, and that some Xs are Ys. Multiply by x, and we have

$$vy = vx,$$

which indifferently implies that some Ys are Xs and some Xs are Ys, being the particular form at which we before arrived.

For convenience of reference the above and some other results have been classified in the annexed Table, the first column of which contains propositions, the second equations, and the third the conditions of final interpretation. It is to be observed, that the auxiliary equations which are given in this column are not independent: they are implied either in the equations of the second column, or in the condition for the interpretation of v. But it has been thought better to write them separately, for greater ease and convenience. And it is further to be borne in mind, that although three different forms are given for the expression of each of the *particular* propositions, everything is really included in the first form.

Table

The class X	x	
The class not-X	$1-x$	

All Xs are Ys }
All Ys are Xs } $x=y$

| All Xs are Ys | $x(1-y)=0$ |
| No Xs are Ys | $xy=0$ |

All Ys are Xs }
Some Xs are Ys } $y=vx$ $vx=$ some Xs
  $v(1-x)=0.$

No Ys are Xs }
Some not-Xs are Ys } $y=v(1-x)$ $v(1-x)=$ some not-Xs
  $vx=0.$

Some Xs are Ys $\begin{cases} v=xy \\ \text{or } vx=vy \\ \text{or } vx(1-y)=0 \end{cases}$ $v=$ some Xs, or some Ys
  $vx=$ some Xs, $vy=$ some Ys
  $v(1-x)=0, v(1-y)=0.$

Some Xs are not Ys $\begin{cases} v=x(1-y) \\ \text{or } vx=v(1-y) \\ \text{or } vxy=0 \end{cases}$ $v=$ some Xs, or some not-Ys
  $vx=$ some Xs, $v(1-y)=$ some not-Ys
  $v(1-x)=0, vy=0.$

THE LAWS OF THOUGHT

CHAPTER II

*Of signs in general, and of the signs appropriate
to the science of logic in particular; also of the laws
to which that class of signs are subject.*

1. That Language is an instrument of human reason, and not merely a medium for the expression of thought, is a truth generally admitted. It is proposed in this chapter to inquire what it is that renders Language thus subservient to the most important of our intellectual faculties. In the various steps of this inquiry we shall be led to consider the constitution of Language, considered as a system adapted to an end or purpose; to investigate its elements; to seek to determine their mutual relation and dependence; and to inquire in what manner they contribute to the attainment of the end to which, as coordinate parts of a system, they have respect.

2. The elements of which all language consists are signs or symbols. Words are signs. Sometimes they are said to represent things; sometimes the operations by which the mind combines together the simple notions of things into complex conceptions; sometimes they express the relations of action, passion or mere quality, which we perceive to exist among the objects of our experience; sometimes the emotions of the perceiving mind. But words, although in this and in other ways they fulfil the office of signs, or representative symbols, are not the only signs which we are capable of employing. Arbitrary marks, which speak only to the eye, and arbitrary sounds or actions, which address themselves to some other sense, are equally of the nature of signs, provided that their representative office is defined and understood. In the mathematical sciences, letters, and the symbols $+$, $-$, $=$, etc., are used as signs, although the term 'sign' is applied to the latter class of symbols, which represent operations or relations, rather than to the former, which represent the elements of number and quantity. As the real import of a sign does not in any way depend upon its particular form or expression, so neither do the laws which determine its use. In the present treatise, however, it is with written signs that we have to do, and it is with reference to these exclusively that the term 'sign' will be employed. The essential properties of signs are enumerated in the following definition.

Definition: A sign is an arbitrary mark, having a fixed interpretation, and susceptible of combination with other signs in subjection to fixed laws dependent upon their mutual interpretation.

3. Let us consider the particulars involved in the above definition separately.

(1) In the first place, a sign is an *arbitrary* mark. It is clearly indifferent what particular word or token we associate with a given idea, provided that the association once made is permanent. The Romans expressed by the word '*civitas*' what we designate by the word 'state'. But both they and we might equally well have employed any other word to represent the same conception. Nothing, indeed, in the nature of Language would prevent us from using a mere letter in the same sense. Were this done, the

laws according to which that letter would require to be used would be essentially the same with the laws which govern the use of '*civitas*' in the Latin, and of 'state' in the English language, so far at least as the use of those words is regulated by any general principles common to all languages alike.

(2) In the second place, it is necessary that each sign should possess, within the limits of the same discourse or process of reasoning, a fixed interpretation. The necessity of this condition is obvious, and seems to be founded in the very nature of the subject. There exists, however, a dispute as to the precise nature of the representative office of words or symbols used as names in the processes of reasoning. By some it is maintained that they represent the conceptions of the mind alone; by others, that they represent things. The question is not of great importance here, as its decision cannot affect the laws according to which signs are employed. I apprehend, however, that the general answer to this and suchlike questions is, that in the process of reasoning, signs stand in the place and fulfil the office of the conceptions and operations of the mind; but that as those conceptions and operations represent things, and the connexions and relations of things, so signs represent things with their connexions and relations; and lastly, that as signs stand in the place of the conceptions and operations of the mind, they are subject to the laws of those conceptions and operations. This view will be more fully elucidated in the next chapter; but it here serves to explain the third of those particulars involved in the definition of a sign, viz. its subjection to fixed laws of combination depending upon the nature of its interpretation.

4. The analysis and classification of those signs by which the operations of reasoning are conducted will be considered in the following Proposition:

Proposition I

All the operations of Language, as an instrument of reasoning, may be conducted by a system of signs composed of the following elements, viz.:

1st. *Literal symbols, as x, y, etc., representing things as subjects of our conceptions.*

2nd. *Signs of operation, as* $+$, $-$, $\times$, *standing for those operations of the mind by which the conceptions of things are combined or resolved so as to form new conceptions involving the same elements.*

3rd. *The sign of identity,* $=$.

And these symbols of Logic are in their use subject to definite laws, partly agreeing with and partly differing from the laws of the corresponding symbols in the science of Algebra.

Let it be assumed as a criterion of the true elements of rational discourse that they should be susceptible of combination in the simplest forms and by the simplest laws, and thus combining should generate all other known and conceivable forms of language. . . .

6. Now, as it has been defined that a sign is an arbitrary mark, it is permissible to replace all signs of the species above described by letters. Let us then agree to represent the class of individuals to which a particular name or description is applicable, by a single letter, as x. If the name is 'men', for instance, let x represent 'all men', or the class 'men'. By a class is usually meant a collection of individuals, to each of which a particular name or description may be applied; but in this work the meaning of the term will be extended so as to include the case in which but a single individual exists, answering to the required name or description, as well as the cases denoted by the terms 'nothing' and 'universe', which as 'classes' should be understood to comprise respectively 'no beings', 'all beings'. Again, if an adjective, as 'good', is employed as a term of description, let us represent by a letter, as y, all things to which the description 'good' is applicable, i.e. 'all good things', or the class 'good things'. Let it further be agreed that by the combination xy shall be represented that class of things to which the names or description represented by x and y are simultaneously applicable. Thus, if x alone stands for 'white things', and y for 'sheep', let xy stand for 'white sheep'; and in like manner, if z stand for 'horned things', and x and y retain their previous interpretations, let xyz represent 'horned white sheep', i.e. that collection of things to which the name 'sheep', and the descriptions 'white' and 'horned' are together applicable.

Let us now consider the laws to which the symbols x, y, etc., used in the above sense, are subject.

7. First, it is evident, that according to the above combinations, the order in which two symbols are written is indifferent. The expressions xy and yx equally represent that class of things to the several members of which the names or descriptions x and y are together applicable. Hence we have,

$$xy = yx. \tag{1}$$

In the case of x representing white things, and y sheep, either of the members of this equation will represent the class of 'white sheep'. There may be a difference as to the order in which the conception is formed, but there is none as to the individual things which are comprehended under it. In like manner if x represents 'estuaries', and y 'rivers', the expressions xy and yx will indifferently represent 'rivers that are estuaries', or 'estuaries that are rivers', the combination in this case being in ordinary language that of two substantives, instead of that of a substantive and adjective as in the previous instance. Let there be a third symbol, as z, representing that class of things to which the term 'navigable' is applicable, and any one of the following expressions,

$$zxy, \ zyx, \ xyz, \text{ etc.}$$

will represent the class of 'navigable rivers that are estuaries'.

If one of the descriptive terms should have some implied reference to another, it is only necessary to include that reference expressly in its stated meaning, in order to render the above remarks still applicable. Thus, if x represent 'wise' and y 'counsellor', we shall have to define whether x implies wisdom in the absolute sense, or only the wisdom of counsel. With such definition the law $xy = yx$ continues to be valid.

We are permitted, therefore, to employ the symbols x, y, z, etc., in the place of the substantives, adjectives and descriptive phrases subject to the rule of interpretation, that any expression in which several of these symbols are written together shall represent all the objects or individuals to which their several meanings are together *applicable, and to the law that the order in which the symbols succeed each other is indifferent.*

As the rule of interpretation has been sufficiently exemplified, I shall deem it unnecessary always to express the subject 'things' in defining the interpretation of a symbol used for an adjective. When I say, let x represent 'good', it will be understood that x only represents 'good' when a subject for that quality is supplied by another symbol, and that, used alone, its interpretation will be 'good things'.

8. Concerning the law above determined, the following observations, which will also be more or less appropriate to certain other laws to be deduced hereafter, may be added.

First, I would remark, that this law is a law of thought, and not, properly speaking, a law of things. Difference in the order of the qualities or attributes of an object, apart from all questions of causation, is a difference in conception merely. The law (1) expresses as a general truth that the same thing may be conceived in different ways, and states the nature of that difference; and it does no more than this.

Secondly, As a law of thought it is actually developed in a law of Language, the product and instrument of thought. Though the tendency of prose writing is towards uniformity, yet even there the order of sequence of adjectives absolute in their meaning, and applied to the same subject, is indifferent, but poetic diction borrows much of its rich diversity from the extension of the same lawful freedom to the substantive also. The language of Milton is peculiarly distinguished by this species of variety. Not only does the substantive often precede the adjectives by which it is qualified, but it is frequently placed in their midst. In the first few lines of the invocation to Light, we meet with such examples as the following:

'*Offspring of heaven first-born.*'
'The rising world of *waters dark and deep.*'
'Bright effluence of *bright essence increate.*'

Now these inverted forms are not simply the fruits of a poetic licence. They are the natural expressions of a freedom sanctioned by the intimate laws of thought, but for reasons of convenience not exercised in the ordinary use of language.

Thirdly, The law expressed by (1) may be characterized by saying that the literal symbols x, y, z are *commutative, like the symbols of Algebra*. In saying this, it is not affirmed that the process of multiplication in Algebra, of which the fundamental law is expressed by the equation

$$xy = yx,$$

possesses in itself any analogy with that process of logical combination which xy has been made to represent above; but only that if the arithmetical and the logical process are expressed in the same manner, their symbolical expressions will be subject to the same formal law. The evidence of that subjection is in the two cases quite distinct.

9. As the combination of two literal symbols in the form xy expresses the whole of that class of objects to which the names or qualities represented by x and y are together applicable, it follows that if the two symbols have exactly the same signification, their combination expresses no more than either of the symbols taken alone would do. In such case we should therefore have

$$xy = x.$$

As y is, however, supposed to have the same meaning as x, we may replace it in the above equation by x, and we thus get

$$xx = x.$$

Now in common Algebra the combination xx is more briefly represented by x^2. Let us adopt the same principle of notation here; for the mode of expressing a particular succession of mental operations is a thing in itself quite as arbitrary as the mode of expressing a single idea or operation (II, 3). In accordance with this notation, then, the above equation assumes the form

$$x^2 = x, \tag{2}$$

and is, in fact, the expression of a second general law of those symbols by which names, qualities, or descriptions, are symbolically represented.

*

Again: If two classes of things, x and y, be identical, that is, if all the members of the one are members of the other, then those members of the one class which possess a given property z will be identical with those members of the other which possess the same property z. Hence if we have the equation

$$x = y;$$

then whatever class or property z may represent, we have also

$$zx = zy.$$

This is formally the same as the algebraic law: If both members of an equation are multiplied by the same quantity, the products are equal.

In like manner it may be shown that if the corresponding members of two equations are multiplied together, the resulting equation is true.

14. Here, however, the analogy of the present system with that of algebra, as commonly stated, appears to stop. Suppose it true that those members of a class x which possess a certain property z are identical with those members of a class y which possess the same property z, it does not follow that the members of the class x universally are identical with the members of the class y. Hence it cannot be inferred from the equation

$$zx = zy,$$

that the equation

$$x = y$$

is also true. In other words, the axiom of algebraists, that both sides of an equation may be divided by the same quantity, has no formal equivalent here. I say no *formal equivalent*, because, in accordance with the general spirit of these inquiries, it is not even sought to determine whether the mental operation which is represented by removing a logical symbol, z, from a combination zx, is in itself analogous with the operation of division in Arithmetic. That mental operation is indeed identical with what is commonly termed Abstraction, and it will hereafter appear that its laws are dependent upon the laws already deduced in this chapter. What has now been shown is, that there does not exist among those laws anything analogous in *form* with a commonly received axiom of Algebra.

But a little consideration will show that even in common algebra that axiom does not possess the generality of those other axioms which have been considered. The deduction of the equation $x=y$ from the equation $zx=zy$ is only valid when it is known that z is not equal to 0. If then the value $z=0$ is supposed to be admissible in the algebraic system, the axiom above stated ceases to be applicable, and the analogy before exemplified remains at least unbroken.

15. However, it is not with the symbols of quantity generally that it is of any importance, except as a matter of speculation, to trace such affinities. We have seen (II, 9) that the symbols of Logic are subject to the special law,

$$x^2 = x.$$

Now of the symbols of Number there are but two, viz. 0 and 1, which are subject to the same formal law. We know that $0^2=0$, and that $1^2=1$; and the equation $x^2=x$, considered as algebraic, has no other roots than 0 and 1. Hence, instead of determining the measure of formal agreement of the symbols of Logic with those of Number generally, it is more immediately suggested to us to compare them with the symbols of quantity *admitting only the* 0 *and* 1. Let us conceive, then, of an Algebra in which the symbols x, y, z, etc. admit indifferently of the values 0 and 1, and of these values alone.

45

JOHN VENN

(1834–1923)

JOHN VENN, English logician, descended from a Devonshire family long distinguished for its erudition, was born at Drypool, Hull, on 4 August, 1834. Representing the eighth generation of his family to study at Cambridge (to be followed by a ninth), he entered Gonville and Caius College in 1853, beginning an association with the College which was to last seventy years. In 1854 he was elected mathematical scholar; he received his degree in January 1857 as Sixth Wrangler and a few months later he was elected Fellow of the College. Ordained a deacon in 1858 and priest in 1859, he served for a short period in parochial work. In 1862 he accepted an appointment as lecturer in moral science at Cambridge. While retaining his sincere religious convictions, Venn resigned his orders in 1883 to devote himself entirely to the study and teaching of logic. In that same year he received his Sc.D. degree at Cambridge and was elected Fellow of the Royal Society. In 1888 he presented to the Cambridge Library the largest private collection of books on logic ever to be brought together. As historian of his College Venn undertook exhaustive researches which culminated in his three-volume publication of the *Biographical History of Gonville and Caius College* (1897). Venn also edited a number of volumes of university archives. Of spare build, even disposition, a devotee of mountain climbing, an excellent linguist, he was loved and admired at home and abroad. Venn died at Cambridge on 4 April, 1923. His son, John Archibald, became president of Queens' College, Cambridge, in 1932.

John Venn wrote three important works which became standard texts almost at once: *The Logic of Chance* (1866), *Symbolic Logic* (1881; second edition, revised and rewritten, 1894) and the *Principles of Empirical Logic* (1889). His *Symbolic Logic* contained the *Venn Diagrams*, a system of overlapping circles or ellipses or other figures, for representing given premises in which shaded regions were used to designate an empty class.

DIAGRAMMATIC REPRESENTATION

Our primary diagram for two terms is thus sketched:

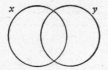

Fig. 45.1.

On the common plan this would represent a *proposition*, and is indeed commonly regarded as standing for the proposition 'some x is y'; though (as was mentioned in the first chapter) it equally involves in addition the two independent propositions 'some x is not y', and 'some y is not x', if we want to express all that it undertakes to tell us. With us, however, it does not as yet represent a proposition at all, but only the framework into which propositions may be fitted; that is, it indicates only the four combinations represented by the letter compounds, xy, $\bar{x}y$, $x\bar{y}$, $\bar{x}\bar{y}$.

Now suppose that we have to reckon with the presence, and consequently with the absence, of a third term z. We just draw a third circle intersecting the above two, thus:

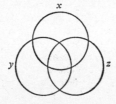

Fig. 45.2.

Each circle is thus cut up into four parts, and each common part of two circles into two parts, so that, including what lies outside of all the three, there are eight compartments. These of course correspond precisely to the eight combinations given by the three literal symbols; viz., xyz, $xy\bar{z}$, $x\bar{y}z$, $x\bar{y}\bar{z}$, $\bar{x}yz$, $\bar{x}y\bar{z}$, $\bar{x}\bar{y}z$, $\bar{x}\bar{y}\bar{z}$. Put a finger upon any compartment, and we have a symbolic name ready provided for it; mention the name, and there can be no doubt as to the compartment thereby referred to.

Both schemes, that of letters and that of areas, agree in their elements being mutually exclusive and collectively exhaustive. No one of the ultimate elements trespasses upon the ground of any other; and, amongst them, they account for all possibilities. Either scheme therefore might be taken as a fair representative of the other.

This process is capable of theoretic extension to any number of terms. The only drawback to its indefinite extension is that with more than three terms we do not find it possible to use such simple figures as circles; for four circles cannot be so drawn as to intersect one another in the way required. With employment of more intricate figures we might go on for ever. All that is requisite is to draw some continuous figure which shall intersect once, and once only, every existing subdivision. The new outline must be so drawn as to cut every one of the previous compartments in two, and so double their number. There is clearly no reason against continuing this process indefinitely.

With four terms in demand the most simple and symmetrical diagram seems to me that produced by making four ellipses intersect one another in the desired manner:

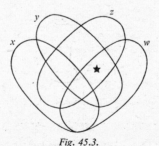

Fig. 45.3.

It is obvious that each component class-figure (say y) is thus divided into eight distinct compartments, producing in all 16 partitions; that these partitions are all different from each other in their composition, and are therefore mutually exclusive; and moreover that they leave nothing unaccounted for, and are therefore collectively exhaustive. And this is all that is required to make them a fitting counterpart of the 16 combinations yielded by x, y, z, w and their negations, in the ordinary tabular statement.

With five terms combined together ellipses fail us, at least in the above simple form. It would not be difficult to sketch out figures of a horse-shoe shape which should answer the purpose, but then any outline which is not very simple and easy to follow fails altogether in its main requirement of being an aid to the eye. What is required is that we should be able to identify any assigned compartment in a moment. Thus it is instantly seen that the compartment marked with an asterisk above is that called $\bar{x}yzw$.

*

What we do, then, is to ascertain what combinations or classes are negatived by any given proposition, and proceed to put some kind of mark against these in the diagram. For this purpose the most effective means is just to shade them out.

For instance the proposition 'all x is y' is interpreted to mean that there is no such class of things in existence as 'x that is not-y', or xy. All that we have to do is to scratch out that subdivision in the two-circle figure,[1] thus:

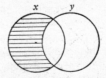

Fig. 45.4.

If we want to represent 'all x is all y', we take this as adding on

1. Other logicians (e.g. Schroder, *Operationskreis*, p. 10; Macfarlane, *Algebra of Logic*, p. 63) have made use of shaded diagrams to direct attention to the compartments under consideration; not, as here, with the view of expressing propositions.

another denial, viz. that of $\bar{x}y$, and proceed to scratch out that division also; thus

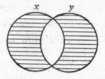

Fig. 45.5.

On the common Eulerian plan we should have to begin with a new figure in each of the two cases respectively, viz. 'all x is y' and 'all y is x'; whereas here we start with the same general outline in each case, merely modifying it in accordance with the varying information given to us.

We postulate at present that every universal proposition may be sufficiently represented by one or more denials, and shall hope to justify this view in its due place. But it will hardly be disputed that every such proposition does in fact negative one or more combinations, and this affords an excellent means of combining two or more propositions together so as to picture their collective import. The first proposition empties out a certain number of compartments. In so far as the next may have covered the same ground it finds its work already done for it, but in so far as it has fresh information to give it displays this by clearing out compartments which the first had left untouched. All that is necessary therefore for a complete diagrammatic illustration is to begin by drawing our figure, as already explained, and then to shade out, or in some way distinguish, the classes which are successively abolished by the various premisses. This will set before the eye, at a glance, the whole import of the propositions collectively.

How widely different this plan is from that of the old-fashioned Eulerian diagrams will be readily seen.[1] One great advantage consists in the ready way in which it lends itself to the representation of successive increments of knowledge as one proposition after another is taken into account, instead of demanding that

1. I have not found any previous attempt to represent propositions on this scheme. Scheffler's elaborate *Naturgesetze* (Part III, on Logic) was apparently published about the same time as my paper in the *Phil. Magazine*.

we should endeavour to represent the net result of them all at a stroke. Our first data abolish, say, such and such classes. This is final, for, as already intimated, all the resultant denials must be regarded as absolute and unconditional: This leaves the field open to any similar accession of knowledge from the next data, and so more classes are swept away. Thus we go on till all the data have had their fire, and the muster-roll at the end will show what classes are, or may be, left surviving. If therefore we simply shade out the compartments in our figure which have thus been successively declared empty, nothing is easier than to continue doing this till all the information furnished by the data is exhausted.

As another very simple illustration of the contrast between the two methods, consider the case of the disjunction, 'All x is either y or z'. It is very seldom even attempted to represent such propositions diagrammatically, (and then, so far as I have seen, only if the alternatives are mutually exclusive), but they are readily enough exhibited when we regard the one in question as merely extinguishing any x that is neither y nor z, thus:

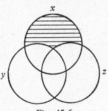

Fig. 45.6.

If to this were added the statement that 'none but the x's are either y or z' we should meet this fresh assertion by the further abolition of $\bar{x}y$ and $\bar{x}z$, and thus obtain:

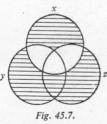

Fig. 45.7.

267

And if, again, we erase the central, or *xyz* compartment, we have then made our alternatives exclusive; i.e. the *x*'s, and they alone, are either *y* or *z* only.

Now if we tried to do this by aid of Eulerian circles we should find at once that we could not do it in the only way in which intricate matters can generally be settled, viz. by breaking them up into details, and taking these step by step, making sure of each as we proceed. The Eulerian figures have to be drawn so as to indicate at once the final outcome of the knowledge furnished. This offers no difficulty in such exceedingly simple cases as those furnished by the various moods of the Syllogism, but it is quite a different matter when we come to handle the complicated results which follow upon the combination of four or five terms. Those who have only looked at the simple diagrams given by Hamilton, Thomson, and most other logicians, in illustration of the Aristotelian Syllogism, have very little conception of the intricate task which would be impressed upon them if they tried, with such resources, to illustrate equations of the type that we must be prepared to take in hand.

As the syllogistic figures are the form of reasoning most familiar to ordinary readers, I will begin with one of these, though they are too simple to serve as effective examples. Take, for instance,

No *Y* is *Z*,

All *X* is *Y*,

∴ No *X* is *Z*.

This would commonly be exhibited thus:

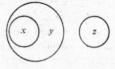

Fig. 45.8.

It is easy enough to do this; for in drawing our circles we have only to attend in two terms at a time, and consequently the relation of X to Z is readily detected; there is not any of that troublesome interconnexion of a number of terms simultaneously with each other which gives rise to the main perplexity in complicated problems. Accordingly such a simple example as this is not a very good one for illustrating the method now proposed; but, in order to mark the distinction, the figure to represent it is given, thus:

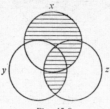

Fig. 45.9.

In this case the one particular relation asked for, viz. that of x to Z, it must be admitted, is not made more obvious on our plan than on the old one. The superiority, if any, in such an example must rather be sought in the completeness of the pictorial information in other respects – as, for instance, in the intimation that, of the four kinds of x which originally had to be taken into consideration, one only, viz. the $xy\bar{z}$, or the 'x that is y but is not z', is left surviving. Similarly with the formal possibilities of y and z: the relative number of these, as compared with the resultant actualities permitted by the data, is detected at a glance.

As a more suitable example consider the following –

$$\begin{cases} \text{All } x \text{ is either } y \text{ and } z, \text{ or not } y, \\ \text{If any } xy \text{ is } z, \text{ then it is } w, \\ \text{No } wx \text{ is } yz; \end{cases}$$

and suppose we are asked to exhibit the relation of x and y to each other. The problem is essentially of the same kind as the syllogistic one; but we certainly could not draw the figures in the off-hand way we did there. Since there are four terms, we sketch the appropriate 4-ellipse figure, and then proceed to analyse the premises in order to see what classes are destroyed by them. The reader will readily see that the first premiss annihilates all 'xy

which is not z', or $xy\bar{z}$; the second destroys 'xyz which is not w', or $xyz\bar{w}$; and the third 'wx which is yz', or $wxyz$. Shade out these three classes, and we see the resultant figure at once, viz.

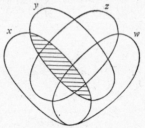

Fig. 45.10.

It is then evident that *all xy* has been thus made away with; that is, x and y must be mutually exclusive, or, as it would commonly be thrown into propositional form, 'No x is y'.

LOGICAL STATEMENTS OR EQUATIONS

Interpretation of equations in general

There are few departments of the Symbolic Logic in which acquired views will have to be more completely abandoned than in reference to the interpretation of our equations. In the common Logic we talk of *the* solution as if there were but one; in fact a plurality of possible answers is considered a fatal defect, so that certain syllogistic figures are rejected on this ground alone.

On the symbolic system all this has, at first sight, to be altered. We must be prepared here for an apparent variety of possible answers. In saying this it is not, of course, implied that conflicting answers could be drawn, but rather that the modes of expression are so various that the same substantial answer can assume a variety of forms.

This distinction rests upon two grounds; firstly, the fact that we put a term and its contradictory (x and $\bar{x}$) on exactly the same footing, whereas the common system seeks always to express itself in positive terms, putting the negation into the predicate. Secondly, there is the obvious difference that whereas but two or three terms are commonly admitted into the former, the latter is prepared to welcome any number.

For instance, take the familiar syllogism, 'all x is y; no z is y; therefore no z is x'. Here it would be said, and very correctly from the customary standpoint, that there is one and only one conclusion possible. Now look at it symbolically: We write the statements in the form $x\bar{y}=0$, $yz=0$. Therefore the full combination of the two may be written $x\bar{y}+yz=0$, and it may be represented in a diagram thus:

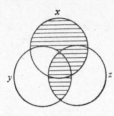

Fig. 45.11.

It will be seen at once, even in such a simple case, what a variety of possible solutions are here open to us. First take the fully complete solutions. These fall into the usual distinction offered by the positive and negative interpretation; that is, according as we enumerate all the abolished classes, or all the possible surviving ones. Thus $x\bar{y}+yz=0$ expanded into its details gives four terms to be destroyed, the remaining four being equated to unity.

$$\begin{cases} x\bar{y}z+x\bar{y}\bar{z}+xyz+\bar{x}yz=0 \\ xy\bar{z}+\bar{x}y\bar{z}+\bar{x}\bar{y}z+\bar{x}\bar{y}\bar{z}=1 \end{cases}$$

These are the complete alternative answers given in detail. The former states, with negative disjunction, that there is nothing which falls into any one of certain four classes; the latter, with affirmative disjunction, that everything does fall into one or other of the remaining four classes.

These ultimate elements we may of course group at will, and thus obtain various simplifications of expression. The former we

271

know will stand as $x\bar{y}+yz=0$, the latter will stand as $y\bar{z}+\bar{x}\bar{y}=1$. They then state respectively that there is nothing which is either yz or $x\bar{y}$, and that everything is either $y\bar{z}$ or $\bar{x}\bar{y}$.

*

Our complete scheme comprises two further modifications on anything here indicated. For we may want to determine not merely x and $\bar{x}$, y and $\bar{y}$, z and $\bar{z}$, but any possible combination or function of these; and this we may want to determine not as here in terms of *all* the remaining elements, but in terms of any selection from amongst these, after elimination of the remaining elements. These extensions will be duly discussed in their proper places.

The above example will serve to show how indefinite is the solution of a logical problem, unless some further indications are given as to the sort of information desired. The sum-total of the facts which are left consistent with the data must necessarily be the same, however they may be expressed. But the various ways of expressing those facts, and still more the various ways of expressing selections and combinations of them, are very numerous. Take, for instance, a slightly more complicated example, such as that indicated in the following figure:

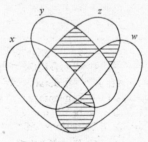

Fig. 45.12.

and observe in what a variety of ways the unshaded portion may be described. The figure represents the results of the data;

$$\begin{cases} \text{All } wx \text{ is } z \\ \text{All } wz \text{ is } x \text{ or } y \quad \text{viz.} \\ \text{All } yz \text{ is } w \text{ or } x \end{cases} \begin{cases} wx\bar{z}=0 \\ wz\bar{x}\bar{y}=0 \\ yz\overline{w}\bar{x}=0. \end{cases}$$

272

One way of course is to say that the surviving classes are all which are not thus obliterated; these being negatively

$$1 - x\bar{z}w - \bar{x}\bar{y}zw - \bar{x}yz\bar{w}$$

or, slightly grouped,

$$1 - x\bar{z}w - \bar{x}z(\bar{y}w + y\bar{w}).$$

Or again, we may express them all positively thus,

$$x(wz + \bar{w}z + \bar{w}\bar{z}) + \bar{x}z(wy + \bar{w}\bar{y}) + \bar{x}\bar{z}$$

or, less completely positive, but briefer,

$$x(1 - w\bar{z}) + \bar{x}z(wy + \bar{w}\bar{y}) + \bar{x}\bar{z}$$

or

$$x(1 - w\bar{z}) + \bar{x}\{z(wy + \bar{w}\bar{y}) + \bar{z}\},$$

each of these symbolic groupings having of course its suitable verbal description. Thus the last may be read 'All x except what is w but not z; and all not-x, provided it be not-z, or, if z, then both or neither w and y'.

The general nature of the problem thus put before us is easily indicated. Suppose there were four terms involved; then our symbolic apparatus provides 2^4 or 16 compartments or possibilities. The data impose material limits upon these possibilities, leaving only a limited number of actualities. That is, they extinguish a certain number and leave only the remainder surviving. In this case out of the 16 original possibilities 12 are left remaining. The full result then of the data is given by enumerating completely either the extinguished compartments or the remaining ones. Either of these enumerations is only possible in one way, provided we give it in full detail. But when we want to group the results, for more convenience, into compendious statements, we see that this can be done in a variety of ways.

The number of different combinations which can be produced increases with the introduction of every fresh class term in a way which taxes the imagination to follow. Thus three terms yield eight subdivisions. From these we might make eight distinct selections of one only; 28 of a pair; 56 of three together, and so on. The total number of distinct groups which can thus be produced is

$$8 + 28 + 56 + 70 + 56 + 28 + 8 + 1 \text{ or } 255.$$

One case, and one only, is excluded necessarily, namely that in which every compartment is erased, for this corresponds to the one formal impossibility of endeavouring to maintain that every one of our exhaustive divisions is unoccupied: this being, as the reader knows, the symbolic generalization of the Law of Excluded Middle. The arithmetical statement of the total number of cases is readily enough written down. Three terms yield eight sub-classes, viz. 2^3; and these eight sub-classes may be taken as above in $2^8 - 1$ ways: viz. $2^{2^3} - 1$ represents the possible varieties before us. Or expressed generally, if there be n terms we can have 2^n classes, and accordingly $2^{2^n} - 1$ distinct groups of these classes. Of course this expression increases with enormous rapidity as n increases. Four terms thus yield 65,535 possibilities in the way of combination of the elements yielded, and so on.

46

CHARLES SANDERS PEIRCE
(1839–1914)

CHARLES SANDERS PEIRCE, one of the most profound and original of American philosophers and mathematicians, founder of the philosophical school of pragmatism, was born at Cambridge, Mass., on 10 September, 1839. The second son of Benjamin Peirce, famous Harvard professor of astronomy and mathematics, he belonged to a family which had been prominent in the affairs of Harvard University for generations. An extraordinarily precocious child, his formal schooling was supplemented by his father's teaching in mathematics and the physical sciences. His early interest in philosophy and logic, on the other hand, met with scant sympathy or approval in the scientific atmosphere of the Peirce household. Charles's training in chemistry, astronomy, geodesy and optics as well as in mathematics left him, as he said, 'saturated through and through with the spirit of the physical sciences'. In 1861 he accepted an appointment as a physicist and mathematician in the United States Coast Survey. Had his activities been restricted to the field of science alone, his accomplishments in this area would have been a sufficient mark of his genius. Peirce's work on astronomical problems and his investigations on the pendulum were especially well received both at home and abroad. Nevertheless, his strong leaning toward philosophy and the sensitivity of his creative powers were never to be overwhelmed. Throughout his life, by means of his personal contacts, lectures and essays, Charles Peirce was at the centre of the development of philosophy in America. James, Royce, Dewey and others frequently acknowledged his influence upon them. Eventually, too, his activities in the newer mathematics instigated the researches undertaken and published by his father (Benjamin Peirce, *Linear Associative Algebra*, 1870).

Peirce retired from the Coast Survey after thirty years of service to devote himself to writing. He left without a pension. Despite a small inheritance and some income derived from his lectures, from the publication of essays, book reviews and numerous contributions to the

Century Dictionary and to the *Dictionary of Philosophy*, he suffered financial hardships during the entire period of his retirement. After his death several hundreds of unpublished manuscripts were found. These came into the care of the Department of Philosophy at Harvard University and have since been edited and published as the *Collected Papers of Charles Sanders Peirce*. The range of subjects on which Peirce wrote was encyclopaedic, including not only logic and philosophy, but also topics in astrophysics, pure mathematics, philosophy of mathematics, probability, philology, criminology, telepathy and optics. All of Peirce's writings were marked by originality of outlook. Intense in his advocacy of the scientific method, vigorously outspoken and unwilling to bridle his violent disapproval of certain popular philosophical beliefs, Peirce failed during his life to win large numbers of disciples. Later scholars were better prepared to evaluate his proposals and his new ideas, presented as they were in the strange but highly significant terms and symbols invented by him. The inspirational and stimulating force of his work has come to enjoy ever greater recognition, and interest in Charles Peirce's vital contributions to the field of logic and scientific method has mounted steadily.

ON THE ALGEBRA OF LOGIC

§3. FORMS OF PROPOSITIONS

173. In place of the two expressions $A-\!\!<B$ and $B-\!\!<A$ taken together we may write $A=B$;[1] in place of the two expressions $A-\!\!<B$ and $\overline{B-\!\!<}A$ taken together we may write $A<B$ or $B>A$;

1. There is a difference of opinion among logicians as to whether $-\!\!<$ or $=$ is the simpler relation. But in my paper on the *Logic of Relatives* (47 n.) I have strictly demonstrated that the preference must be given to $-\!\!<$ in this respect. The term *simpler* has an exact meaning in logic; it means that whose logical depth is smaller; that is, if one conception implies another, but not the reverse, then the latter is said to be the simpler. Now to say that $A=B$ implies that $A-\!\!<B$, but not conversely. *Ergo*, etc. It is to no purpose to reply that $A-\!\!<B$ implies $A=(A$ that is $B)$; it would be equally relevant to say that $A-\!\!<B$ implies $A=A$. Consider an analogous case. Logical sequence is a simpler conception than causal sequence, because every causal sequence is a logical sequence but not every logical sequence is a causal sequence; and it is no reply to this to say that a logical sequence between two facts implies a causal sequence between some two facts whether the same or different.

and in place of the two expressions $A\overline{-<}B$ and $B\overline{-<}A$ taken together [disjunctively] we may write $A\asymp B$.[1]

174. De Morgan, in the remarkable memoir with which he opened his discussion of the syllogism (1846, p. 380)[2] has pointed out that we often carry on reasoning under an implied restriction as to what we shall consider as possible, which restriction, applying to the whole of what is said, need not be expressed. The total of all that we consider possible is called the *universe* of discourse, and may be very limited. One mode of limiting our universe is by considering only what actually occurs, so that everything which does not occur is regarded as impossible.

175. The forms $A-<B$, or A implies B, and $A\overline{-<}B$, or A does not imply B,[3] embrace both hypothetical and categorical pro-

---

The idea that $=$ is a very simple relation is probably due to the fact that the discovery of such a relation teaches us that instead of two objects we have only one, so that it simplifies our conception of the universe. On this account the existence of such a relation is an important fact to learn; in fact, it has the sum of the importances of the two facts of which it is compounded. It frequently happens that it is more convenient to treat the propositions $A-<B$ and $B-<A$ together in their form $A=B$; but it also frequently happens that it is more convenient to treat them separately. Even in geometry we can see that to say that two figures A and B are equal is to say that when they are properly put together A will cover B and B will cover A; and it is generally necessary to examine these facts separately. So, in comparing the numbers of two lots of objects, we set them over against one another, each to each, and observe that for every one of the lot A there is one of the lot B, and for every one of the lot B there is one of the lot A.

In logic, our great object is to analyse all the operations of reason and reduce them to their ultimate elements; and to make a calculus of reasoning is a subsidiary object. Accordingly, it is more philosophical to use the copula $-<$ apart from all considerations of convenience. Besides, this copula is intimately related to our natural logical and metaphysical ideas; and it is one of the chief purposes of logic to show what validity those ideas have. Moreover, it will be seen further on that the more analytical copula does in point of fact give rise to the easiest method of solving problems of logic.

1. i.e. $(A=B)$.

2. 'On the Structure of the Syllogism, and on the Application of the Theory of Probabilities to Questions of Argument and Authority', *Transactions, Cambridge Philosophical Society*, vol. 8, pp. 379–408 (1849). The paper was read and dated 1846.

3. i.e. it is false that $A-<B$.

positions. Thus, to say that all men are mortal is the same as to say that if any man possesses any character whatever then a mortal possesses that character. To say, 'if A, then B' is obviously the same as to say that from A, B follows, logically or extralogically. By thus identifying the relation expressed by the copula with that of illation, we identify the proposition with the inference, and the term with the proposition. This identification, by means of which all that is found true of term, proposition or inference is at once known to be true of all three, is a most important engine of reasoning, which we have gained by beginning with a consideration of the genesis of logic.[1]

§ 4. THE ALGEBRA OF THE COPULA

182. From the identity of the relation expressed by the copula with that of illation, springs an algebra. In the first place, this gives us

$$x -\!\!< x \tag{1}$$

the principle of identity, which is thus seen to express that what we have hitherto believed we continue to believe, in the absence of any reason to the contrary. In the next place, this identification shows that the two inferences

$$
\begin{array}{ccc}
x & & \\
y & \text{and} & x \\
\therefore z & & \therefore y -\!\!< z
\end{array}
\tag{2}
$$

are of the same validity. Hence we have

$$\{x -\!\!< (y -\!\!< z)\} = \{y -\!\!< (x -\!\!< z)\}.[2] \tag{3}$$

183. From (1) we have

$$(x -\!\!< y) -\!\!< (x -\!\!< y),$$

1. In consequence of the identification in question, in $S -\!\!< P$, I speak of S indifferently as *subject*, *antecedent*, or *premiss*, and of P as *predicate*, *consequent*, or *conclusion*.

2. Mr Hugh McColl (*Calculus of Equivalent Statements*, Second Paper, 1878, [*Proceedings, London Mathematical Society*, vol. 9, p. 183 (1877)], makes use of the sign of inclusion several times in the same proposition. He does not, however, give any of the formulae of this section.

whence by (2)

$$x-\!<y \quad x$$
$$\therefore \; y \qquad\qquad (4)$$

is a valid inference.

184. By (4), if x and $x-\!< y$ are true y is true; and if y and $y-\!< z$ are true z is true. Hence, the inference is valid

$$x \quad x-\!<y \quad y-\!<z$$
$$\therefore z.$$

By the principle of (2) this is the same as to say that

$$x-\!<y \quad y-\!<z$$
$$\therefore x-\!<z \qquad\qquad (5)$$

is a valid inference.

A CONTRIBUTION TO
THE PHILOSOPHY OF NOTATION[1]

§1. THREE KINDS OF SIGNS[2]

359. Any character or proposition either concerns one subject, two subjects, or a plurality of subjects. For example, one particle has mass, two particles attract one another, a particle revolves about the line joining two others. A fact concerning two subjects is a dual character or relation; but a relation which is a mere combination of two independent facts concerning the two subjects may be called *degenerate*, just as two lines are called a degenerate conic. In like manner a plural character or conjoint relation is to be called degenerate if it is a mere compound of dual characters.

360. A sign is in a conjoint relation to the thing denoted and

1. *The American Journal of Mathematics*, vol. 7, No. 2, pp. 180–202 (1885); reprinted pp. 1–23.
2. See vol. 2, bk II, for a detailed analysis of signs.

to the mind. If this triple relation is not of a degenerate species, the sign is related to its object only in consequence of a mental association, and depends upon a habit. Such signs are always abstract and general, because habits are general rules to which the organism has become subjected. They are, for the most part, conventional or arbitrary. They include all general words, the main body of speech, and any mode of conveying a judgement. For the sake of brevity I will call them *tokens*.[1]

361. But if the triple relation between the sign, its object, and the mind, is degenerate, then of the three pairs

sign   object
sign   mind
object   mind

two at least are in dual relations which constitute the triple relation. One of the connected pairs must consist of the sign and its object, for if the sign were not related to its object except by the mind thinking of them separately, it would not fulfil the function of a sign at all. Supposing, then, the relation of the sign to its object does not lie in a mental association, there must be a direct dual relation of the sign to its object independent of the mind using the sign. In the second of the three cases just spoken of, this dual relation is not degenerate, and the sign signifies its object solely by virtue of being really connected with it. Of this nature are all natural signs and physical symptoms. I call such a sign an *index*, a pointing finger being the type of the class.

The index asserts nothing; it only says 'There!' It takes hold of our eyes, as it were, and forcibly directs them to a particular object, and there it stops. Demonstrative and relative pronouns are nearly pure indices, because they denote things without describing them; so are the letters on a geometrical diagram, and the subscript numbers which in algebra distinguish one value from another without saying what those values are.

362. The third case is where the dual relation between the sign

1. More frequently called 'symbols'; the word 'token' is later (in 4.537) taken to apply to what in 2.245 is called a 'sinsign'.

and its object is degenerate and consists in a mere resemblance between them. I call a sign which stands for something merely because it resembles it, an *icon*. Icons are so completely substituted for their objects as hardly to be distinguished from them. Such are the diagrams of geometry. A diagram, indeed, so far as it has a general significance, is not a pure icon; but in the middle part of our reasonings we forget that abstractness in great measure, and the diagram is for us the very thing. So in contemplating a painting, there is a moment when we lose the consciousness that it is not the thing, the distinction of the real and the copy disappears, and it is for the moment a pure dream – not any particular existence, and yet not general. At that moment we are contemplating an *icon*.

363. I have taken pains to make my distinction[1] of icons, indices and tokens clear, in order to enunciate this proposition: in a perfect system of logical notation signs of these several kinds must all be employed. Without tokens there would be no generality in the statements, for they are the only general signs; and generality is essential to reasoning. Take, for example, the circles by which Euler represents the relations of terms. They well fulfil the function of icons, but their want of generality and their incompetence to express propositions must have been felt by everybody who has used them.[2] Mr Venn[3] has, therefore, been led to add shading to them; and this shading is a conventional sign of the nature of a token. In algebra, the letters, both quantitative and functional, are of this nature. But tokens alone do not state what is the subject of discourse; and this can, in fact, not be described in general terms; it can only be indicated. The actual world cannot be distinguished from a world of imagination by any description. Hence the need of pronoun and indices, and the more complicated the subject the greater the need of them. The introduction of indices into the algebra of logic is the greatest

1. See *Proceedings, American Academy of Arts and Sciences*, vol. 7, p. 294, May 14, 1867. [1.558.]

2. See 4.356.

3. 'On the Diagrammatic and Mechanical Representations of Propositions and Reasoning'. *Philosophical Magazine*, ser. 5, vol. 10, pp. 1–15 (1880).

merit of Mr Mitchell's system.[1] He writes F_1 to mean the proposition F is true of every object in the universe, and F_u to mean that the same is true of some object.[2] This distinction can only be made in some such way as this. Indices are also required to show in what manner other signs are connected together. With these two kinds of signs alone any proposition can be expressed; but it cannot be reasoned upon, for reasoning consists in the observation that where certain relations subsist certain others are found, and it accordingly requires the exhibition of the relations reasoned within an icon. It has long been a puzzle how it could be that, on the one hand, mathematics is purely deductive in its nature, and draws its conclusions apodictically, while on the other hand, it presents as rich and apparently unending a series of surprising discoveries as any observational science. Various have been the attempts to solve the paradox by breaking down one or other of these assertions, but without success. The truth, however, appears to be that all deductive reasoning, even simple syllogism, involves an element of observation, namely deduction consists in constructing an icon or diagram the relations of whose parts shall present a complete analogy with those of the parts of the object of reasoning, of experimenting upon this image in the imagination, and of observing the result so as to discover unnoticed and hidden relations among the parts. For instance, take the syllogistic formula,

$$\text{All } M \text{ is } P$$
$$S \text{ is } M$$
$$\therefore S \text{ is } P.$$

This is really a diagram of the relations of S, M and P. The fact that the middle term occurs in the two premisses is actually exhibited, and this must be done or the notation will be of no value. As for algebra, the very idea of the art is that it presents formulae which can be manipulated, and that by observing the effects of such manipulation we find properties not to be otherwise discerned. In such manipulation, we are guided by previous

1. *Studies in Logic*, by members of the Johns Hopkins University; Boston, Little, Brown and Co., 1883.
2. Ibid., p. 74.

discoveries which are embodied in general formulae. These are patterns which we have the right to imitate in our procedure, and are the *icons par excellence* of algebra. The letters of applied algebra are usually tokens, but the x, y, z, etc., of a general formula, such as

$$(x+y)z = xz + yz,$$

are blanks to be filled up with tokens, they are indices of tokens. Such a formula might, is it true, be replaced by an abstractly stated rule (say that multiplication is distributive); but no application could be made of such an abstract statement without translating it into a sensible image.

§3. FIRST-INTENTIONAL LOGIC OF RELATIVES

392. The algebra of Boole affords a language by which may be expressed which can be said without speaking of more than one individual at a time. It is true that it can assert that certain characters belong to a whole class, but only such characters as belong to each individual separately. The logic of relatives considers statements involving two and more individuals at once. Indices are here required. Taking, first, a degenerate form of relation, we may write $x_i y_j$ to signify that x is true of the individual i while y is true of the individual j. If z be a relative character z_{ij} will signify that i is in that relation to j. In this way we can express relations of considerable complexity. Thus if

$$1, \quad 2, \quad 3,$$
$$4, \quad 5, \quad 6,$$
$$7, \quad 8, \quad 9,$$

are points in a plane, and l_{123} signifies that 1, 2, and 3 lie on one line, a well-known proposition of geometry[1] may be written

$$l_{159} - < l_{267} - < l_{348} - < l_{147} - < l_{258} - < l_{369} - <$$
$$l_{123} - < l_{456} - < l_{789}.$$

In this notation is involved a *sixth icon*.

1. If the six vertices of a hexagon lie three and three on two straight lines, the three points of intersection of the opposite sides lie on a straight line.

393. We now come to the distinction of *some* and *all*, a distinction which is precisely on a par with that between truth and falsehood; that is, it is descriptive.

All attempts to introduce this distinction in the Boolian algebra were more or less complete failures until Mr Mitchell[1] showed how it was to be effected. His method really consists in making the whole expression of the proposition consist of two parts, a pure Boolian expression referring to an individual and a quantifying part saying what individual this is. Thus, if k means 'he is a king', and h, 'he is happy', the Boolian

$$(\bar{k}+h)$$

means that the individual spoken of is either not a king or is happy. Now, applying the quantification, we may write

$$\text{Any } (\bar{k}+h)$$

to mean that this is true of any individual in the (limited) universe, or

$$\text{Some } (\bar{k}+h)$$

to mean that an individual exists who is either not a king or is happy. So

$$\text{Some } (kh)$$

means some king is happy, and

$$\text{Any } (kh)$$

means every individual is both a king and happy. The rules for the use of this notation are obvious. The two propositions

$$\text{Any } (x) \quad \text{Any } (y)$$

are equivalent to

$$\text{Any } (xy).$$

From the two propositions

$$\text{Any } (x) \quad \text{Some } (y)$$

we may infer

$$\text{Some } (xy).[2]$$

1. Op. cit., p. 79.
2. I will just remark, quite out of order, that the quantification may be made numerical; thus producing the numerically definite inferences of De Morgan and Boole. Suppose at least $\frac{2}{3}$ of the company have white neckties and at least $\frac{3}{4}$ have dress coats. Let w mean 'he has a white necktie', and

Mr Mitchell has also a very interesting and instructive extension of his notation for *some* and *all*, to a two-dimensional universe, that is, to the logic of relatives. Here, in order to render the notation as iconical as possible we may use Σ for *some*, suggesting a sum, and Π for *all*, suggesting a product. Thus $\Sigma_i x_i$ means that x is true of some one of the individuals denoted by i or

$$\Sigma_i x_i = x_i + x_j + x_k + \text{etc.,}[1]$$

or

In the same way, $\Pi_i x_i$ means that x is true of all these individuals, or

$$\Pi_i x_i = x_i x_j x_k, \text{ etc.}[2]$$

If x is a simple relation, $\Pi_i \Pi_j x_{ij}$ means that every i is in this relation to every j, $\Sigma_i \Pi_j x_{ij}$ that some one i is in this relation to every j, $\Pi_j \Sigma^i x_{ij}$ that to every j some i or other is in this relation, $\Sigma_i \Sigma x_{ij}$ that some i is in this relation to some j. It is to be remarked that $\Sigma_i x_j$ and $\Pi_i x_i$ are only *similar* to a sum and a product; they are not strictly of that nature, because the individuals of the universe may be innumerable.

394. At this point, the reader would perhaps not otherwise easily get so good a conception of the notation as by a little

---

d 'he has a dress coat'. Then, the two propositions are

$$\tfrac{2}{3}(w) \text{ and } \tfrac{3}{4}(d).$$

These are to be multiplied together. But we must remember that xy is a mere abbreviation for $\bar{x} + \bar{y}$, and must therefore write

$$\overline{\overline{\tfrac{2}{3}w} + \overline{\tfrac{3}{4}d}}.$$

Now $\overline{\tfrac{2}{3}w}$ is the denial of $\tfrac{2}{3}w$, and this denial may be written $(>\tfrac{1}{3})\bar{w}$, or more than $\tfrac{1}{3}$ of the universe (the company) have not white neckties. So $\tfrac{3}{4}d = (>\tfrac{1}{4})\bar{d}$ the combined premisses thus become

$$\overline{(>\tfrac{1}{3})\bar{w} + (>\tfrac{1}{4})\bar{d}}.$$

Now $(>\tfrac{1}{3})\bar{w} + (>\tfrac{1}{4})\bar{d}$ gives   May be $(\tfrac{1}{3} + \tfrac{1}{4})(\bar{w} + \bar{d})$.

Thus we have   May be $(\tfrac{7}{12})(\bar{w} + \bar{d})$

and this is   (At least $\tfrac{5}{12})(\bar{w} + \bar{d})$, which is the conclusion.

  1. This is the seventh icon?
  2. This is the eighth icon?

practice in translating from ordinary language into this system and back again. Let l_{ij} mean that i is a lover of j, and b_{ij} that i is a benefactor of j. Then

$$\Pi_i \Sigma_j l_{ij} b_{ij}$$

means that everything is at once a lover and a benefactor of something; and

$$\Pi_i \Sigma_j l_{ij} b_{ji}$$

that everything is a lover of a benefactor of itself.

$$\Sigma_i \Sigma_k \Pi_j (l_{ij} + b_{jk})$$

means that these are two persons, one of whom loves everything except benefactors of the other (whether he loves any of these or not is not stated). Let g_i mean that i is a griffin, and c_i that i is a chimera, then

$$\Sigma_i \Pi_j (g_i l_{ij} + \bar{c}_j)$$

means that if there be any chimeras there is some griffin that loves them all; while

$$\Sigma_i \Pi_j g_i (l_{ij} + \bar{c}_j)$$

means that there is a griffin and he loves every chimera that exists (if any exist). On the other hand,

$$\Pi_i \Sigma_i g_{ij} (l_{ij} + \bar{c}_j)$$

means that griffins exist (one, at least), and that one or other of them loves each chimera that may exist; and

$$\Pi_j \Sigma_i^i (g l_{ij} + \bar{c}_j)$$

means that each chimera (if there is any) is loved by some griffin or other.

NOMENCLATURE AND DIVISIONS
OF DYADIC RELATIONS[1]

§1. NOMENCLATURE

573. A dyadic relation proper is either such as can only have place between two subjects of different universes of discourse (as

1. Printed separately in eight pages, c. 1903, apparently intended as the second part of *A Syllabus of Certain Topics of Logic,* published as a supplement to the Lowell Lectures of 1903. See vol. 1, bk II, ch. 1, n.

the membership of a natural person in a corporation), or is such as can subsist between two objects of the same universe. A relation of the former description may be termed a *referential relation*; a relation of the latter description, a *rerelation*.[1]

574. A rerelation may be either such as only subsist between characters or between laws (such as the relation of 'essentially depending on'), or it may be such as can subsist between two existent individual objects. In the former case, it may be termed a *modal* relation (not a good term), in the latter case an *existential relation*. The author's writings on the logic of relations[2] were substantially restricted to existential relations; and the same restriction will be continued in the body of what here follows. A note at the end of this section will treat of modal relations.

575. The number of different species of existential relations for which technical designations are required is so great that it will be best to adopt names for them which shall, by their form, furnish technical definitions of them, in imitation of the nomenclature of chemistry. The following rules will here be used. Any name (for which in this statement of the rules of word-formation we may put *x*), having been adopted for all relations of a given description, the preposition extra (or *ex*, or *e*) will be prefixed to that name ('extra *x*') in order to form a name descriptive of any relation to which the primitive name does not apply; the preposition *contra* will be prefixed (forming 'contra-*x*') to make

1. It is far better to invent a word for a purely technical conception than to use an expression liable to be corrupted by being employed by loose writers. I reduplicate the first syllable of relation to form this word, with little reference to the meaning of the syllable as a preposition. Still, relations of this kind are the only ones that might be asserted of the same relates transposed; and the reduplication of the preposition *re* connotes such transposition.

2. I must, with pain and shame, confess that in my early days I showed myself so little alive to the decencies of science that I presumed to change the name of this branch of logic, a name established by its author and my master, Augustus De Morgan, to 'the logic of relatives'. I consider it my duty to say that this thoughtless act is a bitter reflection to me now, so that young writers may be warned not to prepare themselves similar sources of unhappiness. I am the more sorry, because my designation has come into general use.

a name applicable only to such relations as consist precisely in the non-subsistence of corresponding relations to which the primitive name does apply; the preposition *juxta* will be prefixed so as to bear the sense of *contra-extra*, or (what is the same) *extra-contra*; the preposition *red* (or *re*) will be prefixed to form a name applicable to a relation if, and only if, the correlate of it stands to its relate in a relation to which the primitive name applies, so that, in other words, a 'red-*x*' is a relation the converse of an *x*; the preposition *com* (or *con*, or *co*) will be prefixed to form the general name of any relation which consists in its relate and correlate alike standing in one relation of the primitive kind to one and the same individual correlate; the preposition *ultra* will be prefixed to form a name applicable only to a relation which subsists between any given relate and correlate only in case the former stands in a relation of the primitive kind to some individual to which the latter does not stand in that same relation; the preposition *trans* will be used so as to be equivalent to *contra-red-ultra*, or (what is the same) *recontrultra*, so that *A* will be in a relation '*trans-x*' to *B*, if, and only if, there is an *x*-relation in which *A* stands to whatever individual there may be to which *B* stands in that very same relation; and the preposition *super* to form the name of a relation which is, at once, *ultra* and *trans*, in respect to the very same relation of the primitive kind. Any of these prepositions may be prefixed, in the same sense, occasionally (and where no misunderstanding could result) not only to names of classes of relations and their cognates, but also to relative terms. But it is chiefly the prepositions *com*, *ultra*, *trans* and *super*, that will be so used. For example, taking the relative term 'loves', there will be little occasion to use the first four of the following expressions, especially the first and third, which become almost meaningless, while the last four will often be convenient.

1. *A extra-loves B*; that is, stands in some other relation, whether loving besides or not;

2. *A contra-loves B*; that is, does not love *B*;

3. *A juxta-loves B*; that is, stands in some other relation than than of not loving, whether loving or not;

4. *A reloves B*; that is, is loved by;

5. *A coloves B*; that is, loves something loved by;

6. *A ultraloves B*; that is, loves something not loved by;

7. *A transloves B*; that is, loves whatever may be loved by;

8. *A superloves B*; that is, loves whatever may be loved by and something else.[1]

§2. FIRST SYSTEM OF DIVISIONS

578. An existential dyadic relation may be termed a *lation*[2] to express its possibly subsisting between two existing individuals. . . .

THE SIMPLEST MATHEMATICS

239. The philosophical mathematician, Dr Richard Dedekind,[3] holds mathematics to be a branch of logic. This would not result from my father's definition, which runs, not that mathematics is the science of *drawing* necessary conclusions – which would be deductive logic – but that it is the science which *draws* necessary conclusions. It is evident, and I know as a fact, that he had this distinction in view. At the time when he thought out this definition, he, a mathematician, and I, a logician, held daily discussions about a large subject which interested us both; and

---

1. i.e.

1. extraloves	alb	5. coloves	$(alc)(c\bar{l}b)$
2. contraloves	$-(alb)$	6. ultraloves	$(alc)(c\bar{l}b)$
3. juxtaloves	$-(a\bar{l}b)$	7. transloves	$-[(a\bar{l}c)(c\bar{l}b)]$
4. reloves	$a\bar{l}b$	8. superloves	$\{-[(a\bar{l}c)(c\bar{l}b)][(ald)(d\bar{l}b)]\}$

2. The following schedule may be of aid in this section:

1. $\Sigma_i\Sigma_j r_{ij}$ – lation	11. $\Pi_j\Sigma_i \bar{r}_{ij}$ – extrareperlation
2. $\Sigma_i\Sigma_j \bar{r}_{ij}$ – contralation	12. $\Pi_j\Sigma_i r_{ij}$ – juxtareperlation
3. $\Pi_i\Pi_j \bar{r}_{ij}$ – extralation; $r = 0$	13. $\Sigma_j\Pi_i\Sigma_k r_{ik} \cdot r_{jk}$ – conlation
4. $\Pi_i\Pi_j r_{ij}$ – juxtalation; $r = \infty$	14. $\Sigma_i\Pi_j\Sigma_k r_{ik} \cdot \bar{r}_{jk}$ – ultralation
5. $\Sigma_i\Pi_j r_{ij}$ – perlation	15. $\Pi_i\Sigma_j\Pi_k - (r_{ik} \cdot r_{jk})$ – translation
6. $\Sigma_i\Pi_j \bar{r}_{ij}$ – contraperlation	16. $\Pi_i\Pi_j\Pi_k\Sigma_l - (r_{ik} \cdot r_{il})(r_{il} \cdot \bar{r}_{jl})$
7. $\Pi_i\Sigma_j \bar{r}_{ij}$ – extraperlation	$\qquad$ – superlation
8. $\Pi_i\Sigma_j r_{ij}$ – juxtaperlation	17. $\Sigma_i\Sigma_j r_{ij}\infty$ – essential perlation
9. $\Sigma_j\Pi_i r_{ij}$ – reperlation	18. $\Sigma_i\Sigma_j \bar{r}\infty_{ij}$ – essential reperlation
10. $\Sigma_i\Pi_i \bar{r}_{ij}$ – contrareperlation	19. $\Sigma_i\Sigma_j \bar{r}_{ij}\infty$ – contressentiperlation

$1 = -3$, $5 = -7$, $9 = -11$; $2 = -4$; $6 = -8$; $10 = -12$, $5 -<12$; $9 -<8$; $6 -<11$; $10 -<7$; $17.18 -<3\Psi 4$.

3. *Was sind und was sollen die Zahlen; Vorwort* (1888).

he was struck, as I was, with the contrary nature of his interest and mine in the same propositions. The logician does not care particularly about this or that hypothesis or its consequences, except so far as these things may throw a light upon the nature of reasoning. The mathematician is intensely interested in efficient methods of reasoning, with a view to their possible extension to new problems; but he does not, *qua* mathematician, trouble himself minutely to dissect those parts of this method whose correctness is a matter of course. The different aspects which the algebra of logic will assume for the two men is instructive in this respect. The mathematician asks what value this algebra has as a calculus. Can it be applied to unravelling a complicated question? Will it, at one stroke, produce a remote consequence? The logician does not wish the algebra to have that character. On the contrary, the greater number of distinct logical steps, into which the algebra breaks up an inference, will for him constitute a superiority of it over another which moves more swiftly to its conclusions. He demands that the algebra shall analyse a reasoning into its last elementary steps. Thus, that which is a merit in a logical algebra for one of these students is a demerit in the eyes of the other. The one studies the science of drawing conclusions, the other the science which draws necessary conclusions.

240. But, indeed, the difference between the two sciences is far more than that between two points of view. Mathematics is purely hypothetical: it produces nothing but conditional propositions. Logic, on the contrary, is categorical in its assertions. True, it is not merely, or even mainly, a mere discovery of what really is, like metaphysics. It is a normative science. It thus has a strongly mathematical character, at least in its methodeutic division; for here it analyses the problem of how, with given means, a required end is to be pursued. This is, at most, to say that it has to call in the aid of mathematics; that it has a mathematical branch. But so much may be said of every science. There is a mathematical logic just as there is a mathematical optics and a mathematical economics. Mathematical logic is formal logic. Formal logic, however developed, is mathematics. Formal logic, however, is by no means the whole of logic, or even its

principal part. It is hardly to be reckoned as a part of logic proper. Logic has to define its aim; and in doing so is even more dependent upon ethics, or the philosophy of aims, by far, than it is, in the methodeutic branch, upon mathematics. We shall soon come to understand how a student of ethics might well be tempted to make his science a branch of logic; as, indeed, it pretty nearly was in the mind of Socrates. But this would be no truer a view than the other. Logic depends upon mathematics; still more intimately upon ethics; but its proper concern is with truths beyond the purview of either.

47

GOTTLOB FREGE
(1848–1925)

GOTTLOB FREGE was born in Wismar, Germany, in 1848. Educated
at the Universities of Jena and Gottingen, he received his first appointment at Jena as *Privatdozent* in 1871. He became a professor of mathematics there in 1879. During a long career at Jena, Frege published
several highly significant works, the most important of which are:
Begriffsschrift (1879), *Grundlagen der Arithmetik* (1884), *Funktion und
Begriff* (1891), *Begriff und Gegenstand* (1892), *Sinn und Bedeutung*
(1892), *Grundgesetze I* (1893), *II* (1903). His efforts were directed
towards establishing the principle that mathematics is based on the
general laws of logic. Indeed, he went so far as to take the position that
mathematics is so intimately bound up with logic that no separation of
the two can be effected. In the implementation of this thesis, Frege
pioneered in the construction of a logical symbolism and in the development of symbolical methods of logical analysis. Due in a large
measure to the difficulty of grasping his symbolism as well as to the
novelty of the ideas contained in them, his works were met with a vast
silence. Frege openly expressed his disappointment in seeking mention
of his writings in the technical journals again and again but finding
none whatever. The silence was broken eventually by Bertrand Russell.
However, despite the great extent to which Russell found Frege's views
in agreement with his own, he nevertheless found it possible to derive
from Frege's system, a paradox about the class of all classes which are
not members of themselves. Frege replied to Russell's criticism, but he
failed to convince himself completely of the adequacy of his reply, and
feeling that he had not removed the logical flaw from his work, he
wrote nothing in the field of foundations of arithmetic from that time
on. Later logicians, on the other hand, provided satisfactory replies to
Russell's criticism and also to the objections which had been raised
by Wittgenstein and others to the Frege–Russell definitions. A new
evaluation of Frege's work ensued, in which the rich significance of his

fundamental ideas was recognized. Frege's work proved to be both instructive and inspirational.

The present tendency is to regard Frege as the greatest logician of the nineteenth century, for although approximately a third of his life fell in the twentieth century, his major contributions to logic and mathematics appeared in the nineteenth. In 1948, *Sinn and Bedeutung* was translated into Italian and into English; another English translation of the same work appeared in 1949. *Grundlagen der Arithmetik* was translated into English in 1950 and there have been reprints of this translation since then. Controversies about Frege's work, where they exist, turn not upon the soundness or the value of his ideas but upon the suitability of the technical terms which have been selected (or invented) by translators in their striving to achieve accuracy in their presentation of Frege's meaning. The first to define number in purely logical terms, Frege presented his definition of a number in his *Grund lagen der Arithmetik*, a work in which he made a deliberate effort to use a minimum of symbolism and the simplest possible language.

THE FOUNDATIONS OF ARITHMETIC

translated by J. L. Austin

To obtain the concept of Number, we must fix the sense of a numerical identity.

§ 62. How, then, are numbers to be given to us, if we cannot have any ideas or intuitions of them? Since it is only in the context of a proposition that words have any meaning, our problem becomes this: To define the sense of a proposition in which a number word occurs. That, obviously, leaves us still a very wide choice. But we have already settled that number words are to be understood as standing for self-subsistent objects. And that is enough to give us a class of propositions which must have a sense, namely those which express our recognition of a number as the same again. If we are to use the symbol *a* to signify an object, we must have a criterion for deciding in all cases whether *b* is the same as *a*, even if it is not always in our power to apply this criterion. In our present case, we have to define the sense of the proposition 'the number which belongs to the concept F is the same as that which belongs to the concept G'; that is to say, we must reproduce the content of this proposition in other terms, avoiding the use of the expression 'the Number which belongs to the concept F'. In doing this, we shall be giving a general criterion for the identity of numbers. When we have thus acquired a means of arriving at a determinate number and of recognizing it again as the same, we can assign it a number word as its proper name.

§ 63. Hume[1] long ago mentioned such a means: 'When two numbers are so combined as that the one has always an unit answering to every unit of the other, we pronounce them equal.' This opinion, that numerical equality or identity must be defined in terms of one-one correlation, seems in recent years to have gained widespread acceptance among mathematicians.[2] But it raises at once certain logical doubts and difficulties, which ought not to be passed over without examination.

It is not only among numbers that the relationship of identity is found. From which it seems to follow that we ought not to define it specially for the case of numbers. We should expect the concept of identity to have been fixed first, and that then, from it together with the concept of Number, it must be possible to deduce when Numbers are identical with one another, without there being need for this purpose of a special definition of numerical identity as well.

As against this, it must be noted that for us the concept of Number has not yet been fixed, but is only due to be determined in the light of our definition of numerical identity. Our aim is to construct the content of a judgement which can be taken as an identity such that each side of it is a number. We are therefore proposing not to define identity specially for this case, but to use the concept of identity, taken as already known, as a means for arriving at that which is to be regarded as being identical.

1. Baumann, op. cit., vol. II, p. 565, *Treatise*, bk I, part iii, sect. 1.
2. Cf. E. Schroder, op. cit., pp. 7–8; E. Kossak, *Die Elemente der Arithmetik, Programm des Friedrichs-Werder'schen Gymnasiums*, Berlin, 1872, p. 16; G. Cantor, *Grundlagen einer allgemeinen Mannichfaltigkeitslehre*, Leipzig, 1883.

Admittedly, this seems to be a very odd kind of definition, to which logicians have not yet paid enough attention; but that it is not altogether unheard of, may be shown by a few examples.

§ 64. The judgement 'line *a* is parallel to line *b*', or, using symbols,

$$a \mathbin{//} b,$$

can be taken as an identity. If we do this, we obtain the concept of direction, and say: 'the direction of line *a* is identical with the direction of line *b*'. Thus we replace the symbol $//$ by the more generic symbol $=$, through removing what is specific in the content of the former and dividing it between *a* and *b*. We carve up the content in a way different from the original way, and this yields us a new concept. Often, of course, we conceive of the matter the other way round, and many authorities define parallel lines as lines whose directions are identical. The proposition that 'straight lines parallel to the same straight line are parallel to one another' can then be very conveniently proved by invoking the analogous proposition about things identical with the same thing. Only the trouble is, that this is to reverse the true order of things. For surely everything geometrical must be given originally in intuition. But now I ask whether anyone has an intuition of the direction of a straight line. Of a straight line, certainly; but do we distinguish in our intuition between this straight line and something else, its direction? That is hardly plausible. The concept of direction is only discovered at all as a result of a process of intellectual activity, which takes its start from the intuition. On the other hand, we do have an idea of parallel straight lines. Our convenient proof is only made possible by surreptitiously assuming, in our use of the word 'direction', what was to be proved; for if it were false that 'straight lines parallel to the same straight line are parallel to one another', then we could not transform *a* // *b* into an identity.

We can obtain in a similar way from the parallelism of planes a concept corresponding to that of direction in the case of straight lines; I have seen the name 'orientation'[1] used for this. From geometrical similarity is derived the concept of shape, so

---

1. '*Stellung*'.

that instead of 'the two triangles are similar' we say 'the two triangles are of identical shape' or 'the shape of the one is identical with that of the other'. It is possible to derive yet another concept in this way, to which no name has yet been given, from the collineation of geometrical forms.

§ 65. Now in order to get, for example, from parallelism[1] to the concept of direction, let us try the following definition: The proposition 'line a is parallel to line b' is to mean the same as 'the direction of line a is identical with the direction of line b'.

This definition departs to some extent from normal practice, in that it serves ostensibly to adapt the relation of identity, taken as already known, to a special case, whereas in reality it is designed to introduce the expression 'the direction of line a', which only comes into it incidentally. It is this that gives rise to a second doubt – are we not liable, through using such methods, to become involved in conflict with the well-known laws of identity? Let us see what these are. As analytic truths they should be capable of being derived from the concept itself alone. Now Leibniz's[2] definition is as follows: 'Things are the same as each other without loss of truth.'[3] This I propose to adopt as my own definition of identity. Whether we use 'the same', as Leibniz does, or 'identical', is not of any importance. 'The same' may indeed be thought to refer to complete agreement in all respects, 'identical'[4] only to agreement in this respect or that; but we can adopt a form of expression such that this distinction vanishes. For example, instead of 'the segments are identical in length', we can say 'the length of the segments is identical' or 'the same', and instead of 'the surfaces are identical in colour', 'the colour of the surfaces is identical'. And this is the way in which the word has been used in the examples above. Now, it is actually the case

1. I have chosen to discuss here the case of parallelism, because I can express myself less clumsily and make myself more easily understood. The argument can readily be transferred in essentials to apply to the case of numerical identity.

2. *Non inelegans specimen demonstrandi in abstractis* (Erdmann edn, p. 94).

3. '*Eadem sunt, quorum unum potest substitui alteri salva veritate.*'

4. Still more 'equal' or 'similar', which the German *gleich* can also mean.

that in universal substitutability all the laws of identity are contained.

In order, therefore, to justify our proposed definition of the direction of a line, we should have to show that it is possible, if line *a* is parallel to line *b*, to substitute 'the direction of *b*' everywhere for 'the direction of *a*'. This task is made simpler by the fact that we are being taken initially to know of nothing that can be asserted about the direction of a line except the one thing, that it coincides with the direction of some other line. We should thus have to show only that substitution was possible in an identity of this one type, or in judgement-contents containing such identities as constituent elements.[1] The meaning of any other type of assertion about directions would have first of all to be defined, and in defining it we can make it a rule always to see that it must remain possible to substitute for the direction of any line the direction of any line parallel to it.

§66. But there is still a doubt which may make us suspicious of our proposed definition. In the proposition 'the direction of *a* is identical with the direction of *b*' the direction of *a* plays the part of an object,[2] and our definition affords us a means of recognizing this object as the same again, in case it should happen to crop up in some other guise, say as the direction of *b*. But this means does not provide for all cases. It will not, for instance, decide for us whether England is the same as the direction of the Earth's axis – if I may be forgiven an example which looks nonsensical. Naturally no one is going to confuse England with the direction of the Earth's axis; but that is no thanks to our definition of direction. That says nothing as to

---

1. In a hypothetical judgement, for example, an identity of directions might occur as antecedent or consequent.

2. This is shown by the definite article. A concept is for me that which can be predicate of a singular judgement-content, an object that which can be subject of the same. If in the proposition 'the direction of the axis of the telescope is identical with the direction of the Earth's axis' we take the direction of the axis of the telescope as subject, then the predicate is 'identical with the direction of the Earth's axis'. This is a concept. But the direction of the Earth's axis is only an element in the predicate; it, since it can also be made the subject, is an object.

whether the proposition 'the direction of a is identical with q' should be affirmed or denied, except for the one case where q is given in the form of 'the direction of b'. What we lack is the concept of direction; for if we had that, then we could lay it down that, if q is not a direction, our proposition is to be denied, while if it is a direction, our original definition will decide whether it is to be denied or affirmed. So the temptation is to give as our definition: q is a direction, if there is a line b whose direction is q. But then we have obviously come round in a circle. For in order to make use of this definition, we should have to know already in every case whether the proposition 'q is identical with the direction of b' was to be affirmed or denied.

§ 67. If we were to try saying: q is a direction if it is introduced by means of the definition set out above, then we should be treating the way in which the object q is introduced as a property of q, which it is not. The definition of an object does not, as such, really assert anything about the object, but only lays down the meaning of a symbol. After this has been done, the definition transforms itself into a judgement, which does assert about the object; but now it no longer introduces the object, it is exactly on a level with other assertions made about it. If, moreover, we were to adopt this way out, we should have to be presupposing that an object can only be given in one single way; for otherwise it would not follow, from the fact that q *was* not introduced by means of our definition, that it *could* not have been introduced by means of it. All identities would then amount simply to this, that whatever is given to us in the same way is to be reckoned as the same. This, however, is a principle so obvious and so sterile as not to be worth stating. We could not, in fact, draw from it any conclusion which was not the same as one of our premisses. Why is it, after all, that we are able to make use of identities with such significant results in such divers fields? Surely it is rather because we are able to recognize something as the same again even although it is given in a different way.

§ 68. Seeing that we cannot by these methods obtain any concept of direction with sharp limits to its application, nor therefore, for the same reasons, any satisfactory concept of Number either, let us try another way. If line *a* is parallel to line *b*, then the extension of the concept 'line parallel to line *a*' is identical with the extension of the concept 'line parallel to line *b*'; and conversely, if the extensions of the two concepts just named are identical, then *a* is parallel to *b*. Let us try, therefore, the following type of definition: the direction of line *a* is the extension of the concept 'parallel to line *a*'; the shape of triangle *t* is the extension of the concept 'similar to triangle *t*'.

To apply this to our own case of Number, we must substitute for lines or triangles concepts, and for parallelism or similarity the possibility of correlating one to one the objects which fall under the one concept with those which fall under the other. For brevity, I shall, when this condition is satisfied, speak of the concept *F* being *equal*[1] to the concept *G*; but I must ask that this word be treated as an arbitrarily selected symbol, whose meaning is to be gathered, not from its etymology, but from what is here laid down.

My definition is therefore as follows: the Number which belongs to the concept *F* is the extension[2] of the concept 'equal to the concept *F*'.

§ 69. That this definition is correct will perhaps be hardly evident at first. For do we not think of the extensions of concepts as something quite different from numbers? How we do think

---

1. *Gleichzahlig* – an invented word, literally 'identinumerate' or 'tautarithmic'; but these are too clumsy for constant use. Other translators have used 'equinumerous'; 'equinumerate' would be better. Later writers have used 'similar' in this connexion (but as a predicate of 'class' not of 'concept').

2. I believe that for 'extension of the concept' we could write simply 'concept'. But this would be open to the two objections:

(1) that this contradicts my earlier statement that the individual numbers are objects, as is indicated by the use of the definite article in expressions like 'the number two' and by the impossibility of speaking of ones, twos, etc., in the plural, as also by the fact that the number constitutes only an element in the predicate of a statement of number;

(2) that concepts can have identical extensions without themselves coinciding. I am, as it happens, convinced that both these objections can be met; but to do this would take us too far afield for present purposes. I assume that it is known what the extension of a concept is.

of them emerges clearly from the basic assertions we make about them. These are as follows:

1. that they are identical,
2. that one is wider than the other.

But now the proposition: the extension of the concept 'equal to the concept F' is identical with the extension of the concept 'equal to the concept G' is true if and only if the proposition 'the same number belongs to the concept F as to the concept G' is also true. So that here there is complete agreement.

Certainly we do not say that one number is wider than another, in the sense in which the extension of one concept is wider than that of another; but then it is also quite impossible for a case to occur where the extension of the concept 'equal to the concept F' would be wider than the extension of the concept 'equal to the concept G'. For on the contrary, when all concepts equal to G are also equal to F, then conversely also all concepts equal to F are equal to G. 'Wider' as used here must not, of course, be confused with 'greater' as used of numbers.

Another type of case is, I admit, conceivable, where the extension of the concept 'equal to the concept F' might be wider or less wide than the extension of some other concept, which then could not, on our definition, be a Number; and it is not usual to speak of a Number as wider or less wide than the extension of a concept; but neither is there anything to prevent us speaking in this way, if such a case should ever occur.

Our definition completed and its worth proved.

§ 70. Definitions show their worth by proving fruitful. Those that could just as well be omitted and leave no link missing in the chain of our proofs should be rejected as completely worthless.

Let us try, therefore, whether we can derive from our definition of the Number which belongs to the concept F any of the well-known properties of numbers. We shall confine ourselves here to the simplest.

For this it is necessary to give a rather more precise account still of the term 'equality'. 'Equal' we defined in terms of one-one correlation, and what must now be laid down is how this latter expression is to be understood, since it might easily be supposed that it had something to do with intuition.

We will consider the following example. If a waiter wishes to be certain of laying exactly as many knives on a table as plates, he has no need to count either of them; all he has to do is to lay immediately to the right of every plate a knife, taking care that every knife on the table lies immediately to the right of a plate. Plates and knives are thus correlated one to one, and that by the identical spatial relationship. Now if in the proposition 'a lies immediately to the right of A' we conceive first one and then another object inserted in place of a and again of A, then that part of the content which remains unaltered throughout this process constitutes the essence of the relation. What we need is a generalization of this.

If from a judgement-content which deals with an object a and an object b we subtract a and b, we obtain as remainder a relation-concept which is, accordingly, incomplete at two points. If from the proposition 'the Earth is more massive than the Moon' we subtract the 'Earth', we obtain the concept 'more massive than the Moon'. If, alternatively, we subtract the object, 'the Moon', we get the concept 'less massive than the Earth'. But if we subtract them both at once, then we are left with a relation-concept, which taken by itself has no [assertible] sense any more than a simple concept has: it has always to be completed in order to make up a judgement-content. It can, however, be completed in different ways: instead of Earth and Moon I can put, for example, Sun and Earth, and this *eo ipso* effects the subtraction.

Each individual pair of correlated objects stands to the relation-concept much as an individual object stands to the concept under which it falls – we might call them the subject of the relation-concept. Only here the subject is a composite one. Occasionally, where the relation in question is convertible, this fact achieves verbal recognition, as in the proposition '[Peleus and Thetis were the parents of Achilles'.[1] But not always. For

1. This type of case should not be confused with another, in which the 'and' joins the subjects in appearance only, but in reality joins two propositions.

example, it would scarcely be possible to put the proposition 'the Earth is bigger than the Moon' into other words so as to make 'the Earth and the Moon' appear as a composite subject; the 'and' must always indicate that the two things are being put in some way on a level. However, this does not affect the issue.

The doctrine of relation-concepts is thus, like that of simple concepts, a part of pure logic. What is of concern to logic is not the special content of any particular relation, but only the logical form. And whatever can be asserted of this, is true analytically and known *a priori*. This is as true of relation-concepts as of other concepts.

Just as '*a* falls under the concept *F*' is the general form of a judgement-content which deals with an object *a*, so we can take '*a* stands in the relation ϕ to *b*' as the general form of a judgement-content which deals with an object *a* and an object *b*.

§ 71. If now every object which falls under the concept *F* stands in the relation ϕ to an object falling under the concept *G*, and if to every object which falls under *G* there stands in the relation ϕ an object falling under *F*, then the objects falling under *F* and under *G* are correlated with each other by the relation ϕ.

It may still be asked, what is the meaning of the expression 'every object which falls under *F* stands in the relation ϕ to an object falling under *G*' in the case where no object at all falls under *F*. I understand this expression as follows: the two propositions '*a* falls under *F*' and '*a* does not stand in the relation ϕ to any object falling under *G*' cannot, whatever be signified by *a*, both be true together; so that either the first proposition is false, or the second is, or both are. From this it can be seen that the proposition 'every object which falls under *F* stands in the relation ϕ to an object falling under *G*' is, in the case where there is no object falling under *F*, true; for in that case the first proposition '*a* falls under *F*' is always false, whatever *a* may be. In the same way the proposition 'to every object which falls under *G* there stands in the relation ϕ an object falling under *F*' means that the two propositions '*a* falls under *G*' and 'no object falling under *F* stands to *a* in the relation ϕ' cannot, whatever *a* may be, both be true together.

§ 72. We have thus seen when the objects falling under the concepts F and G are correlated with each other by the relation ϕ. But now in our case, this correlation has to be one-one. By this I understand that the two following propositions both hold good:

1. If d stands in the relation ϕ to a, and if d stands in the relation ϕ to e, then generally, whatever d, a and e may be, a is the same as e.

2. If d stands in the relation ϕ to a, and if b stands in the relation ϕ to a, then generally, whatever d, b and a may be, d is the same as b.

This reduces one-one correlation to purely logical relationships and enables us to give the following definition: the expression 'the concept F is equal to the concept G' is to mean the same as the expression 'there exists a relation ϕ which correlates one to one the objects falling under the concept F with the objects falling under the concept G'.

We now repeat our original definition: the Number which belongs to the concept F is the extension of the concept 'equal to the concept F' and add further: the expression 'n is a Number' is to mean the same as the expression 'there exists a concept such that n is the Number which belongs to it'.

Thus the concept of Number receives its definition, apparently, indeed, in terms of itself, but actually without any fallacy, since 'the Number which belongs to the concept F' has already been defined.

§ 73. Our next aim must be to show that the Number which belongs to the concept F is identical with the Number which belongs to the concept G is the concept F is equal to the concept G. This sounds, of course, like a tautology. But it is not; the meaning of the word 'equal' is not to be inferred from its etymology, but taken to be as I defined it above.

On our definition [of 'the Number which belongs to the concept F'], what has to be shown is that the extension of the concept 'equal to the concept F' is the same as the extension of the concept 'equal to the concept G', if the concept F is equal to the concept G. In other words: it is to be proved that, for F equal to G, the following two propositions hold good universally: if the concept H is equal to the concept F, then it is also equal to the concept G; and if the concept H is equal to the concept G, then it is also equal to the concept F.

The first proposition amounts to this, that there exists a relation which correlates one to one the objects falling under the concept H with those falling under the concept G, if there exists a relation ϕ which correlates one to one the objects falling under the concept F with those falling under the concept G and if there exists also a relation ψ which correlates one to one the objects falling under the concept H with those falling under the concept F. The following arrangement of letters will make this easier to grasp: $H \psi F \phi G$.

Such a relation can in fact be given: it is to be found in the judgement-content 'there exists an object to which c stands in the relation ψ and which stands to b in the relation ϕ', if we subtract from it c and b – take them, that is, as the terms of the relation. It can be shown that this relation is one-one, and that it correlates the objects falling under the concept H with those falling under the concept G.

A similar proof can be given of the second proposition also.[1] And with that, I hope, enough has been indicated of my methods to show that our proofs are not dependent at any point on borrowings from intuition, and that our definitions can be used to some purpose.

§ 74. We can now pass on to the definitions of the individual numbers.

Since nothing falls under the concept 'not identical with itself', I define nought as follows: 0 is the Number which belongs to the concept 'not identical with itself'.

Some may find it shocking that I should speak of a concept in this connexion. They will object, very likely, that it contains a contradiction and is reminiscent of our old friends the square circle and wooden iron. Now I believe that these old friends are not so black as they are painted. To be of any use is, I admit, the last thing we should expect of them; but at the same time, they

---

1. And likewise of the converse: If the number which belongs to the concept F is the same as that which belongs to the concept G, then the concept F is equal to the concept G.

cannot do any harm, if only we do not assume that there is anything which falls under them – and to that we are not committed by merely using them. That a concept contains a contradiction is not always obvious without investigation; but to investigate it we must first possess it and, in logic, treat it just like any other. All that can be demanded of a concept from the point of view of logic and with an eye to rigour of proof is only that the limits to its application should be sharp, that we should be able to decide definitely about every object whether it falls under that concept or not. But this demand is completely satisfied by concepts which, like 'not identical with itself', contain a contradiction; for of every object we know that it does not fall under any such concept.[1]

On my use of the word 'concept', '*a* falls under the concept *F*' is the general form of a judgement-content which deals with an object *a* and permits of the insertion for *a* of anything whatever. And in this sense '*a* falls under the concept "not identical with itself"' has the same meaning as '*a* is not identical with itself' or '*a* is not identical with *a*'.

I could have used for the definition of nought any other concept under which no object falls. But I have made a point of choosing one which can be proved to be such on purely logical grounds; and for this purpose 'not identical with itself' is the most convenient that offers, taking for the definition of 'identical' the one from Leibniz given above [(§ 65)], which is in purely logical terms.

1. The definition of an object in terms of a concept under which it falls is a very different matter. For example, the expression 'the largest proper fraction' has no content, since the definite article claims to refer to a definite object. On the other hand, the concept 'fraction smaller than 1 and such that no fraction smaller than one exceeds it in magnitude' is quite unexceptionable: in order, indeed, to prove that there exists no such fraction, we must make use of just this concept, despite its containing a contradiction. If, however, we wished to use this concept for defining an object falling under it, it would, of course, be necessary first to show two distinct things:

(1) that some object falls under this concept;
(2) that only one object falls under it.

Now since the first of these propositions, not to mention the second, is false, it follows that the expression 'the largest proper fraction' is senseless.

§ 75. Now it must be possible to prove, by means of what has already been laid down, that every concept under which no object falls is equal to every other concept under which no object falls, and to them alone; from which it follows that 0 is the Number which belongs to any such concept, and that no object falls under any concept if the number which belongs to that concept is 0.

If we assume that no object falls under either the concept F or the concept G, then in order to prove them equal we have to find a relation ϕ which satisfies the following conditions: every object which falls under F stands in the relation ϕ to an object which falls under G; and to every object which falls under G there stands in the relation ϕ an object falling under F.

In view of what has been said above [(§ 71)] on the meaning of these expressions, it follows, on our assumption [that no object falls under either concept], that these conditions are satisfied by every relation whatsoever, and therefore among others by identity, which is moreover a one-one relation; for it meets both the requirements laid down [in § 72] above.

If, to take the other case, some object, say a, does fall under G, but still none falls under F, then the two propositions 'a falls under G' and 'no object falling under F stands to a in the relation ϕ' are both true together for every relation ϕ; for the first is made true by our first assumption and the second by our second assumption. If, that is, there exists no object falling under F, then a fortiori there exists no object falling under F which stands to a in any relation whatsoever. There exists, therefore, no relation by which the objects falling under F can be correlated with those falling under G so as to satisfy our definition [of equality], and accordingly the concepts F and G are unequal.

§ 76. I now propose to define the relation in which every two adjacent members of the series of natural numbers stand to each other. The proposition: 'there exists a concept F, and an object falling under it x, such that the Number which belongs to the concept F is n and the Number which belongs to the concept "falling under F but not identical with x" is m' is to mean the same as 'n follows in the series of natural numbers directly after m'.

I avoid the expression '*n* is *the* Number following next after *m*', because the use of the definite article cannot be justified until we have first proved two propositions. For the same reason I do not yet say at this point '*n*=*m*+1', for to use the symbol=is likewise to designate (*m*+1) an object.

§ 77. Now in order to arrive at the number 1, we have first of all to show that there is something which follows in the series of natural numbers directly after 0.

Let us consider the concept – or, if you prefer it, the predicate – 'identical with 0'. Under this falls the number 0. But under the concept 'identical with 0 but not identical with 0', on the other hand, no object falls, so that 0 is the Number which belongs to this concept. We have, therefore, a concept 'identical with 0' and an object falling under it 0, of which the following propositions hold true: the Number which belongs to the concept 'identical with 0' is identical with the Number which belongs to the concept 'identical with 0'; the Number which belongs to the concept 'identical with 0 but not identical with 0' is 0.

Therefore, on our definition [§ 76], the Number which belongs to the concept 'identical with 0' follows in the series of natural numbers directly after 0.

Now if we give the following definition: 1 is the Number which belongs to the concept 'identical with 0', we can then put the preceding conclusion thus: 1 follows in the series of natural numbers directly after 0.

It is perhaps worth pointing out that our definition of the number 1 does not presuppose, for its objective legitimacy, any matter of observed fact.[1] It is easy to get confused over this, seeing that certain subjective conditions must be satisfied if we are to be able to arrive at the definition, and that sense experiences are what prompt us to frame it.[2] All this, however, may be perfectly correct, without the propositions so arrived at ceasing to be *a priori*. One such condition is, for example, that blood of the right quality must circulate in the brain in sufficient volume – at least so far as we know; but the truth of our last proposition does not depend on this; it still holds, even if the

1. Non-general proposition.
2. Cf. B. Erdmann, *Die Axiome der Geometrie*, p. 164.

circulation stops; and even if all rational beings were to take to hibernating and fall asleep simultaneously, our proposition would not be, say, cancelled for the duration, but would remain quite unaffected. For a proposition to be true is just not the same thing as for it to be thought.

§ 78. I proceed to give here a list of several propositions to be proved by means of our definitions. The reader will easily see for himself in outline how this can be done.

1. If a follows in the series of natural numbers directly after 0, then a is $=1$.

2. If 1 is the Number which belongs to a concept, then there exists an object which falls under that concept.

3. If 1 is the Number which belongs to a concept F; then, if the object x falls under the concept F and if y falls under the concept F, x is $=y$; that is, x is the same as y.

4. If an object falls under the concept F, and it can be inferred generally from the propositions that x falls under the concept F and that y falls under the concept F that x is $=y$, then 1 is the number which belongs to the concept F.

5. The relation of m to n which is established by the proposition: 'n follows in the series of natural numbers directly after m' is a one-one relation.

6. Every Number except 0 follows in the series of natural numbers directly after a Number.

MODERN ALGEBRA

48

GABRIEL CRAMER
(1704–52)

GABRIEL CRAMER, an eminent Swiss mathematician, widely known among students of mathematics for his rule for solving systems of equations by determinants, was born in Geneva on 31 July, 1704. He belonged to an ancient Holstein family known first in Strassburg, and then in Geneva, where his father and grandfather were beloved physicians. Cramer was educated at the University of Geneva at a time when that institution was renowned for its great teachers. When he was barely twenty years old, his capacity was well recognized at the University, and in 1724 he was given an appointment there as a professor of mathematics. In 1727 he took a two-year leave for travel during which time he made the acquaintance of Jean Bernoulli in Basel and thereby began his long and fruitful association with the Bernoulli family. Returning to Geneva after travelling to England, Holland and France, Cramer resumed his professional duties. A skilful researcher in mathematics, Cramer wrote many memoirs in the development of topics in mathematics, and his writings appeared in various scientific journals. Gabriel Cramer was the first outstanding scholar who set aside his own investigations for the purpose of seriously undertaking the laborious and often thankless work of editing and publishing the writings of others. His publications, such as the work of Jean Bernoulli, and his two volume edition of the correspondence between Leibniz and Jean Bernoulli remain invaluable sources for historical research in the field of mathematics.

Cramer's greatest mathematical work, *Introduction to the Analysis of Algebraic Curves* (1750), containing the famous 'Cramer's Rule' mentioned above, was written while his health was failing. At this time, a severe fall suffered when his carriage accidently overturned rendered his condition worse. Seeking aid toward his recovery, he went to the south of France late in 1751. However, he survived only a short time and died in Bagnols near Nîmes on 4 January, 1752.

INTRODUCTION TO THE
ANALYSIS OF ALGEBRAIC CURVES

translated by Henrietta O. Midonick

CHAPTER III

The number of coefficients a, b, c, d, e, etc., in every general equation is the same as the number of its terms. But the number of these coefficients can be diminished by one because the second member of each equation is zero and consequently the entire first member can be divided by any one of the coefficients. If, for example, we divide the entire first member by the coefficient of the highest power of y or x, then after the division this term will have the number one, exactly, as its coefficient.

Thus, dividing the general equation of the first order
$$a + by + cx = 0$$
by c, we reduce it to $\dfrac{a}{c} + \dfrac{b}{c} y + x = 0$

where, in addition to unity, the coefficient of x, there are but two coefficients $\dfrac{a}{c}$ and $\dfrac{b}{c}$.

Accordingly, if v is an exponent of any order whatsoever, the number of coefficients of the general equation of this order will be $\frac{1}{2} vv + \frac{3}{2} v$, which is the sum of the arithmetical progression $2 + 3 + 4 + 5 +$ etc. where the difference is 1, the first term 2 and the number of terms v.[1]

From this it follows that a curve of order v can always be passed through $\frac{1}{2} vv + \frac{3}{2} v$ given points, that is, any curve of order v is determined and its equation given when there are $\frac{1}{2} vv + \frac{3}{2} v$ fixed points through which it must pass.

Thus a curve of the first order is determined by two given points; a curve of the second order by 5; one of the third order by 9; one of the fourth by 14; one of the fifth by 20, and so on.

Only one example is required for the demonstration.[2]

---

1. Stirling, *Lineae Tertii Ordinis*, etc. p. 3 ff.
2. Stirling, *Lineae Tertii Ordinis*, etc. p. 69. Newton, *Arithmetica Universalis*. Problem LXI.

Let A, B, C, D and E be five given points through which a curve of the second order is to be passed. Draw any two lines FG and FH through a point F. Take F as the origin and the two lines FG and FH as the axes. Then draw Aa, Bb, Cc, Dd and Ee, the ordinates of the given points. These ordinates as well as the

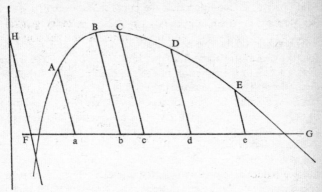

Fig. 48.1.

abscissas Fa, Fb, Fc, Fd and Fe will be given values. Denote Aa by a, Bb by b, Cc by c, Dd by d, Ee by e and Fa by α, Fb by β, Fc by γ, Fd by δ, Fe by ϵ, and take

$$A + By + Cx + Dyy + Exy + xx = 0$$

as the equation of the second order which must pass through the given points, A, B, C, D and E. The problem is thus reduced to finding the values of the five unknown coefficients, A, B, C, D, E. Now, to do this, we have five equations. For, since the curve passes through the point A, Fa (α) must be the abscissa corresponding to the ordinate aA (a). Hence, if α is a value of x, a is the corresponding value of y. Substituting α for x and a for y in the equation $A + By + Cx + Dyy + Exy + xx = 0$, the condition that the curve must pass through A is expressed by the equation

$$A + Ba + C\alpha + Daa + Ea\alpha + \alpha\alpha = 0.$$

The condition that the curve must pass through B similarly yields

$$A + Bb + C\beta + Dbb + Eb\beta + \beta\beta = 0.$$

The condition that the curve must pass through C yields

$$A + Bc + C\gamma + Dcc + Ec\gamma + \gamma\gamma = 0.$$

The condition that the curve must pass through D yields

$$A + Bd + C\delta + Ddd + Ed\delta + \delta\delta = 0.$$

And finally, the condition that the curve pass through E yields

$$A + Be + C\epsilon + Dee + Ee\epsilon + \epsilon\epsilon = 0.$$

The values of the five coefficients, A, B, C, D and E can be found by means of these five equations; and $A + By + Cx + Dyy + Exy + xx = 0$, the equation of the required curve will thus be determined. The calculations involved in the solution would in fact be very long;[1] but it is not necessary to carry out the solution in order to convince oneself that it is always possible to pass a curve of the second order through five given points, A, B, C, D and E, and in general, to pass a curve of order v through $\frac{1}{2} vv + \frac{3}{2} v$ given points. It will suffice merely to observe that every point provides an equation and that one can determine as many coefficients as there are equations. Thus, $\frac{1}{2} vv + \frac{3}{2} v$ points determine $\frac{1}{2} vv + \frac{3}{2} v$ coefficients. This means that all the coefficients contained in the general equation of a curve of order v (see preceding section) can be determined.

As the unknowns A, B, C, D and E in these equations never go above the first degree, the solution of the problem will always be possible and no exception or limitation can be introduced through imaginary roots. For the coefficients are determined without any necessity whatsoever for the extraction of a root, the sole operation by which imaginary roots can be introduced into a calculation. However, it may happen that some of the coefficients will be zero. In that case the equation of the curve will contain a smaller number of terms. Or it may happen that some of the coefficients may prove to be infinite and consequently the terms affected by these coefficients would alone constitute the whole equation, the other terms vanishing in comparison

1. The science of algebra has given us the means of shortening the calculations. I believe I have found a rule for this purpose, which is very easy and general, given any number of equations and a like number of unknowns, none of which is above the first degree. It will be found in *Appendix* No. 1.

with them. Or some of the coefficients may remain undetermined and in that case it will be possible to pass an infinity of curves of the same order through the given points.

If finding the actual equation of the curve passing through a number of given points is required, the calculations may be shortened by taking one of the points, A, for example, as the origin (Figure 48.2). Since the abscissa and ordinate are zero at this point, both α and a will equal zero. Hence the first of the

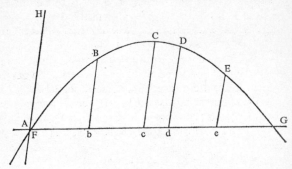

Fig. 48.2.

five equations above reduces to $A = 0$, and the other four equations become

$$Bb + C\beta + Dbb + E\beta b + \beta\beta = 0$$
$$Bc + C\gamma + Dcc + Ec\gamma + \gamma\gamma = 0$$
$$Bd + C\delta + Ddd + Ed\delta + \delta\delta = 0$$
$$Be + C\varepsilon + Dee + Ee\varepsilon + \varepsilon\varepsilon = 0$$

from which we derive

$$= \frac{\beta\gamma de(\beta-\gamma)(\delta e-d\varepsilon)-\beta c\delta e(\beta-\delta)(\gamma e-c\varepsilon)+\beta cd\varepsilon(\beta-\varepsilon)(\gamma d-c\delta)-b\gamma\delta e(\gamma-\delta)(\beta e-b\varepsilon)}{+b\gamma d\varepsilon(\gamma-\varepsilon)(\beta d-b\delta)-bc\delta\varepsilon(\delta-\varepsilon)(\beta c-b\gamma)}{(\beta c-b\gamma)(\delta e-d\varepsilon)(bc+de)-(\beta d-b\delta)(\gamma e-c\varepsilon)(bd+ce)+(\beta e-b\varepsilon)(\gamma d-c\delta)(be+cd)}$$

$$= \frac{(\beta\beta c-b\gamma\gamma)(\delta e-d\varepsilon)de-(\beta\beta d-b\delta\delta)(\gamma e-c\varepsilon)ce+(\beta\beta e-b\varepsilon\varepsilon)(\gamma d-c\delta)cd+(\gamma\gamma d-c\delta\delta)(\beta e-b\varepsilon)be}{-(\gamma\gamma e-c\varepsilon\varepsilon)(\beta d-b\delta)bd+(\delta\delta e-c\varepsilon\varepsilon)(\beta c-b\gamma)bc}{(\beta c-b\gamma)(\delta e-d\varepsilon)(bc+de)-(\beta d-b\delta)(\gamma e-c\varepsilon)(bd+ce)+(\beta e-b\varepsilon)(\gamma d-c\delta)(be+cd)}$$

$$= \frac{(\beta c-b\gamma)(\delta e-d\varepsilon)(\beta\gamma+\delta e)-(\beta d-b\delta)(\gamma e-c\varepsilon)(\beta\delta+\gamma\varepsilon)+(\beta e-b\varepsilon)(\gamma d-c\delta)(\beta\varepsilon+\gamma\delta)}{(\beta c-b\gamma)(\delta e-d\varepsilon)(bc+de)-(\beta d-b\delta)(\gamma e-c\varepsilon)(bd+ce)+(\beta e-b\varepsilon)(\gamma d-c\delta)(be+cd)}$$

$$= \frac{(\beta c-b\gamma)(\delta e-d\varepsilon)(\beta c+b\gamma+\delta e+d\varepsilon)-(\beta d-b\delta)(\gamma e-c\varepsilon)(\beta d+b\delta+\gamma e+c\varepsilon)}{+(\beta e-b\varepsilon)(\gamma d-c\delta)(\beta e+b\varepsilon+\gamma d+c\delta)}{(\beta c-b\gamma)(\delta e-d\varepsilon)(bc+de)-(\beta d-b\delta)(\gamma e-c\varepsilon)(bd+ce)+(\beta e-b\varepsilon)(\gamma d-c\delta)(be+cd)}$$

But the work can be considerably shortened again by drawing lines (Figure 48.3) from A to each of the given points B and E,

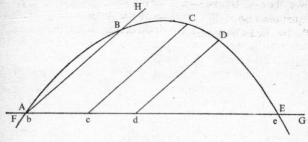

Fig. 48.3.

taking AB as the axis of the ordinates and AE as the axis of the abscissas. Then with the abscissa (β) of zero value, corresponding to the ordinate AB (b) and the ordinate (e) of zero value, corresponding to the abscissa AE (ϵ), the values of B, C, D and E are reduced to

$$B = -\frac{b\gamma\delta[d(\epsilon-\gamma)-(\epsilon-\delta)c]}{cd[\gamma(b-d)-(b-c)\delta]}$$

$$C = -\frac{cd\epsilon[\gamma(b-d)-(b-c)\delta]}{cd[\gamma(b-d)-(b-c)\delta]} = -\epsilon$$

$$D = +\frac{\gamma\delta[d(\epsilon-\gamma)-(\epsilon-\delta)c]}{cd[\gamma(b-d)-(b-c)\delta]} = -\frac{B}{b}$$

$$E = +\frac{c\delta(b-c)(\epsilon-\delta)-\gamma d(b-d)(\epsilon-\gamma)}{cd[\gamma(b-d)-(b-c)\delta]}$$

APPENDIX: ON THE ELIMINATION OF UNKNOWNS

If a problem involves several unknowns whose relations are so complicated, that several equations are required to express them, the values of the unknowns may be found by eliminating all of the unknowns except one. The single remaining unknown combined with the known quantities will yield a *final equation*. If the problem is determinate, the *final equation* will produce as a first result the value of the aforesaid unknown, and then by means of it, the values of all the remaining unknowns.

Algebra provides infallible rules for the success of this procedure, provided that one has the patience to follow them. However, the calculations become extremely long when the number of equations and unknowns is excessively large. . . . The purpose of the two items following is to remove these disadvantages.

No. I.

Let z, y, x, v, etc., represent several unknowns. Take as many equations as there are unknowns; let the equations involving the unknowns, be:

$$A_1 = Z_1 z + Y_1 y + X_1 x + V_1 v + \text{etc.},$$
$$A_2 = Z_2 z + Y_2 y + X_2 x + V_2 v + \text{etc.},$$
$$A_3 = Z_3 z + Y_3 y + X_3 x + V_3 v + \text{etc.},$$
$$A_4 = Z_4 z + Y_4 y + X_4 x + V_4 v + \text{etc.},$$
$$\text{etc.,}$$

where the letters A_1, A_2, A_3, A_4, etc., denote the first member (assumed to be known) of the first, second, third, fourth, etc., equations, respectively, and not the powers of A, as ordinarily. Similarly, Z_1, Z_2, etc., are the coefficients of z in the first, second, etc., equations, Y_1, Y_2, etc., are the coefficients of y, X_1, X_2, etc., are the coefficients of x, and V_1, V_2, etc., of v in the first, second, etc., equations, and so on.

Granting this notation, if there is but one equation, and the only unknown is z, we will have $z = \dfrac{A_1}{Z_1}$. If there are two equations and two unknowns, z and y, we will find

$$z = \frac{A_1 Y_2 - A_2 Y_1}{Z_1 Y_2 - Z_2 Y_1}, \text{ and } y = \frac{Z_1 A_2 - Z_2 A_1}{Z_1 Y_2 - Z_2 Y_1}.$$

If there are three equations and three unknowns z, y, and x, we will find

$$z = \frac{A_1 Y_2 X_3 - A_1 Y_3 X_2 - A_2 Y_1 X_3 + A_2 Y_3 X_1 + A_3 Y_1 X_2 - A_3 Y_2 X_1}{Z_1 Y_2 X_3 - Z_1 Y_3 X_2 - Z_2 Y_1 X_3 + Z_2 Y_3 X_1 + Z_3 Y_1 X_2 - Z_3 Y_2 X_1}$$

$$y = \frac{Z_1 A_2 X_3 - Z_1 A_3 X_2 - Z_2 A_1 X_3 + Z_2 A_3 X_1 + Z_3 A_1 X_2 - Z_3 A_2 X_1}{Z_1 Y_2 X_3 - Z_1 Y_3 X_2 - Z_2 Y_1 X_3 + Z_2 Y_3 X_1 + Z_3 Y_1 X_2 - Z_3 Y_2 X_1}$$

$$x = \frac{Z_1 Y_2 A_3 - Z_1 Y_3 A_2 - Z_2 Y_1 A_3 + Z_2 Y_3 A_1 + Z_3 Y_1 A_2 - Z_3 Y_2 A_1}{Z_1 Y_2 X_3 - Z_1 Y_3 X_2 - Z_2 Y_1 X_3 + Z_2 Y_3 X_1 + Z_3 Y_1 X_2 - Z_3 Y_2 X_1}$$

An inspection of these formulas yields the following general rule: If there are n equations and n unknowns, the value of each unknown will be found by forming n fractions each having the same denominator; the number of terms in the denominator will be the same as the number of different possible arrangements of n distinct objects, taken n at a time; each term of the denominator will be composed of the letters $ZYXV$, etc., always written in the same order, but with the first n natural numbers as exponents (actually merely indices written as superscripts) distributed among the letters in every possible way. Thus, if there are three unknowns, the denominator will consist of $(1 \times 2 \times 3 =)$ 6 terms, each composed of the three letters ZYX to which the sets of indices 123, 132, 213, 231, 312, 321 are successively applied. The sign $+$ or $-$ is assigned to each term in accordance with the following rule: If, within the same term, any index is followed by a smaller index, either immediately or with other indices intervening, I shall call this an alteration or an inversion. The number of inversions must be counted for each term. If the number of inversions in a term is even or zero, the sign of the term will be $+$; if the number of inversions in a term is odd, the sign of the term will be $-$. For example in the term $Z_1 Y_2 V_3$, there is no inversion. Hence the sign of this term is $+$. The sign of the term $Z_3 Y_1 X_2$ is also $+$, because there are two inversions in the order of its indices, 3 preceding 1, and 3 preceding 2. But the term $Z_3 Y_2 X_4$, which has three inversions, 3 preceding 2, 3 preceding 1, and 2 preceding 1, takes the sign $-$.

The denominator common to the expression for the value of each unknown having been thus formed, the value of z will be obtained by assigning to this denominator a numerator formed by changing Z to A in every term of the denominator. The value of y is likewise a fraction which has the same denominator, its numerator being the quantity obtained by changing Y to A in every term of the denominator. The values of the other unknowns are found by a similar procedure.

Generally speaking, the problem is determinate. But there may be some special cases where it is indeterminate and others where it becomes impossible. This occurs when the denominator common to the unknowns proves to be equal to zero, as, for example, if there are only two equations, the denominator $Z_1 Y_2 - Z_2 Y_1$ $=0$ or if there are three unknowns, the denominator

$$Z_1 Y_2 Z_3 - Z_1 Y_3 X_2 - Z_2 Y_1 X_3 + Z_2 Y_3 X_1 + Z_3 Y_1 X_2 - Z_3 Y_2 X_1 = 0$$

and so on. Now, if the quantities A_1, A_2, A_3, etc., are such that the numerators are also zero, the problem is indeterminate. For the fractions which must yield the values of the unknowns are then $\frac{0}{0}$ and indeterminate. But if the quantities A_1, A_2, A_3, etc., are such that while the denominator common to all unknowns is equal to zero, all the numerators, or some of them are not zero, then the problem is impossible, or at least, the quantities required for the solution of the problem are either all infinite or some of them are. For example, if we have two equations, $2 = 3z - 2y$ and $5 = 6z - 4y$, it will be found that $z = \frac{2}{0}$ and $y = \frac{3}{0}$. Hence z and y are infinite magnitudes which are in the ratio of 2 to 3. Solving for the unknowns by the ordinary methods, we should stumble into the absurd statement $\frac{2}{3} = \frac{5}{6}$. For the first equation gives $z = \frac{2}{3}y + \frac{2}{3}$ and the second gives $z = \frac{4}{6}y + \frac{5}{6}$. Hence, $\frac{2}{3}y + \frac{2}{3} = \frac{4}{6}y + \frac{5}{6}$, which is absurd if z and y are finite quantities. However, if they are infinite, it may be said without absurdity that $z = \frac{2}{3}y + \frac{2}{3}$ and at the same time, that $z = \frac{2}{3}y + \frac{5}{6}$, because the finite quantities $\frac{2}{3}$ and $\frac{5}{6}$ are insignificant in comparison with the infinite quantities z and $\frac{2}{3}y$, and the two equations $z = \frac{2}{3}y + \frac{2}{3}$ and $z = \frac{2}{3}y + \frac{5}{6}$ both reduce to $z = \frac{2}{3}y$ which contains no contradiction.

49

CASPAR WESSEL
(1745–1818)

THE first clear, scientific exposition of the modern geometrical method of representing $\sqrt{-1}$ was given in a paper entitled *On the Analytical Representation of Direction* before the Royal Academy of Sciences and Letters of Denmark in 1797 by a Norwegian surveyor, Caspar Wessel. The treatise contained a complete development of the laws governing operations with directed line segments as representations of numbers in the form $a+b\sqrt{-1}$ and their applications, as well as a partial theory of rotation.

Caspar Wessel was born in Jonsrud, Norway, on 8 June, 1745. His father was a pastor in that town and Caspar was one of a bustling household of thirteen children. He received an excellent education, begun at Jonsrud and continued in 1757 at the high school in Christiania. In 1763 Wessel went to Copenhagen for further study. The following year he was engaged by the Danish Academy of Sciences as an assistant in the preparation of a map of Denmark. He remained in the employ of the Academy from that time, uninterruptedly until 1805. Possessing broad interests and a versatile turn of mind, Wessel also studied Roman law, passing an examination in this field in 1778. Wessel received many marks of appreciation of his distinguished services in Denmark. For a special assignment performed for the Academy after his retirement, he was awarded the Academy's silver medal and a complete set of its memoirs. He was made Knight of the Danebrog in 1815. The high place Wessel held in the estimation of his contemporaries was sustained in 1819 when many of the maps of Denmark were declared out of date, while Wessel's trigonometric calculations were retained, being specifically excluded from the general discord.

It speaks well for the Academy that they received Wessel's paper sympathetically, since he was neither a member of the Academy nor was he considered a mathematician. Sponsored and encouraged by Tetens, Councillor-of-state, Wessel's presentation of his work in 1797 was followed by its publication in 1798 and its appearance in Vol. V of the Memoirs of the Academy in 1799. Written in Danish, it failed to achieve wide accessibility to the mathematicians of other countries with the result that this excellent and significant work did not become generally known until a French translation of it was published in 1897.

———

ON THE ANALYTICAL REPRESENTATION OF DIRECTION

AN ATTEMPT[1]

APPLIED CHIEFLY TO THE SOLUTION OF PLANE AND SPHERICAL POLYGONS (BY CASPAR WESSEL, SURVEYOR)

translated from the Danish by Professor Martin A. Nordgaard, St Olaf College, Northfield, Minnesota

This present attempt deals with the question, how may we represent direction analytically; that is, how shall we express right lines so that in a single equation involving one unknown line and others known, both the length and the direction of the unknown line may be expressed.

To help answer this question I base my work on two propositions which to me seem undeniable. The first one is: changes in direction which can be effected by algebraic operations shall be indicated by their signs. And the second: direction is not a subject for algebra except in so far as it can be changed by algebraic operations. But since these cannot change direction (at least, as commonly explained) except to its opposite, that is, from positive to negative, or *vice versa*, these two are the only directions it should be possible to designate, by present methods; for the other

---

1. In recent histories of mathematics, there have come about very misleading translations into English of Wessel's title word '*forsog*' as 'essay on, etc.'. This possibility comes from the word '*essai*' used in the French translation of Wessel's memoir, the French word meaning both an attempt or endeavour, and a treatise (essay). Wessel's word '*forsog*' can only mean *attempt* or *endeavour*.

directions the problem should be unsolvable. And I suppose this is the reason no one has taken up the matter.[1] It has undoubtedly been considered impermissible to change anything in the accepted explanation of these operations.

And to this we do not object so long as the explanation deals only with quantities in general. But when in certain cases the nature of the quantities dealt with seems to call for more precise definitions of these operations and these can be used to advantage, it ought not to be considered impermissible to offer modifications. For as we pass from arithmetic to geometric analysis, or from operations with abstract numbers to those with right lines, we meet with quantities that have the same relations to one another as numbers, surely; but they also have many more. If we now give these operations a wider meaning, and do not as hitherto limit their use to right lines of the same or opposite direction; but if we extend somewhat our hitherto narrow concept of them so that it becomes applicable not only to the same cases as before, but also to infinitely many more; I say, if we take this liberty, but do not violate the accepted rules of operations, we shall not contravene the first law of numbers. We only extend it, adapt it to the nature of the quantities considered, and observe the rule of method which demands that we by degrees make a difficult principle intelligible.

It is not an unreasonable demand that operations used in geometry be taken in a wider meaning than that given to them in arithmetic. And one will readily admit that in this way it should be possible to produce an infinite number of variations in the directions of lines. Doing this we shall accomplish, as will be proved later, not only that all impossible operations can be avoided – and we shall have light on the paradoxical statement that at times the possible must be tried by impossible means – but also that the direction of all lines in the same plane can be expressed as analytically as their lengths without burdening the mind with new signs or new rules. There is no question that the general validity of geometric propositions is frequently seen with greater ease if direction can be indicated analytically and governed

1. Unless it be Magister Gilbert, in Halle, whose prize memoir on *Calculus Situs* possibly contains an explanation of this subject.

by algebraic rules than when it is represented by a figure, and that only in certain cases. Therefore it seems not only permissible, but actually profitable, to make use of operations that apply to other lines than the equal (those of the same direction) and the opposite. On that account my aim in the following chapters will be:

1. First, to define the rules for such operations;

2. Next, to demonstrate their application when the lines are in the same plane, by two examples;

3. To define the direction of lines lying in different planes by a new method of operation, which is not algebraic;

4. By means of this method to solve plane and spherical polygons;

5. Finally to derive in the same manner the ordinary formulae of spherical trigonometry.

These will be the chief topics of this treatise. The occasion for its being was my seeking a method whereby I could avoid the impossible operations; and when I had found this, I applied it to convince myself of the universality of certain well-known formulae. The Honourable Mr Tetens, Councillor-of-State, was kind enough to read through these first investigations. It is due to the encouragement, counsel and guidance of this distinguished savant that this paper is minus some of its first imperfections and that it has been deemed worthy to be included among the publications of the Royal Academy.

A METHOD WHEREBY FROM GIVEN RIGHT LINES TO FORM OTHER RIGHT LINES BY ALGEBRAIC OPERATIONS: AND HOW TO DESIGNATE THEIR DIRECTIONS AND SIGNS

Certain homogeneous quantities have the property that if they are placed together, they increase or diminish one another only as increments or decrements.

There are others which in the same situation effect changes in one another in innumerable other ways. To this class belong right lines.

Thus the distance of a point from a plane may be changed in

innumerable ways by the point describing a more or less inclined right line outside the plane.

For, if this line is perpendicular to the axis of the plane, that is, if the path of the point makes a right angle with the axis, the point remains in a plane parallel to the given plane, and its path has no effect on its distance from the plane.

If the described line is indirect, that is, if it makes an oblique angle with the axis of the plane, it will add to or subtract from the distance by a length less than its own; it can increase or diminish the distance in innumerable ways.

If it is direct, that is, in line with the distance, it will increase or diminish the same by its whole length; in the first case it is positive, in the second, negative.

Thus, all the right lines which can be described by a point are, in respect to their effects upon the distance of a given point from a plane outside the point, either direct or indirect or perpendicular[1] according as they add to or subtract from the distance the whole, a part, or nothing, of their own lengths.

Since a quantity is called absolute if its value is given as immediate and not in relation to another quantity, we may in the preceding definitions call the distance the absolute line; and the share of the relative line in lengthening or shortening the absolute line may be called the 'effect' of the relative line.

There are other quantities besides right lines among which such relations exist. It would therefore not be a valueless task to explain these relations in general, and to incorporate their general concept in an explanation on operations. But I have accepted the advice of men of judgement, that in this paper both the nature of the contents and plainness of exposition demand that the reader be not burdened here with concepts so abstract. I shall consequently make use of geometric explanation only. These follow.

§ 1

Two right lines are added if we unite them in such a way that the second line begins where the first one ends, and then pass a right

1. 'Indifferent' would be a more fitting name were it not so unfamiliar to our ears.

line from the first to the last point of the united lines. This line is the sum of the united lines.

For example, if a point moves forward three feet and backward two feet, the sum of these two paths is not the first three and the last two feet combined; the sum is one foot forward. For this path, described by the same point, gives the same effect as both the other paths.

Similarly, if one side of a triangle extends from a to b and the other from b to c, the third one from a to c shall be called the sum. We shall represent it by $ab+bc$, so that ac and $ab+bc$ have the same meaning; or $ac=ab+bc=-ba+bc$, if ba is the opposite of ab. If the added lines are direct, this definition is in complete agreement with the one ordinarily given. If they are indirect, we do not contravene the analogy by calling a right line the sum of two other right lines united, as it gives the same effect as these. Nor is the meaning I have attached to the symbol $+$ so very unusual; for in the expression $ab+\dfrac{ba}{2}=\frac{1}{2}ab$ it is seen that $\dfrac{ba}{2}$ is not a part of the sum. We may therefore set $ab+bc=ac$ without, on that account, thinking of bc as a part of ac; $ab+bc$ is only the symbol representing ac.

§ 2

If we wish to add more than two right lines we follow the same procedure. They are united by attaching the terminal point of the first to the initial point of the second and the terminal point of this one to the initial point of the third, etc. Then we pass a right line from the point where the first one begins to the point where the last one ends; and this we call their sum.

The order in which these lines are taken is immaterial; for no matter where a point describes a right line within three planes at right angles to one another, this line has the same effect on the distances of the point from each of the planes. Consequently any one of the added lines contributes equally much to the determination of the position of the last point of the sum whether it have first, last, or any other place in the sequence. Consequently, too, the order in the addition of right lines is immaterial. The sum will always be the same; for the first point is supposed to be given and the last point always assumes the same position.

So that in this case, too, the sum may be represented by the added lines connected with one another by the symbol+. In a quadrilateral, for example, if the first side is drawn from a to b, the second from b to c, the third from c to d, but the fourth from a to d, then we may write: $ad = ab + bc + cd$.

§ 3

If the sum of several lengths, breadths and heights is equal to zero, then is the sum of the lengths, the sum of the breadths, and the sum of the heights each equal to zero.

§ 4

It shall be possible in every case to form the product of two right lines from one of its factors in the same manner as the other factor is formed from the positive or absolute line set equal to unity. That is:

Firstly, the factors shall have such a direction that they both can be placed in the same plane with the positive unit.

Secondly, as regards length, the product shall be to one factor as the other factor is to the unit. And,

Finally, if we give the positive unit, the factors, and the product a common origin, the product shall, as regards its direction, lie in the plane of the unit and the factors and diverge from the one factor as many degrees, and on the same side, as the other factor diverges from the unit, so that the direction angle of the product, or its divergence from the positive unit, becomes equal to the sum of the direction angles of the factors.

§ 5

Let $+1$ designate the positive rectilinear unit and $+\epsilon$ a certain other unit perpendicular to the positive unit and having the same origin; then the direction angle of $+1$ will be equal to $0°$, that of -1 to $180°$, that of $+\epsilon$ to $90°$, and that of $-\epsilon$ to $-90°$ or $270°$. By the rule that the direction angle of the product shall equal the sum of the angles of the factors, we have: $(+1)(+1) = +1$; $(+1)(-1) = -1$; $(-1)(-1) = +1$; $(+1)(+\epsilon) = +\epsilon$; $(+1)(-\epsilon) = -\epsilon$; $(-1)(+\epsilon) = -\epsilon$; $(-1)(-\epsilon) = +\epsilon$; $(+\epsilon)(+\epsilon) = -1$; $(+\epsilon)(-\epsilon) = +1$; $(-\epsilon)(-\epsilon) = -1$.

From this it is seen that ϵ is equal to $\sqrt{-1}$; and the divergence

326

of the product is determined such that not any of the common rules of operation are contravened.

§ 6

The cosine of a circle arc beginning at the terminal point of the radius $+1$ is that part of the radius, or of its opposite, which begins at the centre and ends in the perpendicular dropped from the terminal point of the arc. The sine of the arc is drawn perpendicular to the cosine from its end point to the end point of the arc.

Thus, according to § 5, the sine of a right angle is equal to $\sqrt{-1}$. Set $\sqrt{-1} = \epsilon$. Let v be any angle, and let sin v represent a right line of the same length as the sine of the angle v, positive, if the measure of the angle terminates in the first semi-circumference, but negative, if in the second. Then it follows from §§ 4 and 5 that ϵ sin v expresses the sine of the angle v in respect of both direction and extent. . . .

§ 7

In agreement with §§ 1 and 6, the radius which begins at the centre and diverges from the absolute or positive unit by angle v is equal to cos $v + \epsilon$ sin v. But, according to § 4, the product of the two factors, of which one diverges from the unit by angle v and the other by angle u, shall diverge from the unit by angle $v + u$. So that if the right line cos $v + \epsilon$ sin v is multiplied by the right line cos $u + \epsilon$ sin u, the product is a right line whose direction angle is $v + u$. Therefore, by §§ 1 and 6, we may represent the product by cos $(v + u) + \epsilon$ sin $(v + u)$.

§ 8

The product (cos $v + \epsilon$ sin v) (cos $u + \epsilon$ sin u), or cos $(v + u) + \epsilon$ sin $(v + u)$, can be expressed in still another way, namely, by adding into one sum the partial products that result when each of the added lines whose sum constitutes one factor is multiplied by each of those whose sum constitutes the other. Thus, if we use the known trigonometric formulae.

$$\cos (v + u) = \cos v \cos u - \sin v \sin u,$$
$$\sin (v + u) = \cos v \sin u + \cos u \sin v,$$

we shall have this form:

$$(\cos v + \epsilon \sin v)(\cos u + \epsilon \sin u) = \cos v \cos u - \sin v \sin u$$
$$+ \epsilon (\cos v \sin u + \cos u \sin v).$$

For the above two formulae can be shown, without great difficulty, to hold good for all cases – be one or both of the angles acute or obtuse, positive or negative. In consequence, the propositions derived from these two formulae also possess universality.

§ 9

By § 7 $\cos v + \epsilon \sin v$ is the radius of a circle whose length is equal to unity and whose divergence from $\cos 0°$ is the angle v. It follows that $r \cos v + r\epsilon \sin v$ represents a right line whose length is r and whose direction angle is v. For if the sides of a right-angled triangle increase in length r times, the hypotenuse increases r times; but the angle remains the same. However, by § 1, the sum of the sides is equal to the hypotenuse; hence,

$$r \cos v + r\epsilon \sin v = r(\cos v + \epsilon \sin v).$$

This is therefore a general expression for every right line which lies in the same plane with the lines $\cos 0°$ and $\epsilon \sin 90°$, has the length r, and diverges from $\cos 0°$ by v degrees.

§ 10

If a, b, c denote direct lines of any length, positive or negative, and the two indirect lines $a + \epsilon b$ and $c + \epsilon d$ lie in the same plane with the absolute unit, their product can be found, even when their divergences from the absolute unit are unknown. For we need only to multiply each of the added lines that constitute one sum by each of the lines of the other and add these products; this sum is the required product both in respect to extent and direction: so that $(a + \epsilon b)(c + \epsilon d) = ac - bd + \epsilon(ad + bc)$.

Proof: Let the length of the line $a + \epsilon b$ be A, and its divergence from the absolute unit be v degrees, also let the length of $c + \epsilon d$ be C, and its divergence be u. Then, by § 9, $a + \epsilon b = A \cos v + A\epsilon \sin v$, and $c + \epsilon d = C \cos u + C\epsilon \sin u$. Thus $a = A \cos v$, $b = A \sin v$, $c = C \cos u$, $d = C \sin u$ (§ 3). But, by § 4, $(a + \epsilon b)(c + \epsilon d) = AC [\cos (v + u) + \epsilon \sin (v + u)] = AC [\cos v \cos u - \sin v \sin u + \epsilon (\cos v \sin u + \cos u \sin v)]$ (§ 8). Consequently, if instead of

328

$AC \cos v \cos u$ we write ac, and for $AC \sin v \sin u$ write bd, etc., we shall derive the relation we set out to prove.

It follows that, although the added lines of the sum are not all direct, we need make no exception in the known rule on which the theory of equations and the theory of integral functions and their simple divisors are based, namely, that if two sums are to be multiplied, then must each of the added quantities in one be multiplied by each of the added quantities in the other. It is, therefore, certain that if an equation deals with right lines and its root has the form $a + \epsilon b$, then an indirect line is represented. Now, if we should want to multiply together right lines which do not both lie in the same plane with the absolute unit, this rule would have to be put aside. That is the reason why the multiplication of such lines is omitted here. Another way of representing changes of direction is taken up later, in §§ 24–35.

The quotient multiplied by the divisor shall equal the dividend. We need no proof that these lines must lie in the same plane with the absolute unit, as that follows directly from the definition in § 4. It is easily seen also that the quotient must diverge from the absolute unit by angle $v - u$, if the dividend diverges from the same unit by angle v and the divisor by angle u.

Suppose, for example, that we are to divide $A (\cos v + \epsilon \sin v)$ by $B (\cos u + \epsilon \sin u)$. The quotient is

$$\frac{A}{B}[\cos (v - u) + \epsilon \sin (v - u)] \quad \text{since}$$

$$\frac{A}{B}[\cos (v - u) + \epsilon \sin (v - u)] \times B (\cos u + \epsilon \sin u) = A (\cos v + \epsilon \sin v),$$

by § 7. That is, since $\frac{A}{B}[\cos (v - u) + \epsilon \sin (v - u)]$ multiplied by the divisor $B (\cos u + \epsilon \sin u)$ equals the dividend $A (\cos v + \epsilon \sin v)$, then $\frac{A}{B}[\cos (v - u) + \epsilon \sin (v - u)]$ must be that required quotient. . . .

50

CARL FRIEDRICH GAUSS
(1777–1855)

CARL FRIEDRICH GAUSS (originally Johann Carl Friedrich Gauss),
astronomer, physicist and greatest mathematician of his time, was
born in Brunswick, Germany, on 23 April, 1777. He was the son of
Gerhard Gauss, a day labourer and small contractor who was himself
the son of a peasant. His mother, of sturdy stock, lived to the age of
ninety-seven. Gerhard Gauss had the usual parental desire to have his
sons join him in his trade, but he was thwarted in this by the extremely
early and startling evidences of Carl's unique genius. Speaking of the
experiences of his youth, Gauss used to say that he could reckon almost
before he had learned to talk. When he was scarcely three years old, as
he watched his father at work on the computation of weekly wages and
the like, one day, the talented child corrected an error in one of his
father's accounts. When he was ten years old, he was familiar with the
binomial theorem and the theory of infinite series. At his first school in
Brunswick, Gauss attracted the attention of a gifted young mathematical assistant, Bartels, under whose sympathetic and understanding
tutelage he made phenomenal strides in mathematics. When Gauss
had reached the age of fourteen, Bartels arranged an audience for him
with Charles William Ferdinand, the reigning Duke of Brunswick.
The Duke, impressed by Gauss's personality and potentialities, agreed
to sponsor his education, and in 1792 Gauss was sent to the Collegium
Carolinum (at first against his father's wishes). In 1795 he enrolled in
the University of Göttingen, where he came under the influence of
Kastner. His three years at Göttingen were extraordinarily productive
of original work. Gauss stated in his diary that so many new ideas
came crowding into his mind before he was twenty that he could record
only a fraction of them. By 1796 mathematics had become his favourite
field of study. 'Mathematics,' he would say, 'is the queen of the
sciences, and arithmetic is the queen of mathematics.' By 1798 most of
the work for his famous *Disquisitiones Arithmeticae* (published 1801)
had been done. Gauss spent that year at the University of Helmstadt
where he had access to a fine library and where he met the professor

of mathematics, Johann F. Pfaff, who was to be his lifelong friend. He received his doctor's degree at Helmstadt in 1799 and soon thereafter added investigations in astronomy, geodesy and physics to his mathematical researches. In 1807, while he was still a private teacher in Brunswick, the fame of his work brought him an offer of a professorial chair in the St Petersberg Academy. However, he was immediately offered a double appointment at home, as the first director of the new Göttingen Observatory and professor of mathematics at the university. Accepting the offer at Göttingen, Gauss's life thereafter was dedicated to observations, study, writing and teaching. The teaching phase of his work, unfortunately, was not always the happiest. 'This winter,' Gauss wrote to his friend, Bessel, 'I am giving two courses of lectures to three students, of which one is only moderately prepared, the other less than moderately, and the third lacks both preparation and ability. Such are the onera of a mathematical profession.'[1] Nevertheless, Gauss was exceedingly generous in his expression of appreciation of mathematical ability wherever it genuinely existed, and among the distinguished mathematicians who were his pupils, we find great names such as Dirichlet and Riemann.

Gauss remained a man of simple tastes throughout his life. It is said that he never wore any of the numerous decorations that were granted to him. He was fortunate in the enjoyment of good health, and despite the strain of night observations he rarely required a physician. Gauss's later years continued to be marked by his exceptional mental vigour and originality. His interest in languages was undiminished. At the age of sixty he studied and mastered the Russian language. At the age of seventy he added a new proof to those he had already composed for the fundamental theorem of algebra. He remained active until a few months before his death on 23 February, 1855.

Combining an abundant inventiveness and a zeal for perfection of form, Gauss enriched and influenced almost every field of pure and applied mathematics. His motto, '*Pauca sed matura*', applied to all his writings. Fundamental theorems in many branches of analysis were introduced in *Disquisitiones Arithmeticae* (1801) and in his *Superficies Curvas* (1827). The theory of numbers as a separate, systematic branch of mathematics dates from the publication of his *Disquisitiones*. His *Theoria Motus* (1809) was a landmark in the application of mathematics to celestial mechanics. Every work Gauss produced was an event in the history of science.

---

1. R. E. Moritz, *Memorabilia Mathematica*, no. 974.

ON THE CONGRUENCE OF NUMBERS

translated from the Latin by Ralph G. Archibald
from Disquisitiones Arithmeticae (1801)

FIRST SECTION
CONCERNING CONGRUENCE OF NUMBERS IN GENERAL

Congruent Numbers, Moduli, Residues, and Non-residues

1

If a number a divides the difference of the numbers b and c, b and c are said to be *congruent with respect to* a; but if not, *incongruent*. We call a the *modulus*. In the former case, each of the numbers b and c is called a *residue* of the other, but in the latter case, a *non-residue*.

These notions apply to all integral numbers both positive and negative,[1] but not to fractions. For example, -9 and $+16$ are congruent with respect to the modulus 5; -7 is a residue of $+15$ with respect to the modulus 11, but a non-residue with respect to the modulus 3. Now, since every number divides zero, every number must be regarded as congruent to itself with respect to all moduli.

2

If k denotes an indeterminate integral number, all residues of a given number a with respect to the modulus m are contained in the formula $a + km$. The easier of the propositions which we shall give can be readily demonstrated from this standpoint; but anyone will just as easily perceive their truth at sight.

We shall denote in future the congruence of two numbers by this sign, $\equiv$, and adjoin the modulus in parentheses when necessary. For example, $-16 \equiv 9 \pmod 5$, $-7 \equiv 15 \pmod{11}$.[2]

1. Obviously, the modulus is always to be taken *absolutely* – that is, without any sign.

2. We have adopted this sign on account of the great analogy which exists between an equality and a congruence. For the same reason Legendre, in memoirs which will later be frequently quoted, retained the sign of equality itself for a congruence. We hesitated to follow this notation lest it introduce an ambiguity.

3

Theorem: *If there be given the m consecutive integral numbers a, a+1, a+2, . . ., a+m−1, and another integral number A, then some one of the former will be congruent to this number A with respect to the modulus m; and, in fact, there will be only one such number.*

If, for instance, $\frac{a-A}{m}$ is an integer, we shall have $a \equiv A$; but if it is fractional, let k be the integer immediately greater (or, when it is negative, immediately *smaller* if no regard is paid to sign). Then $A+km$ will fall between a and $a+m$, and will therefore be the number desired. Now, it is evident that all the quotients $\frac{a-A}{m}, \frac{a+1-A}{m}, \frac{a+2-A}{m}$, etc., are situated between $k-1$ and $k+1$. Therefore not more than one can be integral.

Least Residues

4

Every number, then, will have a residue not only in the sequence 0, 1, 2, . . ., $m-1$, but also in the sequence 0, −1, −2, . . ., −(m−1). We shall call these *least residues*. Now, it is evident that, unless 0 is a residue, there will always be two: one *positive*, the other *negative*. If they are of different magnitudes, one of them will be less than $\frac{m}{2}$; but if they are of the same magnitude, each will equal $\frac{m}{2}$ when no regard is paid to sign. From this it is evident that any number has a residue not exceeding half the modulus. This residue is called the *absolute minimum*.

For example, with respect to the modulus 5, −13 has the positive least residue 2, which at the same time is the absolute minimum, and has −3 as the negative least residue. With respect to the modulus 7, +5 is its own positive least residue, −2 is the negative least residue and at the same time the absolute minimum.

Elementary Propositions Concerning Congruences

5

From the notions just established we may derive the following obvious properties of congruent numbers.

The numbers which are congruent with respect to a composite modulus, will certainly be congruent with respect to any one of its divisors.

If several numbers are congruent to the same number with respect to the same modulus, they will be congruent among themselves (with respect to the same modulus).

The same identity of moduli is to be understood in what follows.

Congruent numbers have the same least residues, incongruent numbers different least residues.

6

If the numbers A, B, C, etc., and the numbers a, b, c, etc., are congruent each to each with respect to any modulus, that is, if

$$A \equiv a, \ B \equiv b, \text{ etc.,}$$

then we shall have

$$A + B + C + \text{etc.,} \ \equiv a + b + c + \text{etc.}$$

If $A \equiv a$ and $B \equiv b$, we shall have $A - B \equiv a - b$.

7

If $A \equiv a$, we shall also have $kA \equiv ka$.

If k is a positive number, this is merely a particular case of the proposition of the preceding article when we place $A = B = C$ etc., and $a = b = c$ etc. If k is negative, $-k$ will be positive. Then $-kA \equiv -ka$, and consequently $kA \equiv ka$.

If $A \equiv a$ and $B \equiv b$, we shall have $AB \equiv ab$. For, $AB \equiv Ab \equiv ba$.

8

If the numbers A, B, C, etc., and the numbers a, b, c, etc., are congruent each to each, that is, if $A \equiv a$, $B \equiv b$, etc., the products of *the numbers of each set will be congruent; that is, ABC etc., ≡abc etc.*

From the preceding article, $AB \equiv ab$, and for the same reason $ABC \equiv abc$; in a like manner we can consider as many factors as desired.

If we take all the numbers A, B, C, etc., equal, and also the corresponding numbers a, b, c, etc., we obtain this theorem:

If $A \equiv a$ and if k is a positive integer, we shall have $A^k \equiv a^k$.

9

Let X be a function of the indeterminate x, of the form

$$Ax^a + Bx^b + Cx^c + \text{etc.},$$

where A, B, C, etc., denote any integral numbers, and a, b, c, etc., non-negative integral numbers. If, now, to the indeterminate x there be assigned values which are congruent with respect to any stated modulus, the resulting values of the function X will then be congruent.

Let f and g be two congruent values of x. Then by the preceding articles $f^a \equiv g^a$ and $Af^a \equiv Ag^a$; in the same way $Bf^b \equiv Bg^b$, etc. Hence

$$Af^a + Bf^b + Cf^c + \text{etc.}, \equiv Ag^a + Bg^b + Cg^c + \text{etc.} \quad \text{Q.E.D.}$$

It is easily seen, too, how this theorem can be extended to functions of several indeterminates.

10

If, therefore, all consecutive integral numbers are substituted for x, and if the values of the function X are reduced to least residues, these residues will constitute a sequence in which the same terms repeat after an interval of m terms (m denoting the modulus); or, in other words, this sequence will be formed by a *period of m terms* repeated indefinitely. Let, for example, $X = x^3 - 8x + 6$ and $m = 5$. Then for $x = 0$, 1, 2, 3, etc., the values of X give the positive least residues, 1, 4, 3, 4, 3, 1, 4, etc., where the first five, namely, 1, 4, 3, 4, 3, are repeated without end. And furthermore, if the sequence is continued backwards, that is, if negative values are assigned to x, the same period occurs in the inverse order. It is therefore evident that terms different from those constituting the period cannot occur in the sequence.

In this example, then, X can be neither $\equiv 0$ nor $\equiv 2 \pmod 5$, and can still less be $=0$ or $=2$. Whence it follows that the equations $x^3-8x+6=0$ and $x^3-8x+4=0$ cannot be solved in integral numbers, and therefore, as we know, cannot be solved in rational numbers. It is obviously true in general that, if it is impossible to satisfy the congruence $X\equiv 0$ with respect to some particular modulus, then the equation $X=0$ has no rational root when X is a function of the unknown x, of the form

$$x^n+Ax^{n-1}+Bx^{n-2}+\text{etc.},+N,$$

where A, B, C, etc., are integers and n is a positive integer. (It is well known that all algebraic equations can be brought to this form.) This criterion, though presented here in a natural manner, will be treated at greater length in Section VIII. From this brief indication, some idea, no doubt, can be formed regarding the utility of these researches.

THEORIA MOTUS

Theory of the Motion of the Heavenly Bodies Moving About
the Sun in Conic Sections

translated by Charles Henry Davis

PREFACE

To determine the orbit of a heavenly body, without any hypothetical assumption, from observations not embracing a great
period of time, and not allowing a selection with a view to the
application of special methods, was almost wholly neglected up
to the beginning of the present century; or, at least, not treated
by any one in a manner worthy of its importance, since it assuredly
commended itself to mathematicians by its difficulty and elegance,
even if its great utility in practice were not apparent. An opinion
had universally prevailed that a complete determination from
observations embracing a short interval of time was impossible –
an ill-founded opinion – for it is now clearly shown that the orbit
of a heavenly body may be determined quite nearly from good

observations embracing only a few days; and this without any hypothetical assumption.

Some ideas occurred to me in the month of September of the year 1801, engaged at the time on a very different subject, which seemed to point to the solution of the great problem of which I have spoken. Under such circumstances we not unfrequently, for fear of being too much led away by an attractive investigation, suffer the associations of ideas, which, more attentively considered, might have proved most fruitful in results, to be lost from neglect. And the same fate might have befallen these conceptions had they not happily occurred at the most propitious moment for their preservation and encouragement that could have been selected. For just about this time the report of the new planet, discovered on the first day of January of that year with the telescope at Palermo, was the subject of universal conversation; and soon afterwards the observations made by that distinguished astronomer Piazzi from the above date to the eleventh of February were published. Nowhere in the annals of astronomy do we meet with so great an opportunity, and a greater one could hardly be imagined, for showing most strikingly, the value of this problem, than in this crisis and urgent necessity, when all hope of discovering in the heavens this planetary atom, among innumerable small stars after the lapse of nearly a year, rested solely upon a sufficiently approximate knowledge of its orbit to be based upon these very few observations. Could I ever have found a more seasonable opportunity to test the practical value of my conceptions, than now in employing them for the determination of the orbit of the planet Ceres, which during these forty-one days had described a geocentric arc of only three degrees, and after the lapse of a year must be looked for in a region of the heavens very remote from that in which it was last seen? This first application of the method was made in the month of October, 1801, and the first clear night, when the planet was sought for[1] as directed by the numbers deduced from it, restored the fugitive to observation. Three other new planets, subsequently discovered, furnished new opportunities for examining and verifying the efficiency and generality of the method.

1. By de Zach, 7 December, 1801.

Several astronomers wished me to publish the methods employed in these calculations immediately after the second discovery of Ceres; but many things – other occupations, the desire of treating the subject more fully at some subsequent period, and, especially, the hope that a further prosecution of this investigation would raise various parts of the solution to a greater degree of generality, simplicity and elegance – prevented my complying at the time with these friendly solicitations. I was not disappointed in this expectation, and have no cause to regret the delay. For, the methods first employed have undergone so many and such great changes, that scarcely any trace of resemblance remains between the method in which the orbit of Ceres was first computed, and the form given in this work. Although it would be foreign to my purpose, to narrate in detail all the steps by which these investigations have been gradually perfected, still, in several instances, particularly when the problem was one of more importance than usual, I have thought that the earlier methods ought not to be wholly suppressed. But in this work, besides the solutions of the principal problems, I have given many things which, during the long time I have been engaged upon the motions of the heavenly bodies in conic sections, struck me as worthy of attention, either on account of their analytical elegance, or more especially on account of their practical utility. But in every case I have devoted greater care both to the subjects and methods which are peculiar to myself, touching lightly and so far only as the connection seemed to require, on those previously known.

FIRST BOOK

General Relations Between Those Quantities by Which the Motions of Heavenly Bodies About the Sun Are Defined

First Section. Relations pertaining simply to position in the orbit

1

In this work we shall consider the motions of the heavenly bodies so far only as they are controlled by the attractive force of the sun. All the secondary planets are therefore excluded from our plan, the perturbations which the primary planets exert upon each other are excluded, as is also all motion of rotation. We regard the moving bodies themselves as mathematical points, and we assume that all motions are performed in obedience to the following laws, which are to be received as the basis of all discussion in this work.

I. The motion of every heavenly body takes place in the same fixed plane in which the centre of the sun is situated.

II. The path described by a body is a conic section having its focus in the centre of the sun.

III. The motion in this path is such that the areas of the spaces described about the sun in different intervals of time are proportional to those intervals. Accordingly, if the times and spaces are expressed in numbers, any space whatever divided by the time in which it is described gives a constant quotient.

IV. For different bodies moving about the sun, the squares of these quotients are in the compound ratio of the parameters of their orbits, and of the sum of the masses of the sun and the moving bodies.

Denoting, therefore, the parameter of the orbit in which the body moves by $2p$, the mass of this body by μ (the mass of the sun being put $=1$), the area it describes about the sun in the time t by $\frac{1}{2}g$, then $\dfrac{g}{t\sqrt{p}\sqrt{(1+\mu)}}$ will be a constant for all heavenly bodies. Since then it is of no importance which body we use for determining this number, we will derive it from the motion of the earth, the mean distance of which from the sun we shall adopt for the unit of distance; the mean solar day will always be our unit of time. Denoting, moreover, by π the ratio of the circumference of the circle to the diameter, the area of the entire ellipse described by the earth will evidently be $\pi\sqrt{p}$, which must therefore be put $=\frac{1}{2}g$, if by t is understood the sidereal year; whence, our constant becomes $=\dfrac{2\pi}{t\sqrt{(1+\pi)}}$. In order to ascertain the numerical value of this constant, hereafter to be denoted by k, we will put, according to the latest determination, the sidereal year or $t=365\cdot2563835$, the mass of the earth, or $\mu=\dfrac{1}{354710}=$ $0\cdot0000028192$, whence results

log 2π	.	.	.	.	.	0·7981798684
Compl. log t	.	.	.	.	7·4374021852
Compl. log. $\sqrt{(1+\pi)}$	.	.	.	9·9999993878

log k	.	.	.	.	.	8·2355814414
$k=$					0·01720209895

2

The laws above stated differ from those discovered by our own Kepler in no other respect than this, that they are given in a form applicable to all kinds of conic sections, and that the action of moving body on the sun, on which depends the factor $\sqrt{(1+\pi)}$, is taken into account. If we regard these laws as phenomena derived from innumerable and indubitable observations, geometry shows what action ought in consequence to be exerted upon bodies moving about the sun, in order that these phenomena may be continually produced. In this way it is found that the action of the sun upon the bodies moving about it is exerted just as if an attractive force, the intensity of which is reciprocally proportional to the square of the distance, should urge the bodies towards the centre of the sun. If now, on the other hand, we set out with the assumption of such an attractive force, the phenomena are deduced from it as necessary consequences. It is sufficient here merely to have recited these laws, the connexion of which with the principle of gravitation it will be the less necessary to dwell upon in this place, since several authors subsequently to the eminent Newton have treated this subject, and among them the illustrious La Place, in that most perfect work the *Mécanique Céleste*, in such a manner as to leave nothing further to be desired.

3

Inquiries into the motions of the heavenly bodies, so far as they take place in conic sections, by no means demand a complete theory of this class of curves; but a single general equation rather on which all others can be based, will answer our purpose. And it appears to be particularly advantageous to select that one to which, while investigating the curve described according to the law of attraction, we are conducted as a characteristic equation. If we determine any place of a body in its orbit by the distances x, y, from two right lines drawn in the plane of the orbit intersecting each other at right angles in the centre of the sun, that is, in one of the foci of the curve, and further, if we denote the distance of the body from the sun by r (always positive), we shall have between r, x, y the linear equation $r + \alpha x + \beta y = \gamma$, in which α, β, γ represent constant quantities, γ being from the nature of the case always positive. By changing the position of the right lines to which x, y, are referred, this position being essentially arbitrary, provided only the lines continue to intersect each other at right angles, the form of the equation and also the value of γ will not be changed, but the values of α and β will vary, and it is plain that the position may be so determined that β shall become $=0$, and α, at least, not negative. In this way by putting for α, γ, respectively e, p, our equation takes the form $r + ex = p$. The right line to which the distances y are referred in this case is called the *line of apsides*, p is the *semi-parameter*, e the *eccentricity*; finally the conic section is distinguished by the name of *ellipse*, *parabola* or *hyperbola*, according as e is less than unity, equal to unity, or greater than unity.

It is readily perceived that the position of the line of apsides would be fully determined by the conditions mentioned, with the exception of the single case where both α and β were $=0$; in which case r is always $=p$, whatever the right lines to which x, y are referred. Accordingly, since we have $e=0$, the curve (which will be a circle) is according to our definition to be assigned to the class of ellipses, but it has this peculiarity, that the position of the apsides remains wholly arbitrary, if indeed we choose to extend that idea to such a case.

4

Instead of the distance x let us introduce the angle v, contained between the line of apsides and a straight line drawn from the sun to the place of the body (*the radius vector*), and this angle may commence at that part of the line of apsides at which the distances x are positive, and may be supposed to increase in the direction of the motion of the body. In this way we have $x = r \cos v$, and thus our formula becomes $r = \dfrac{p}{1+e \cos v}$, from which immediately result the following conclusions:

I. For $v=0$, the value of the radius vector r becomes a minimum, that is, $= \dfrac{p}{1+e}$: this point is called the perihelion.

II. For opposite values of v, there are corresponding equal values of r; consequently the line of apsides divides the conic section into two equal parts.

III. In the *ellipse*, v increases continuously from $v=0$, until it attains its maximum value, $\dfrac{p}{1-e}$, in *aphelion*, corresponding to $v=180°$; after aphelion, it decreases in the same manner as it had increased, until it reaches the perihelion, corresponding to $v=360°$. That portion of the line of apsides terminated at one extremity by the perihelion and at the other by the aphelion is called the *major axis*; hence the semi-axis major, called also the *mean distance*, $= \dfrac{p}{1-ee}$; the distance of the middle point of the axis (*the centre of the ellipse*) from the focus will be $\dfrac{ep}{1-ee} = ea$, denoting by a the semi-axis major.

IV. On the other hand, the aphelion in its proper sense is wanting in the parabola, but r is increased indefinitely as v approaches $+180°$, or $-180°$. For $v = \pm 180°$ the value of r becomes infinite, which shows that the curve is not cut by the line of apsides at a point opposite the perihelion. Wherefore, we cannot, with strict propriety of language, speak of the major axis or of the centre of the curve; but by an extension of the formulae found in the ellipse, according to the established usage of analysis, an infinite value is assigned to the major axis, and the centre of the curve is placed at an infinite distance from the focus.

V. In the hyperbola, lastly, v is confined within still narrower limits, in fact between $v = -(180 - \Psi)$, and $v = +(180 - \Psi)$, denoting by Ψ the angle of which the cosine $= \dfrac{1}{e}$. For whilst v

approaches these limits, r increases to infinity; if, in fact, one of these two limits should be taken for v, the value of r would result infinite, which shows that the hyperbola is not cut at all by a right line inclined to the line of apsides above or below by an angle $180° - \psi$. For the values thus excluded, that is to say, from $180° - \psi$ to $180° + \psi$, our formula assigns to r a negative value. The right line inclined by such an angle to the line of apsides does not indeed cut the hyperbola, but if produced reversely, meets the other branch of the hyperbola, which, as is known, is wholly separated from the first branch and is convex towards that focus, in which the sun is situated. But in our investigation, which, as we have already said, rests upon the assumption that r is taken positive, we shall pay no regard to that other branch of the hyperbola in which no heavenly body could move, except one on which the sun should, according to the same laws, exert not an attractive but a repulsive force. Accordingly, the aphelion does not exist, properly speaking, in the hyperbola also; that point of the reverse branch which lies in the line of apsides, and which corresponds to the values $v = 180°$, $r = -\dfrac{p}{e-1}$, might be considered as analogous to the aphelion. If now, we choose after the manner of the ellipse to call the value of the expression $\dfrac{p}{1-ee}$, even here where it becomes negative, the semi-axis major of the hyperbola, then this quantity indicates the distance of the point just mentioned from the perihelion, and at the same time the position opposite to that which occurs in the ellipse. In the same way $\dfrac{ep}{1-ee}$, that is, the distance from the focus to the middle point between these two points (the centre of the hyperbola), here obtains a negative value on account of its opposite direction.

5

We call the angle v the *true anomaly* of the moving body, which, in the parabola is confined within the limits $-180°$ and $+180°$, in the hyperbola between $-(180° - \psi)$ and $+(180° - \psi)$, but which in the ellipse runs through the whole circle in periods constantly renewed. Hitherto, the greater number of astronomers have been accustomed to count the true anomaly in the ellipse not from the perihelion but from the aphelion, contrary to the analogy of the parabola and hyperbola, where, as the aphelion is wanting, it is necessary to begin from the perihelion: we have the less hesitation in restoring the analogy among all classes of conic sections, that the most recent French astronomers have by their example led the way.

It is frequently expedient to change a little the form of the expression $r = \dfrac{p}{1 + e \cos v}$; the following forms will be especially observed:

$$r = \frac{p}{1 + e - 2e \sin^2 \tfrac{1}{2}v} = \frac{p}{1 - e + 2e \cos^2 \tfrac{1}{2}v}$$

$$r = \frac{p}{(1 + e) \cos^2 \tfrac{1}{2}v + (1 - e) \sin^2 \tfrac{1}{2}v}.$$

Accordingly, we have in the parabola

$$r = \frac{p}{2 \cos^2 \tfrac{1}{2}v};$$

in the hyperbola the following expression is particularly convenient,

$$r = \frac{p \cos \psi}{2 \cos \tfrac{1}{2}(v + \psi) \cos \tfrac{1}{2}(v - \psi)}$$

THE FOUNDATIONS OF MATHEMATICS

translated by G. Waldo Dunnington

The subject of mathematics includes all extensive magnitudes (those in which parts can be conceived); intensive magnitudes (all non-extensive magnitudes) insofar as they are dependent on the extensives. To the former class of magnitudes belong: space or geometrical magnitudes which include lines, surfaces, solids and angles; to the latter: velocity, density, rigidity, pitch and timbre of tones, intensity of tones and of light, probability, etc.

A magnitude in itself cannot become the subject of a scientific investigation; mathematics considers magnitudes only in reference to each other. The relationship of magnitudes to each other which they have only in so far as they are magnitudes, is called an arithmetical relationship. In geometrical magnitudes there occurs a relation in respect to position and this is called a geometric relationship. It is clear that geometric magnitudes can also have arithmetical relationships to each other.

Mathematics really teaches general truths which concern the relations of magnitudes and the purpose of it is to present magnitudes which have known relationships to known magnitudes or to magnitudes known to these, i.e. to make a presentation of them possible. But now we can have a presentation of a magnitude in a twofold manner, either by direct perception (a direct presentation), or by comparison with others, by direct perception of given magnitudes (indirect presentation). Accordingly the duty of the mathematician is either really to present the magnitude sought or to indicate the way one proceeds from the presentation of a magnitude directly given to the presentation of the magnitude sought (arithmetical presentation). This latter occurs by means of numbers, which show how many times one must repeatedly present the directly given magnitude[1] to get a presentation of the magnitude sought. That magnitude one calls unity and the process measuring.

These various relations of magnitudes and the various modes of presenting magnitudes are the foundations of both major disciplines of mathematics. Arithmetic regards magnitudes in arithmetical relations and presents them arithmetically; geometry regards magnitudes in geometric relations and presents them geometrically. To present geometrically magnitudes which have arithmetical relations, which was so customary among the ancients, is no longer the custom at present, otherwise one would have to regard this as a part of geometry. On the contrary, one applies the arithmetical mode of presentation most frequently to magnitudes in geometrical relationships, e.g. in trigonometry, also in the theory of curves, which are regarded as geometric disciplines. That moderns have thus preferred the arithmetical presentation rather than the geometrical, does not occur without reason, especially since our method of counting (the base ten) is so much easier than that of the ancients.

1. Occasionally, also, how many times one must conceive a part of the same as repeated, which then gives the idea of the broken number (fraction).

Since a great difference can occur among the arithmetical relations of magnitudes to each other, the parts of the arithmetical sciences are of a very different nature. Most important is the circumstance of whether in this relationship the concept of the infinite must be presupposed or not; the first case belongs to the calculation of the infinite, or higher mathematics, the latter to common or lower mathematics.

*

1. What is the essential condition that a combination of concepts may be thought of as referring to a *quantity*?

2. Everything becomes much simpler if one at first turns from infinite divisibility and merely considers discrete quantities. As, e.g. in biquadratic residues the points as objects, the passages from one to another, i.e. relationships, as quantities, where the meaning of $a+bi-c-di$ is at once clear.

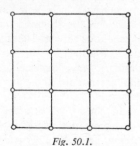

Fig. 50.1.

3. Mathematics is thus in the most general sense the science of relationships. Relationship predicates two things and is then called simple relationships, etc.

4. Points in a line are the general presentation of things, where each thing has a relationship of inequality only to two.

If a point can have a relationship to more than two things, then the representation of that is the condition of points in a plane, which are connected by lines. However, if an investigation is to be possible here, then it can concern only those points which are in a mutual relationship to three, and where there is a relationship between the relationships.

GENERAL INVESTIGATIONS OF
CURVED SURFACES OF 1825 AND 1827

translated by James C. Morehead and Adam M. Hiltereitel

GAUSS'S ABSTRACT OF THE 'DISQUISITIONES
GENERALES CIRCA SUPERFICIES CURVAS' PRESENTED
TO THE ROYAL SOCIETY OF GÖTTINGEN[1]

Although geometers have given much attention to general investigations of curved surfaces and their results cover a significant portion of the domain of higher geometry, this subject is still so far from being exhausted, that it can well be said that, up to this time, but a small portion of an exceedingly fruitful field has been cultivated. Through the solution of the problem, to find all representations of a given surface upon another in which the smallest elements remain unchanged, the author sought some years ago to give a new phase to this study. The purpose of the present discussion is further to open up other new points of view and to develop some of the new truths which thus become accessible. We shall here give an account of those things which can be made intelligible in a few words. But we wish to remark at the outset that the new theorems as well as the presentations of new ideas, if the greatest generality is to be attained, are still partly in need of some limitations or closer determinations, which must be omitted here.

In researches in which an infinity of directions of straight lines in space is concerned, it is advantageous to represent these directions by means of those points upon a fixed sphere, which are the end points of the radii drawn parallel to the lines. The centre and the radius of this *auxiliary sphere* are here quite arbitrary. The radius may be taken equal to unity. This procedure agrees fundamentally with that which is constantly employed in astronomy, where all directions are referred to a fictitious celestial sphere of infinite radius. Spherical trigonometry and certain other theorems, to which the author has added a new one of frequent application, then serve for the solution of

1. *Göttingen gelehrte Anzeigen*, no. 177, 1827, pp. 1761–8.

the problems which the comparison of the various directions involved can present.

If we represent the direction of the normal at each point of the curved surface by the corresponding point of the sphere, determined as above indicated, namely, in this way, to every point on the surface, let a point on the sphere correspond; then, generally speaking, to every line on the curved surface will correspond a line on the sphere, and to every part of the former surface will correspond a part of the latter. The less this part differs from a plane, the smaller will be the corresponding part on the sphere. It is, therefore, a very natural idea to use as the measure of the total curvature, which is to be assigned to a part of the curved surface, the area of the corresponding part of the sphere. For this reason the author calls this area the *integral curvature* of the corresponding part of the curved surface. Besides the magnitude of the part, there is also at the same time its *position* to be considered. And this position may be in the two parts similar or inverse, quite independently of the relation of their magnitudes. The two cases can be distinguished by the positive or negative sign of the total curvature. This distinction has, however, a definite meaning only when the figures are regarded as upon definite sides of the two surfaces. The author regards the figure in the case of the sphere on the outside, and in the case of the curved surface on that side upon which we consider the normals erected. It follows then that the positive sign is taken in the case of convexo-convex or concavo-concave surfaces (which are not essentially different), and the negative in the case of concavo-convex surfaces. If the part of the curved surface in question consists of parts of these different sorts, still closer definition is necessary, which must be omitted here.

The comparison of the areas of two corresponding parts of the curved surface and of the sphere leads now (in the same manner as, e.g. from the comparison of volume and mass springs the idea of density) to a new idea. The author designates as *measure of curvature* at a point of the curved surface the value of the fraction whose denominator is the area of the infinitely small part of the curved surface at this point and whose numerator is the area of the corresponding part of the surface of the auxiliary sphere, or the integral curvature of that element. It is clear that, according to the idea of the author, integral curvature and measure of curvature in the case of curved surfaces are analogous to what, in the case of curved lines, are called respectively amplitude and curvature simply. He hesitates to apply to curved surfaces the latter expressions, which have been accepted more from custom than on account of fitness. Moreover, less depends upon the choice of words than upon this, that their introduction shall be justified by pregnant theorems.

The solution of the problem, to find the measure of curvature at any point of a curved surface, appears in different forms according to the manner in which the nature of the curved surface is given. When the points in space, in general, are distinguished by three rectangular coordinates, the simplest method is to express one coordinate as a function of the other two. In this way we obtain the simplest expression for the measure of curvature. But, at the same time, there arises a remarkable relation between this measure of curvature and the curvatures of the curves formed by the intersections of the curved surface with planes normal to it. Euler, as is well known, first showed that two of these cutting planes which intersect each other at right angles have this property, that in one is found the greatest and in the other the smallest radius of curvature; or, more correctly, that in them the two extreme curvatures are found. It will follow then from the above mentioned expression for the measure of curvature that this will be equal to a fraction whose numerator is unity and whose denominator is the product of the extreme radii of curvature. The expression for the measure of curvature will be less simple, if the nature of the curved surface is determined by an equation in x, y, z. And it will become still more complex, if the nature of the curved surface is given so that x, y, z are expressed in the form of functions of two new variables p, q. In this last case the expression involves fifteen elements, namely, the partial differential coefficients of the first and second orders of x, y, z with respect to p and q. But it is less important in itself than for the reason that it facilitates the transition to another expression, which must be classed with the most remarkable theorems of this study. If the nature of the curved surface be expressed by this method, the general expression for any linear element upon it, or for $\sqrt{(dx^2+dy^2+dz^2)}$, has the form $\sqrt{(E\,dp^2+2F\,dp\,.\,dq+G\,dq^2)}$, where E, F, G are again functions of p and q. The new expression for the measure of curvature mentioned above contains merely these magnitudes and their partial differential coefficients of the first and second order. Therefore we notice that, in order to determine the measure of curvature, it is necessary to know only the general expression for a linear element; the expressions for the coordinates x, y, z are not required. A direct result from this is the remarkable theorem: If a curved surface, or a part of it, can be developed upon another surface, the measure of curvature at every point remains unchanged after the development. In particular, it follows from this further: Upon a curved surface that can be developed upon a plane, the measure of curvature is everywhere equal to zero. From this we derive at once the characteristic equation of surfaces developable upon a plane, namely,

$$\frac{\partial^2 z}{\partial x^2}\cdot\frac{\partial^2 z}{\partial y^2}-\left\{\frac{\partial^2 z}{\partial x\,.\,\partial y}\right\}^2=0,$$

when z is regarded as a function of x and y. This equation has been known for some time, but according to the author's judgement it has not been established previously with the necessary rigour.

These theorems lead to the consideration of the theory of curved surfaces from a new point of view, where a wide and still wholly uncultivated field is open to investigation. If we consider surfaces not as boundaries of bodies, but as bodies of which one dimension vanishes, and if at the same time we conceive them as flexible but not extensible, we see that two essentially different relations must be distinguished, namely, on the one hand, those that presuppose a definite form of the surface in space; on the other hand, those that are independent of the various forms which the surface may assume. This discussion is concerned with the latter. In accordance with what has been said, the measure of curvature belongs to this case. But it is easily seen that the consideration of figures constructed upon the surface, their angles, their areas and their integral curvatures, the joining of the points by means of shortest lines, and the like, also belong to this case. All such investigations must start from this, that the very nature of the curved surface is given by means of the expression of any linear element in the form $\sqrt{(Edp^2 + 2Fdp \cdot dq + Gdq^2)}$. The author has embodied in the present treatise a portion of his investigations in this field, made several years ago, while he limits himself to such as are not too remote for an introduction, and may, to some extent, be generally helpful in many further investigations. In our abstract, we must limit ourselves still more, and be content with citing only a few of them as types. The following theorems may serve for this purpose.

If upon a curved surface a system of infinitely many shortest lines of equal lengths be drawn from one initial point, then will the line going through the end points of these shortest lines cut each of them at right angles. If at every point of an arbitrary line on a curved surface shortest lines of equal lengths be drawn at right angles to this line, then will all these shortest lines be perpendicular also to the line which joins their other end points. Both these theorems, of which the latter can be regarded as a generalization of the former, will be demonstrated both analytically and by simple geometrical considerations. *The excess of the sum of the angles of a triangle formed by shortest lines over two right angles is equal to the total curvature of the triangle.* It will be assumed here that that angle (57° 17′ 45″) to which an arc equal to the radius of the sphere corresponds will be taken as the unit for the angles and that for the unit of total curvature will be taken a part of the spherical surface, the area of which is a square whose side is equal to the radius of the sphere. Evidently we can express this important theorem thus also: the excess over two right angles of the angles of a triangle formed by shortest lines is to eight right angles as the part of the surface of the auxiliary sphere, which corresponds to it as its integral curvature, is to the whole surface of the sphere. In general, the excess over $2n - 4$ right angles of the angles of a polygon of n sides, if these are shortest lines, will be equal to the integral curvature of the polygon.

The general investigations developed in this treatise will, in the conclusion, be applied to the theory of triangles of shortest lines, of which we shall introduce only a couple of important theorems. If a, b, c be the sides of such a triangle (they will be regarded as magnitudes of the first order); A, B, C the angles opposite; α, β, γ

the measures of curvature at the angular points; σ the area of the triangle, then, to magnitudes of the fourth order, $\frac{1}{3}(\alpha+\beta+\gamma)\sigma$ is the excess of the sum $A+B+C$ over two right angles. Further, with the same degree of exactness, the angles of a plane rectilinear triangle whose sides are a, b, c, are respectively

$$A-\tfrac{1}{12}(2\alpha+\beta+\gamma)\sigma$$
$$B-\tfrac{1}{12}(\alpha+2\beta+\gamma)\sigma$$
$$C-\tfrac{1}{12}(\alpha+\beta+2\gamma)\sigma.$$

We see immediately that this last theorem is a generalization of the familiar theorem first established by Legendre. By means of this theorem we obtain the angles of a plane triangle, correct to magnitudes of the fourth order, if we diminish each angle of the corresponding spherical triangle by one-third of the spherical excess. In the case of non-spherical surfaces, we must apply unequal reductions to the angles, and this inequality, generally speaking, is a magnitude of the third order. However, even if the whole surface differs only a little from the spherical form, it will still involve also a factor denoting the degree of the deviation from the spherical form. It is unquestionably important for the higher geodesy that we be able to calculate the inequalities of those reductions and thereby obtain the thorough conviction that, for all measurable triangles on the surface of the earth, they are to be regarded as quite insensible. So it is, for example, in the case of the greatest triangle of the triangulation carried out by the author. The greatest side of this triangle is almost fifteen geographical[1] miles, and the excess of the sum of its three angles over two right angles amounts almost to fifteen seconds. The three reductions of the angles of the plane triangle are $4''\cdot95113$, $4''\cdot95104$, $4''\cdot95131$. Besides, the author also developed the missing terms of the fourth order in the above expressions. Those for the sphere possess a very simple form. However, in the case of measurable triangles upon the earth's surface, they are quite insensible. And in the example here introduced they would have diminished the first reduction by only two units in the fifth decimal place and increased the third by the same amount.

1. This German geographical mile is four minutes of arc at the equator, namely, 7·42 kilometres, and is equal to about 4·6 English statute miles. [Translators.]

51

BENJAMIN PEIRCE
(1809–80)

BENJAMIN PEIRCE, professor of astronomy and mathematics at Harvard University and leading mathematician of his day, was born at Salem, Mass., on 4 April, 1809. His father, who had been a member first of the lower house and then of the upper house of the state legislature, was Librarian of Harvard University from 1826 to 1831, and the author (posthumously) of the History of Harvard University from its inception to the period of the Revolution. As an undergraduate, Benjamin came under the influence of Nathaniel Bowditch. His deep affection for Professor Bowditch continued undiminished in later years and it was to him that Peirce dedicated the greatest and most extensive of his texts, his *Analytic Mechanics*. 'To the cherished and revered memory,' he wrote, 'of my master in science, Nathaniel Bowditch, the father of American geometry.' Before graduating from Harvard, Peirce had already made a firm choice of a career, the teaching of mathematics. He was appointed tutor in mathematics at Harvard at the same time as Charles W. Eliot, and the two young mathematics tutors joined forces in securing the innovation of written final examinations in mathematics in the place of oral examinations which had formerly been held in the presence of visiting committees of the Board of Overseers. The practice of holding written examinations soon spread to other departments at Harvard and to other universities. Peirce was the first professor of mathematics who consistently encouraged his students to engage in mathematical research. Through the great series of text-books which he wrote, Benjamin Peirce exerted a profound and lasting influence upon the teaching of mathematics in America.

Vastly more than teaching and text-books filled Peirce's forty-nine active years at Harvard. In 1847 he was one of a committee of five to plan a programme for the organization of the Smithsonian Institution. In that same year he received the degree of LL.D. from the University of North Carolina. Thereafter not a year went by without his participation in some vitally important activity or without his receiving academic recognition from learned societies and universities both at home and abroad. An office of the American Nautical Almanac was established at Harvard so that without interrupting his duties as professor he was able to serve the Almanac as consulting astronomer (1849–67). He was Director of the longitude determination for the United States Coast Survey from 1832 to 1867. He served as Superintendent of the Coast Survey from 1867 to 1874, and as the consulting geometer from 1874 to 1880. In 1870 he personally conducted an expedition to Sicily to observe the eclipse of the sun. He was instrumental in the establishment of the Dudley Observatory at Albany in 1855 and he was largely responsible for the institution of an observatory at Harvard. Possessing a huge fund of personal dynamism, Benjamin Peirce was active and influential in numerous scientific associations, some of which he helped to found. He was president of the American Association for the Advancement of Science in 1853. He was one of the fifty incorporators of the National Academy of Sciences and the chairman of the mathematics and physics class in 1864.

Each of Benjamin Peirce's four sons achieved professional distinction. James Mills Peirce, the eldest, was for forty years professor of mathematics at Harvard. Benjamin Mills Peirce was a mining engineer. Herbert Henry Peirce was a diplomat. Charles Sanders Peirce, the least successful financially of the entire family, has come to be regarded as the greatest and most original of American philosophers. The researches which led Benjamin Peirce to the publication of his memoir, *Linear Associative Algebra*, were undertaken by him upon the urging of his son, Charles. Stimulated by Charles's lively interest in the new mathematics being developed on a level of generality and application heretofore unknown, Benjamin Peirce wrote the memoir which he described as a 'philosophic study of the laws of algebraic operation', developed from fundamental principles, no reference being made to other branches of mathematics. Introducing the work in 1870, Benjamin Peirce stated, 'This work has been the pleasantest mathematical effort of my life. In no other have I seemed to myself to have received so full a reward for my mental labour in the novelty and breadth of the results.' The method employed by Benjamin Peirce in his *Linear Associative Algebra* was extended and perfected by later writers, notably by H. E. Hawkes who wrote on hypercomplex numbers in seven units.

LINEAR ASSOCIATIVE ALGEBRA

A Memoir read before the National Academy of Sciences
in Washington, 1870

by Benjamin Peirce,
with Notes and Addenda, by C. S. Peirce,
Son of the Author

1. Mathematics is the science which draws necessary conclusions.

This definition of mathematics is wider than that which is ordinarily given, and by which its range is limited to quantitative research. The ordinary definition, like those of other sciences, is objective; whereas this is subjective. Recent investigations, of which quaternions is the most noteworthy instance, make it manifest that the old definition is too restricted. The sphere of mathematics is here extended, in accordance with the derivation of its name, to all demonstrative research, so as to include all knowledge strictly capable of dogmatic teaching. Mathematics is not the discoverer of laws, for it is not induction; neither is it the framer of theories, for it is not hypothesis; but it is the judge over both, and it is the arbiter to which each must refer its claims; and neither law can rule nor theory explain without the sanction of mathematics. It deduces from a law all its consequences, and develops them into the suitable form for comparison with observation, and thereby measures the strength of the argument from observation in favour of a proposed law or of a proposed form of application of a law.

Mathematics, under this definition, belongs to every inquiry, moral as well as physical. Even the rules of logic, by which it is rigidly bound, could not be deduced without its aid. The laws of argument admit of simple statement, but they must be curiously transposed before they can be applied to the living speech and verified by observation. In its pure and simple form the syllogism cannot be directly compared with all experience, or it would not have required an Aristotle to discover it. It must be transmuted into all the possible shapes in which reasoning loves to clothe itself. The transmutation is the mathematical process in the establishment of the law. Of some sciences, it is so large a portion that they have been quite abandoned to the mathematician – which may not have been altogether to the advantage of philosophy. Such is the case with geometry and analytic mechanics. But in many other sciences, as in all those of mental philosophy and most of the branches of natural history, the deductions are so immediate and of such simple construction, that it is of no practical use to separate the mathematical portion and subject it to isolated discussion.

2. The branches of mathematics are as various as the sciences to which they belong, and each subject of physical inquiry has its appropriate mathematics. In every form of material manifestation, there is a corresponding form of human thought, so that the human mind is as wide in its range of thought as the physical universe in which it thinks. The two are wonderfully matched. But where there is a great diversity of physical appearance, there is often a close resemblance in the processes of deduction. It is important, therefore, to separate the intellectual work from the external form. Symbols must be adopted which may serve for the embodiment of forms of argument, without being trammelled by the conditions of external representation or special interpretation. The words of common language are usually unfit for this purpose, so that other symbols must be adopted, and mathematics treated by such symbols is called *algebra*. Algebra, then, is formal mathematics.

3. All relations are either qualitative or quantitative. Qualitative relations can be considered by themselves without regard to quantity. The algebra of such inquiries may be called logical algebra, of which a fine example is given by Boole.

Quantitative relations may also be considered by themselves without regard to quality. They belong to arithmetic, and the corresponding algebra is the common or arithmetical algebra.

In all other algebras both relations must be combined, and the algebra must conform to the character of the relations.

4. The symbols of an algebra, with the laws of combination, constitute its *language*; the methods of using the symbols in the drawing of inferences is its *art*; and their interpretation is its *scientific application*. This three-fold analysis of algebra is

adopted from President Hill, of Harvard University, and is made the basis of a division into books.

BOOK I THE LANGUAGE OF ALGEBRA[1]

5. The language of algebra has its alphabet, vocabulary and grammar.

6. The symbols of algebra are of two kinds: one class represent its fundamental conceptions and may be called its *letters*, and the other represent the relations or modes of combinations of the letters and are called *the signs*.

7. The *alphabet* of an algebra consists of its letters; the *vocabulary* defines its signs and the elementary combinations of its letters; and the *grammar* gives the rules of composition by which the letters and signs are united into a complete and consistent system.

The Alphabet

8. Algebras may be distinguished from each other by the number of their independent fundamental conceptions, or of the letters of their alphabet. Thus an algebra which has only one letter in its alphabet is a *single* algebra; one which has two letters is a *double* algebra; one of three letters a *triple* algebra; one of four letters a *quadruple* algebra, and so on.

This artificial divisions of the algebras is cold and uninstructive like the artificial Linnean system of botany. But it is useful in a preliminary investigation of algebras, until a sufficient variety is obtained to afford the material for a natural classification.

Each fundamental conception may be called a *unit*; and thus each unit has its corresponding letter, and the two words, unit and letter, may often be used indiscriminately in place of each other, when it cannot cause confusion.

9. The present investigation, not usually extending beyond the sextuple algebra, limits the demand of the algebra for the most part to six letters; and the six letters, i, j, k, l, m and n, will be restricted to this use except in special cases.

1. Only this book was ever written. [C.S.P.]

10. *For any given letter another may be substituted*, provided a new letter represents a combination of the original letters of which the replaced letter is a necessary component.

For example, any combination of two letters, which is entirely dependent for its value upon both of its components, such as their sum, difference or product, may be substituted for either of them.

This *principle of the substitution of letters* is radically important, and is a leading element of originality in the present investigation; and without it such an investigation would have been impossible. It enables the geometer to analyse an algebra, reduce it to its simplest and characteristic forms, and compare it with other algebras. It involves in its principle a corresponding substitution of *units* of which it is in reality the formal representative.

There is, however, no danger in working with the symbols, irrespective of the ideas attached to them, and the consideration of the change of the original conceptions may be safely reserved for the *book of interpretation*.

11. In making the substitution of letters, the original letter will be preserved with the distinction of a subscript number.

Thus, for the letter i there may successively be substituted i_1, i_2, i_3, etc. In the final forms, the subscript numbers can be omitted, and they may be omitted at any period of the investigation, when it will not produce confusion.

It will be practically found that these subscript numbers need scarcely ever be written. They pass through the mind, as a sure ideal protection from erroneous substitution, but disappear from the writing with the same facility with which those evanescent chemical compounds, which are essential to the theory of transformation, escape the eye of the observer.

12. A *pure* algebra is one in which every letter is connected by some indissoluble relation with every other letter.

13. When the letters of an algebra can be separated into two groups, which are mutually independent, it is a *mixed algebra*. It is mixed even when there are letters common to the two groups. Were an algebra employed for the simultaneous discussion of distinct classes of phenomena, such as those of sound and light, and were the peculiar units of each class to have their appropriate letters, but were there no recognized dependence of the phenomena upon each other, so that the phenomena of each class might have been submitted to independent research, the one algebra would be actually a mixture of two algebras, one appropriate to sound, the other to light.

It may be further observed that when, in such a case as this, the component algebras are identical in form, they are reduced to the case of one algebra with two diverse interpretations.

The Vocabulary

14. Letters which are not appropriated to the alphabet of the algebra[1] may be used in any convenient sense. But it is well to employ *the small letters* for expressions of common algebra, and *the capital letters* for those of the algebra under discussion.

There must, however, be exceptions to this notation; thus the letter D will denote the derivative of an expression to which it is applied, and Σ the summation of cognate expressions, and other exceptions will be mentioned as they occur. Greek letters will generally be reserved for angular and functional notation.

15. The three symbols J, ∂, and G will be adopted with the signification

$$J = \sqrt{-1}$$

∂ = the ratio of circumference to diameter of circle

$$= 3 \cdot 1415926536$$

G = the base of Naperian logarithms = $2 \cdot 7182818285$, which gives the mysterious formula

$$J^{-J} = \sqrt{G^{\partial}} = 4 \cdot 810477381.$$

16. All the signs of common algebra will be adopted; but any signification will be permitted them which is not inconsistent with their use in common algebra; so that if by any process an expression to which they refer is reduced to one of common algebra, they must resume their ordinary signification.

17. The sign =, which is called that of equality, is used in its ordinary sense to denote that the two expressions which it

---

1. See §9.

separates are the same whole, although they represent different combinations of parts.

18. The signs $>$ and $<$ which are those of inequality, and denote 'more than' or 'less than' in quantity, will be used to denote the relations of a whole to its part, so that the symbol which denotes the part shall be at the vertex of the angle, and that which denotes the whole at its opening. This involves the proposition that the smaller of the quantities is included in the class expressed by the larger. Thus

$$B < A \quad \text{or} \quad A > B$$

denotes that A is a whole of which B is a part, so that all B is A.[1]

If the usual algebra had originated in qualitative, instead of quantitative, investigations, the use of the symbols might easily have been reversed; for it seems that all conceptions involved in A must also be involved in B, so that B is more than A in the sense that it involves more ideas.

The combined expression

$$B > C < A$$

denotes that there are quantities expressed by C which belong to the class A and also to the class B. It implies, therefore, that some B is A and that some A is B.[2] The intermediate C might be omitted if this were the only proposition intended to be expressed, and we might write

$$B > \quad < A.$$

In like manner the combined expression

$$B < C > A$$

denotes that there is a class which includes both A and B,[3] which proposition might be written

$$B < \quad > A.$$

19. A vertical mark drawn through either of the preceding signs reverses its signification. Thus

$$A \neq B$$

---

1. The formula in the text implies, also, that some A is not B. [C. S. P.]
2. This, of course, supposes that C does not vanish. [C. S. P.]
3. The universe will be such a class unless A or B is the universe. [C. S. P.]

denotes that B and A are essentially different wholes;

$$A \not> B \quad \text{or} \quad B \not< A$$

denotes that all B is not A,[1] so that if they have only quantitative relations, they must bear to each other the relation of

$$A = B \quad \text{or} \quad A < B.$$

20. The sign $+$ is called *plus* in common algebra and denotes *addition*. It may be retained with the same name, and the process which it indicates may be called addition. In the simplest cases it expresses a mere mixture, in which the elements preserve their mutual independence. If the elements cannot be mixed without mutual action and a consequent change of constitution, the mere union is still expressed by the sign of addition, although some other symbol is required to express the character of the mixture as a peculiar compound having properties different from its elements. It is obvious from the simplicity of the union recognized in this sign, that the order of the admixture of the elements cannot affect it; so that it may be assumed that

$$A + B = B + A$$

and

$$(A + B) + C = A + (B + C) = A + B + C.$$

21. The sign $-$ is called *minus* in common algebra, and denotes *subtraction*. Retaining the same name, the process is to be regarded as the reverse of addition; so that if an expression is first added and then subtracted, or the reverse, it disappears from the result; or, in algebraic phrase, it is *cancelled*. This gives the equations

$$A + B - B = A - B + B = A$$

and

$$B - B = 0.$$

The sign minus is called the negative sign in ordinary algebra, and any term preceded by it may be united with it, and the combination may be called a *negative term*. This use will be adopted into all the algebras, with the provision that the derivation of the word negative must not transmit its interpretation.

22. The sign $\times$ may be adopted from ordinary algebra with the

1. The general interpretation is rather that either A and B are identical or that some B is not A. [C. S. P.]

name of the sign of *multiplication*, but without reference to the meaning of the process. The result of multiplication is to be called the *product*. The terms which are combined by the sign of multiplication may be called *factors*; the factor which precedes the sign being distinguished as the *multiplier*, and that which follows it being the *multiplicand*. The words multiplier, multiplicand and product, may also be conveniently replaced by the terms adopted by Hamilton, of *facient*, *faciend* and *factum*. Thus the equation of the product is multiplier × multiplicand = product; *or* facient × faciend = factum. When letters are used, the sign of multiplication can be *omitted* as in ordinary algebra.

23. When an expression used as a factor in certain combinations gives a product which vanishes, it may be called in those combinations a *nilfactor*. Where as the multiplier it produces vanishing products it is *nilfacient*, but where it is the multiplicand of such a product it is *nilfaciend*.

24. When an expression used as a factor in certain combinations overpowers the other factors and is itself the product, it may be called an *idemfactor*. When in the production of such a result it is the multiplier, it is *idemfacient*, but when it is the multiplicand it is *idemfaciend*.

25. When an expression raised to the square or any higher power vanishes, it may be called *nilpotent*; but when, raised to a square or higher power, it gives itself as the result, it may be called *idempotent*.

The defining equation of nilpotent and idempotent expressions are respectively $A^n = 0$, and $A^n = A$; but with reference to idempotent expressions, it will always be assumed that they are of the form

$$A^2 = A,$$

unless it be otherwise distinctly stated.

26. *Division* is the reverse of multiplication, by which its results are verified. It is the process for obtaining one of the factors of a given product when the other factor is given. It is important to distinguish the position of the given factor, whether it is facient or faciend. This can be readily indicated by combining the sign of multiplication, and placing it before or after the given factor just

as it stands in the product. Thus when the multiplier is the given factor, the correct equation of division is

$$\text{quotient} = \frac{\text{dividend}}{\text{divisor} \times}$$

and the equation of verification is divisor × quotient = dividend. But when the multiplicand is the given factor, the equation of division is

$$\text{quotient} = \frac{\text{dividend}}{\times \text{ divisor}}$$

and the equation of verification is quotient × divisor = dividend.

27. Exponents may be introduced just as in ordinary algebra, and they may even be permitted to assume the forms of the algebra under discussion. There seems to be no necessary restriction to giving them even a wider range and introducing into one algebra the exponents from another. Other signs will be defined when they are needed.

The definition of the fundamental operations is an essential part of the vocabulary, but as it is subject to the rules of grammar which may be adopted, it must be reserved for special investigation in the different algebras.

The Grammar

28. Quantity enters as a form of thought into every inference. It is always implied in the syllogism. It may not, however, be the direct object of inquiry; so that there may be logical and chemical algebras into which it only enters accidentally, agreeably to § 1. But where it is recognized, it should be received in its most general form and in all its variety. The algebra is otherwise unnecessarily restricted, and cannot enjoy the benefit of the most fruitful forms of philosophical discussion. But while it is thus introduced as a part of the formal algebra, it is *subject to every degree and kind of limitation in its interpretation.*

The free introduction of quantity into an algebra does not even involve the reception of its unit as one of the independent units of the algebra. But it is probable that without such a unit, no algebra is adapted to useful investigation. It is so admitted into quaternions, and its admission seems to have misled some philosophers into the opinion that quaternions is a triple and not a quadruple algebra. This will be the more evident from the form in which quaternions first present themselves in the present investigation, and in which the unit of quantity is not distinctly recognizable without a transmutation of the form.

29. The introduction of quantity into an algebra naturally carries with it, not only the notation of ordinary algebra, but likewise many of the rules to which it is subject. Thus, when a quantity is a factor of a product, it has the same influence whether it be facient or faciend, so that with the notation of § 14, there is the equation

$$Aa = aA,$$

and in such a product, the quantity a may be called the *coefficient*.

In like manner, terms which only differ in their coefficients, may be added by adding their coefficients; thus,

$$(a \pm b)A = aA \pm bA = Aa \pm Ab = A(a \pm b).$$

30. The exceeding simplicity of the conception of an equation involves the identity of the equations

$$A = B \quad \text{and} \quad B = A$$

and the substitution of B for A in every expression, so that

$$MA \pm C = MB \pm C,$$

or that, *the members of an equation may be mutually transposed or simultaneously increased or decreased or multiplied or divided by equal expressions.*

31. How far the principle of § 16 limits the extent within which the ordinary symbols may be used, cannot easily be decided. But it suggests limitations which may be adopted during the present discussion, and leave an ample field for curious investigation.

The distributive principle of multiplication may be adopted; namely, the principle that the product of an algebraic sum of factors into or by a common factor, is equal to the corresponding algebraic sum of the individual products of the various factors into or by the common factor; and it is expressed by the equations

$$(A \pm B)C = AB \pm BC.$$
$$C(A \pm B) = CA \pm CB.$$

32. *The associative principle of multiplication* may be adopted; namely, that the product of successive multiplications is not affected by the order in which the multiplications are performed, provided there is no change in the relative position of the factors; and it is expressed by the equations

$$ABC = (AB)C = A(BC).$$

This is quite an important limitation, and the algebras which are subject to it will be called *associative*.

33. The principle that the value of a product is not affected by the relative position of the factors is called *the commutative principle*, and is expressed by the equation

$$AB = BA.$$

This principle is *not* adopted in the present investigation.

34. An algebra in which every expression is reducible to the form of an algebraic sum of terms, each of which consists of a single *letter* with a quantitative coefficient, is called *a linear* algebra. Such are all the algebras of the present investigation.

35. Wherever there is a limited number of independent conceptions, a linear algebra may be adopted. For a combination which was not reducible to such an algebraic sum as those of linear algebra, would be to that extent independent of the original conceptions, and would be an independent conception additional to those which were assumed to constitute the elements of the algebra.

36. An algebra in which there can be complete interchange of its independent units, without changing the formulae of combination, is a *completely symmetrical algebra*; and one in which there may be a partial interchange of its units is *partially symmetrical*. But the term symmetrical should not be applied, unless the interchange is more extensive than that involved in the distributive and commutative principles. An algebra in which the interchange is effected in a certain order which returns into itself is a *cyclic algebra*.

Thus, quaternions is a cyclic algebra, because in any of its fundamental equations, such as

$$i^2 = -1$$
$$ij = -ji = k$$
$$ijk = -1$$

there can be an interchange of the letters in the order i, j, k, i, each letter being changed into that which follows it. The double algebra in which

$$i^2 = i, \quad ij = i$$
$$j^2 = j, \quad ji = j$$

is cyclic because the letters are interchangeable in the order i, j, i. But neither of these algebras is commutative.

37. When an algebra can be reduced to a form in which all the letters are expressed as powers of some one of them, it may be called a *potential algebra*. If the powers are all squares, it may be called *quadratic*; if they are cubes, it may be called *cubic*; and similarly in other cases.

Linear Associative Algebra

38. *All the expressions of an algebra are distributive, whenever the distributive principle extends to all the letters of the alphabet.*

For it is obvious that in the equation

$$(i+j)(k+l) = ik + jk + il + jl$$

each letter can be multiplied by an integer, which gives the form

$$(ai+bj)(ck+dl) = acik + bcjk + adil + bdjl,$$

in which a, b, c and d are integers. The integers can have the ratios of any four real numbers, so that by simple division they can be reduced to such real numbers. Other similar equations can also be formed by writing for a and b, a_1 and b_1, or for c and d, c_1 and d_1, or by making both these substitutions simultaneous. If then the two first of these new equations are multiplied by J and the last by -1; the sum of the four equations will be the same as that which would be obtained by substituting for a, b, c and d, $a+Ja_1, b+Jb_1, c+Jc_1$ and $d+Jd_1$. Hence a, b, c and d may be any numbers, real or imaginary, and in general whatever mixtures

A, B, C and D may represent of the original units under the form of an algebraic sum of the letters i, j, k, etc. we shall have

$$(A+B)(C+D)=AC+BC+AD+BD,$$

which is the complete expression of the distributive principle.

39. *An algebra is associative whenever the associative principle extends to all the letters of its alphabet.*

For if

$$A=\Sigma(ai)=ai+a_1j+a_2k+\text{etc.}$$
$$B=\Sigma(bi)=bi+b_1j+b_2k+\text{etc.}$$
$$C=\Sigma(ci)=ci+c_1j+c_2k+\text{etc.}$$

it is obvious that

$$AB=\Sigma(ab_1ij)$$
$$BC=\Sigma(bc_1ij)$$
$$(AB)C=\Sigma(ab_1c_2ijk)=A(BC)=ABC$$

which is the general expression of the associative principle.

52

JAMES JOSEPH SYLVESTER
(1814–97)

JAMES JOSEPH SYLVESTER, one of the foremost mathematicians of the nineteenth century, an innovator in the theory of numbers, analysis, differential equations and higher algebra, was born in London, England, on 13 September 1814. His mathematical genius was early in evidence and long lived. At the age of fifteen he won a prize of five hundred dollars for his solution of a problem proposed to him by the Directors of the Lotteries Contractors of the United States. At the age of eighty-two, his talent, zeal and vision undimmed, he was creating mathematics in number theory.

At the age of fourteen, Sylvester studied for a semester under Professor De Morgan in the University of London. He then entered the Royal Institution at Liverpool where he was so outstanding in mathematics that a special class was created for him. About two years later, he entered St John's College, Cambridge. Sylvester graduated at St John's College in 1837 as Second Wrangler. Nevertheless, he was denied a degree because, as a Jew, he was unable to pass the prescribed test of faith. Later, he was granted a degree at the University of Dublin, and in 1871 he also received his degree at Cambridge. At the unusually early age of twenty-five, Sylvester was elected Fellow of the Royal Society. After some fruitless attempts to secure a teaching position, Sylvester worked for a period of ten years as an actuary for an insurance company in London. During this time he also studied law, and he was called to the Bar in 1850. Sylvester and Cayley met through their activities connected with the study of law. Their lifelong friendship grew out of their common interest in mathematics and the mutual admiration and respect of each for the extraordinary capacities of the other. Sylvester's interest in pure mathematics having been rekindled through his discussions with Cayley, he applied for and received an appointment as professor of mathematics at the Royal Military Academy at Woolwich. Some sixteen years later, he was retired. 'Superannuated' at Woolwich at the age of fifty-six, Sylvester was actually on the threshold of many more years of vital activity. In the next few years he wrote on a variety of topics, including problems of link-motion. He invented a skew pantograph, and he published a pamphlet, *The Laws of Verse*, on the construction of poetry. In 1875 he became the first professor of mathematics at the new Johns Hopkins University at Baltimore, Md, where for six years he was surrounded by numerous hard-working and happy students. In 1876 he founded and became the editor of the *American Journal of Mathematics*. In 1883, at the age of seventy, he returned to England to fill the Savilian chair of geometry at Oxford University.

A vigorous and stimulating personality, Sylvester was an accomplished linguist with considerable talent in poetry and music. An eloquent speaker and writer, he is known to have held an audience spellbound for an hour and a half merely with the material introductory to the main text of his lecture. His mathematical papers were published in scientific periodicals all over the world. He was prolific in the creation of new mathematics and of new terminology as it was needed. He introduced almost all of the terminology in the theory of invariants. Terms in common use today such as invariant, covariant, Hessian, contravariant, combinant, commutant, con-comitant, are but a few of those due to him. In his generous appreciation of the work of others, in his philosophical view of mathematics as an unlimited field of endeavour for the imagination and for invention, Sylvester encouraged his students, colleagues and readers, imparting to them his boundless enthusiasm, his method, and his power.

PRESIDENTIAL ADDRESS TO SECTION 'A' OF THE BRITISH ASSOCIATION[1]

From The Collected Papers of James Joseph Sylvester, Vol. II

It is said of a great party leader and orator in the House of Lords that, when lately requested to make a speech at some religious or charitable (at all events a non-political) meeting, he declined to do

1. The Address was also reprinted by the Author in a volume issued by Longmans, Green and Co., London, 1870, of which the earlier portion deals with the Laws of Verse. Some additional notes to the Address there given are reproduced at the end of this volume.

so on the ground that he could not speak unless he saw an adversary before him – somebody to attack or reply to. In obedience to a somewhat similar combative instinct, I set to myself the task of considering certain recent utterances of a most distinguished member of this Association, one whom I no less respect for his honesty and public spirit than I admire for his genius and eloquence, but from whose opinions on a subject which he has not studied I feel constrained to differ. Goethe has said

> *Verstaendige Leute kannst du irren sehn*
> *In Sachen, naemlich, die sie nicht verstehn.*[1]

I have no doubt that had my distinguished friend, the probable President-elect of the next Meeting of the Association applied his uncommon powers of reasoning, induction, comparison, observation and invention to the study of mathematical science he would have become as great a mathematician as he is now a biologist; indeed he has given public evidence of his ability to grapple with the practical side of certain mathematical questions; but he has not made a study of mathematical science as such, and the eminence of his position and the weight justly attaching to his name render it only the more imperative that any assertions proceeding from such a quarter, which may appear to me erroneous, or so expressed as to be conducive to error, should not remain unchallenged or be passed over in silence.

He says 'mathematical training is almost purely deductive. The mathematician starts with a few simple propositions, the proof of which is so obvious that they are called self-evident, and the rest of his work consists of subtle deductions from them. The teaching of languages, at any rate as ordinarily practised, is of the same general nature – authority and tradition furnish the data, and the mental operations are deductive.' It would seem from the above somewhat singularly juxtaposed paragraphs that, according to Professor Huxley, the business of the mathematical student is from a limited number of propositions (bottled up and labelled ready for future use) to deduce any required result by a process of the same general nature as a student of language employs in

1. Understanding people you may see erring – in those things, to wit, which they do not understand.

declining and conjugating his nouns and verbs – that to make out a mathematical proposition and to construe or parse a sentence are equivalent or identical mental operations. Such an opinion scarcely seems to need serious refutation. The passage is taken from an article in *Macmillan's Magazine* for June last, entitled 'Scientific Education – Notes of an After-dinner Speech', and I cannot but think would have been couched in more guarded terms by my distinguished friend had his speech been made *before* dinner instead of *after*.

The notion that mathematical truth rests on the narrow basis of a limited number of elementary propositions from which all others are to be derived by a process of logical inference and verbal deduction, has been stated still more strongly and explicitly by the same eminent writer in an article of even date with the preceding in the *Fortnightly Review*, where we are told that 'Mathematics is that study which knows nothing of observation, nothing of experiment, nothing of induction, nothing of causation.' I think no statement could have been made more opposite to the undoubted facts of the case, that mathematical analysis is constantly invoking the aid of new principles, new ideas and new methods, not capable of being defined by any form of words, but springing direct from the inherent powers and activity of the human mind, and from continually renewed introspection of that inner world of thought of which the phenomena are as varied and require as close attention to discern as those of the outer physical world (to which the inner one in each individual man may, I think, be conceived to stand in somewhat the same general relation of correspondence as a shadow to the object from which it is projected, or as the hollow palm of one hand to the closed fist which it grasps of the other), that it is unceasingly calling forth the faculties of observation and comparison, that one of its principal weapons is induction, that it has frequent recourse to experimental trial and verification, and that it affords a boundless scope for the exercise of the highest efforts of imagination and invention.

. . . Were it not unbecoming to dilate on one's personal experience, I could tell a story of almost romantic interest about my own latest researches in a field where Geometry, Algebra and the Theory of Numbers melt in a surprising manner into one another, like sunset tints or the colours of the dying dolphin, 'the last still loveliest' (a sketch of which has just appeared in the *Proceedings of the London Mathematical Society*[1]), which would very strikingly illustrate how much observation, divination, induction, experimental trial and verification, causation, too (if that means, as I suppose it must, mounting from phenomena to their reasons or causes of being), have to do with the work of the mathematician. In the face of these facts, which every analyst in this room or out of it can vouch for out of its own knowledge and personal experience, how can it be maintained, in the words of Professor Huxley, who, in this instance, is speaking of the sciences as they are in themselves and without any reference to scholastic discipline, that Mathematics 'is that study which knows nothing of observation, nothing of induction, nothing of experiment, nothing of causation'?

I, of course, am not so absurd as to maintain that the habit of observation of external nature will be best or in any degree cultivated by the study of mathematics, at all events as that study is at present conducted; and no one can desire more earnestly than myself to see natural and experimental science introduced into our schools as a primary and indispensable branch of education: I think that that study and mathematical culture should go on hand in hand together, and that they would greatly influence each other for their mutual good. I should rejoice to see mathematics taught with that life and animation which the presence and example of her young and buoyant sister could not fail to impart, short roads preferred to long ones, Euclid honourably shelved or buried 'deeper than did ever plummet sound' out of the schoolboy's reach, morphology introduced into the elements of Algebra – projection, correlation and motion accepted as aids to geometry – the mind of the student quickened and elevated and his faith awakened by early initiation into the ruling ideas of polarity, continuity, infinity and familiarization with the doctrine of the imaginary and inconceivable.

It is this living interest in the subject which is so wanting in our traditional and medieval modes of teaching. In France, Germany

1. Under the title of 'Outline Trace of the Theory of Reducible Cyclodes'.

and Italy, everywhere where I have been on the Continent, mind acts direct on mind in a manner unknown to the frozen formality of our academic institutions, schools of thought and centres of real intellectual cooperation exist; the relation of master and pupil is acknowledged as a spiritual and a lifelong tie, connecting successive generations of great thinkers with each other in an unbroken chain, just in the same way as we read, in the catalogue of our French Exhibition, or of the Salon at Paris, of this man or that being the pupil of one great painter or sculptor and the master of another. When followed out in this spirit, there is no study in the world which brings into more harmonious action all the faculties of the mind than the one of which I stand here as the humble representative, there is none other which prepares so many agreeable surprises for its followers, more wonderful than the changes in the transformation-scene of a pantomime, or, like this, seems to raise them, by successive steps of initiation, to higher and higher states of conscious intellectual being.

This accounts, I believe, for the extraordinary longevity of all the greatest masters of the Analytical art, the Dii Majores of the mathematical Pantheon. Leibnitz lived to the age of 70; Euler to 76; Lagrange to 77; Laplace to 78; Gauss to 78; Newton, the crown and glory of his race, to 85; Archimedes, the nearest akin, probably, to Newton in genius, was 75, and might have lived on to be 100, for aught we can guess to the contrary, when he was slain by the impatient and illmannered sergeant, sent to bring him before the Roman general, in the full vigour of his faculties, and in the very act of working out a problem; Pythagoras, in whose school, I believe, the word mathematician (used, however, in a somewhat wider than its present sense) originated, the second founder of geometry, the inventor of the matchless theorem which goes by his name, the precognizer of the undoubtedly mis-called Copernican theory, the discoverer of the regular solids and the musical canon, who stands at the very apex of this pyramid of fame, (if we may credit the tradition) after spending 22 years studying in Egypt, and 12 in Babylon, opened school when 56 or 57 years old in Magna Graecia, married a young wife when past 60, and died, carrying on his work with energy unspent to the last, at the age of 99. The mathematician lives long and lives young; the wings of his soul do not early drop off, nor do its pores become clogged with the earthy particles blown from the dusty highways of vulgar life.

... Time was when all the parts of the subject were dissevered, when algebra, geometry and arithmetic either lived apart or kept up cold relations of acquaintance confined to occasional calls upon one another; but that is now at an end; they are drawn together and are constantly becoming more and more intimately related and connected by a thousand fresh ties, and we may confidently look forward to a time when they shall form but one body with one soul. Geometry formerly was the chief borrower from arithmetic and algebra, but it has since repaid its obligations with abundant usury; and if I were asked to name, in one word, the pole-star round which the mathematical firmament revolves, the central idea which pervades as a hidden spirit the whole corpus of mathematical doctrine, I should point to Continuity as contained in our notions of space, and say, it is this, it is this! Space is the *Grand Continuum* from which, as from an inexhaustible reservoir, all the fertilizing ideas of modern analysis are derived; and as Brindley, the engineer, once allowed before a parliamentary committee that, in his opinion, rivers were made to feed navigable canals, I feel almost tempted to say that one principal reason for the existence of space, or at least one principal function which it discharges, is that of feeding mathematical invention. Everybody knows what a wonderful influence geometry has exercised in the hands of Cauchy, Puiseux, Riemann, and his followers Clebsch, Gordan, and others, over the very form and presentment of the modern calculus, and how it has come to pass that the tracing of curves, which was once to be regarded as a puerile amusement, or at best useful only to the architect or decorator, is now entitled to take rank as a high philosophical exercise, inasmuch as every new curve or surface, or other circumscription of space is capable of being regarded as the embodiment of some specific organized system of continuity.

APPENDIX

Induction and analogy are the special characteristics of modern mathematics, in which theorems have given place to theories, and no truth is regarded otherwise than as a link in an infinite chain. *'Omne exit in infinitum'* is their favourite motto and accepted axiom. No mathematician now-a-days sets any store on the discovery of isolated theorems, except as affording hints of an unsuspected new sphere of thought, like meteorites detached from some undiscovered planetary orb of speculation. The form, as well as matter, of mathematical science, as must be the case in any true living organic science, is in a constant state of flux, and the position of its centre of gravity is liable to continual change. At different periods in its history defined, with more or less accuracy, as the science of number or quantity, or extension or operation or arrangement, it appears at present to be passing through a phase in which the development of the notion of Continuity plays the leading part. In exemplification of the generalizing tendency of modern mathematics, take so simple a fact as that of two straight lines or two planes being incapable of including 'a space'. When analysed this statement will be found to resolve itself into the assertion that if two out of the four triads that can be formed with four points lie respectively *in directo*, the same must be true of the remaining two triads; and that if two of the five tetrads that can be formed with five points lie respectively *in plano*, the remaining three tetrads (subject to a certain obvious exception) must each do the same. This, at least, is one way of arriving at the notion of an unlimited rectilinear and planar schema of points. The two statements above made, translated into the language of determinants, immediately suggest as their generalized expression my great 'Homaloidal Law', which affirms that the vanishing of a certain specifiable number of minor determinants of a given order of any matrix (i.e. rectangular array of quantities) implies the simultaneous evanescence of all the rest of that order. I made (*inter alia*) a beautiful application of this law (which is, I believe, recorded in Mr Spottiswoode's valuable treatise on Determinants, but where besides I know not) to the establishment of the well-known relations, wrung out with so much difficulty by Euler, between the cosines of the nine angles, which two sets of rectangular axes in space make with one another. This is done by contriving and constructing a matrix such that the six known equations connecting the nine cosines taken both ways in sets of threes shall be expressed by the evanescence of six of its minors; the simultaneous evanescence of the remaining minors given by the Homaloidal Law will then be found to express the relations in question (which, Euler has put on record, it drove him almost to despair to obtain), but which are thus obtained by a simple process of inspection and reading off, without any labour whatever. The fact that such a law, containing in a latent form so much refined algebra, and capable of such interesting immediate applications, should present itself to the *observation* merely as the extended expression of the ground of the possibility of our most elementary and seemingly intuitive conceptions concerning the right line and plane, has often filled me with amazement to reflect upon.

ADDRESS ON COMMEMORATION DAY AT JOHNS HOPKINS UNIVERSITY, 1877

It is with unaffected feelings of diffidence that I present myself before you, for, save on rare and exceptional occasions, it has not been my wont to make my voice heard in public assemblies. I know, indeed, and can conceive of no pursuit so antagonistic to the cultivation of the oratorical faculty – that faculty so prevalent in this country that the possession of it is not regarded as a gift but the want of it as a defect – as the study of Mathematics. An eloquent mathematician must, from the nature of things, ever remain as rare a phenomenon as a talking fish, and it is certain that the more anyone gives himself up to the study of oratorical effect the less will he find himself in a fit state of mind to mathematicize. It is the constant aim of the mathematician to reduce all his expressions to their lowest terms, to retrench every superfluous word and phrase, and to condense the Maximum of meaning into the Minimum of language. He has to turn his eye ever inwards, to see everything in its dryest light, to train and inure himself to a habit of internal and impersonal reflection and elaboration of abstract thought, which makes it most difficult for him to touch or enlarge upon any of those themes which appeal to the emotional nature of his fellow-men. When called upon to speak in public he feels as a man might do who has passed all his life in peering through a microscope, and is suddenly called upon to take charge of an astronomical observatory. He has to go out of himself, as it were, and change the habitual focus of his vision.

One of our great English judges observed on some occasion, when he was outvoted by his brethren on the bench (or, perchance, it may have been the twelfth outstanding juryman, who protested that never before in his life had he been shut up with eleven other such obstinate men) that 'opinions ought to count by weight rather than by number', and so I would say that the good done by a university is to be estimated not so much by the mere number of its members as by the spirit which actuates and the work that is done by them. When I hear, as I have heard, of members of this University, only hoping to be enabled to keep body and soul together in order that they may continue to enjoy the advantages which it affords, it may be for a decade of years to come; when I find classes diligently attending lectures on the most abstruse branches of scholarship and science, remote from all the avenues which lead to fortune or public recognition; when I observe the earnestness with which our younger members address themselves to the studies of the place, and the absence of all manifestations of disorder or levity, without the necessity for the exercise of any external restraint, it seems to me that this establishment, even in its cradle, better responds to what its name should import, more fully embodies the true idea of a university, than if its halls and lecture-rooms swarmed with hundreds of idle and indifferent students, or with students, diligent, indeed, but working not from a pure love of knowledge, not even for the chaplet of olive, or the laurel crown, but for high places in examinations, for marks, as we say in England, the counters or vouchers to enable their fortunate possessor to draw large stakes out of the pool of sinecure fellowships or lucrative civil appointments.

At this moment I happen to be engaged in a research of fascinating interest to myself, and which, if the day only responds to the promise of its dawn, will meet, I believe, a sympathetic response from the Professors of our divine Algebraical art wherever scattered through the world.

There are things called Algebraical Forms. Professor Cayley calls them Quantics. These are not, properly speaking, Geometrical Forms, although capable, to some extent, of being embodied in them, but rather schemes of processes, or of operations for forming, for calling into existence, as it were, Algebraic quantities.

To every such Quantic is associated an infinite variety of other forms that may be regarded as engendered from and floating, like an atmosphere, around it – but infinite in number as are these derived existences, these emanations from the parent form, it is found that they admit of being obtained by composition, by mixture, so to say, of a certain limited number of fundamental forms, standard rays, as they might be termed in the Algebraic Spectrum of the Quantic to which they belong. And, as it is a leading pursuit of the Physicists of the present day to ascertain the fixed lines in the spectrum of every chemical substance, so it is the aim and object of a great school of mathematicians to make out the fundamental derived forms, the Covariants and Invariants, as they are called, of these Quantics.

This is the kind of investigation in which I have for the last month or two been immersed, and which I entertain great hopes of bringing to a successful issue. Why do I mention it here? It is to illustrate my opinion as to the invaluable aid of teaching to the teacher, in throwing him back upon his own thoughts and leading him to evolve new results from ideas that would have otherwise remained passive or dormant in his mind.

But for the persistence of a student of this University in urging upon me his desire to study with me the modern Algebra I should never have been led into this investigation; and the new facts and principles which I have discovered in regard to it (important facts, I believe), would, so far as I am concerned, have remained still hidden in the womb of time. In vain I represented to this inquisitive student that he would do better to take up some other subject lying less off the beaten track of study, such as the higher parts of the Calculus or Elliptic Functions, or the theory of Substitutions, or I wot not what besides. He stuck with perfect respectfulness, but with invincible pertinacity, to his point. He would have the New Algebra (Heaven knows where he had heard about it, for it is almost unknown in this continent), that or nothing. I was obliged to yield, and what was the consequence? In trying to throw light upon an obscure explanation in our textbook, my brain took fire, I plunged with re-quickened zeal into a subject which I had for years abandoned, and found food for thoughts which have engaged my attention for a considerable time past, and will probably occupy all my powers of contemplation advantageously for several months to come.

I remember, too, how, in like manner, when a very young professor, fresh from the University of Cambridge, in the act of teaching a private pupil the simpler parts of Algebra, I discovered the principle now generally adopted into the higher text-books, which goes by the name of the 'Dialytic Method of Elimination'. So much for the reaction of the student on the teacher.[1] May the

___

1. Not to speak of professor on professor. Thus it was in order to be able to meet the threatened interrogatories of my valued colleague, the irrepressible Mr Rowland, that I was led, on my return passage to England last summer, to look into Professor Clerk Maxwell's extremely valuable, but ill-digested and somewhat unduly pretentious treatise on Electricity and Magnetism, which led to my theory of the Bipotential, and to my writing the paper published in the *Philosophical Magazine* for October last, which ought to have the effect of causing the author to rewrite one of his leading chapters on Statical Electricity.

I have at present a class of from eight to ten students attending my lectures on the Modern Higher Algebra. One of them, a young engineer, engaged from eight in the morning to six at night in the duties of his office, with an interval of an hour and a half for his dinner or lectures, has furnished me with the best proof, and the best expressed, I have ever seen of what I call the Law of Concomitant Interchange, applicable to permutation systems, i.e. the law which affirms that every complete set of permuted elements may be separated into two parts, or if we like to say so, be presented in the form of a diptych with two precisely similar Alae, such that a single interchange between any two elements is accompanied with a total interchange between the two Alae. This is the theorem which lies at the basis of the great theory of simple equations, which every school-boy is supposed to understand, but which was not really made out until a bevy of great Mathematicians, including Leibnitz, Laplace and Lagrange, had turned their attention to the subject. Jacobi, I have read somewhere, used to say that if he at all excelled other mathematicians, it was chiefly due to his greater facility in manipulating simple equations that he owed it. The same Jacobi, who, I remember, visited our English Cambridge, and so much relished the Trinity audit ale which he drank there, and who once being asked whether he was brother to time never come when the two offices of teaching and research-ing shall be sundered in this University! So long as man remains a gregarious and sociable being, he cannot cut himself off from the gratification of the instinct of imparting what he is learning, of propagating through others the ideas and impressions seething in his own brain, without stunting and atrophying his moral nature and drying up the surest sources of his future intellectual replenishment.

I should be sorry to suppose that I was to be left in sole possession of so vast a field as is occupied by modern mathe-matics. Mathematics is not a book confined within a cover and bound between brazen clasps, whose contents it needs only patience to ransack; it is not a mine, whose treasures may take long to reduce into possession, but which fill only a limited number of veins and lodes; it is not a soil, whose fertility can be exhausted by the yield of successive harvests; it is not a continent or an ocean, whose area can be mapped out and its contour de-fined: it is limitless as that space which it finds too narrow for its aspirations; its possibilities are as infinite as the worlds which are forever crowding in and multiplying upon the astronomer's gaze; it is as incapable of being restricted within assigned boundaries or being reduced to definitions of permanent validity, as the con-sciousness, the life, which seems to slumber in each nomad, in every atom of matter, in each leaf and bud and cell, and is for-ever ready to burst forth into new forms of vegetable and animal existence.

the eminent physicist, Professor Jacobi, of St Petersburg, replied: 'Quite the contrary – he is my brother.' And *apropos* of the zeal of the student in question, let me mention for the benefit of my English friends, I have been agreeably surprised to find how widely diffused a spirit there exists in this country of disinterested love of learning. Out of Italy, especially Tuscany, where my friend Enrico Betti, as I had the opportunity of observing, and in his own country too, where no man is supposed to be a prophet, the neighbourhood of Pistoja, as a Professor is more influential, more honoured and courted than he could be if he were a rich Marquis, I believe there is no nation in the world where ability with character counts for so much, and the mere possession of wealth (in spite of all that we hear about the Almighty dollar), for so little as in America, with exception it may be of certain of the Trans-Atlantic cities, which are really only colonies and emporiums for the trading classes of Europe.

53

ARTHUR CAYLEY
(1821–95)

ARTHUR CAYLEY, a prolific writer in mathematics, through whose fertile imagination and genius, new and vital branches of modern mathematics were created, was born at Richmond, Surrey, England, on 16 August, 1821. His family was of old origin, dating back to William the Conqueror. In 1829, his father, a merchant in the Russian trade, on retiring from business, returned to England and thus Cayley's early education was begun in England. Indications of his mathematical genius were so impressive that Cayley, who had been intended for a place in his father's former business, was sent instead to Cambridge. He was a distinguished student from the first. In 1842 he was chosen Senior Wrangler without the *viva voce* tests which were a customary part of the Tripos. Cayley was called to the Bar in 1848, but his chief interest continued to be in mathematics and he always reserved a substantial portion of his time for study and research in this field. In the fourteen years during which he practised law, Cayley wrote between two hundred and three hundred mathematical papers. Many of his most famous memoirs belong to this period. These include memoirs on quantics, contributions to the theory of symmetric functions of the roots of an equation, calculations connected with planetary and lunar theories, reports on theoretical dynamics and his important work on matrices. Although he was the author of only one book, *A Treatise on Elliptic Functions*, Cayley published more than eight hundred papers which appeared in a steady stream throughout his life.

In 1863 Cayley was elected to the newly constituted Sadlerian Professorship of Pure Mathematics at Cambridge. He married Susan Moline that same year and settled in Cambridge to a life of serenity, devoting himself to mathematical research and the quiet round of activity in the University. Cayley was highly respected for his fairmindedness and the soundness of his judgement. He was frequently called upon by the University and the scientific societies for his opinion on legal matters. In 1881 he accepted an invitation to lecture at Johns

Hopkins University where his friend Sylvester was professor of mathematics. Apart from this interlude, his tenure at Cambridge was uninterrupted until his retirement in 1892. His health failed gradually and he died on 26 January, 1895.

Numerous honours were conferred upon Cayley at home and on the continent by universities and academies of science, in recognition of his many fundamental contributions to modern pure mathematics. Cayley belonged to the group of mathematicians who, working from a small number of fundamental concepts, developed modern algebraic and geometrical systems. Cayley is credited with the invention of an n-dimensional geometry. He was the founder, with Sylvester, of the theory of invariants. The theory of matrices, extensively developed within the last hundred years, was originated by Cayley in his famous paper, *A Memoir on the Theory of Matrices*, published in the *Philosophical Transactions of the Royal Society of London* in 1858. Almost every branch of pure mathematics was enriched by his contributions.

A MEMOIR ON
THE THEORY OF MATRICES

From the Philosophical Transactions of the Royal Society of London *vol. cxlviii. for the year 1858, pp. 17–37. Received 10 December, 1857, Read 14 January, 1858*

The term matrix might be used in a more general sense, but in the present memoir I consider only square and rectangular matrices, and the term matrix used without qualification is to be understood as meaning a square matrix; in this restricted sense, a set of quantities arranged in the form of a square, e.g.

$$\left(\begin{array}{ccc} a, & b, & c \\ a', & b', & c' \\ a'', & b'', & c'' \end{array} \right)$$

is said to be a matrix. The notion of such a matrix arises naturally from an abbreviated notation for a set of linear equations, viz. the equations

$$X = ax + by + cz,$$
$$Y = a'x + b'y + c'z,$$
$$Z = a''x + b''y + c''z,$$

may be more simply represented by

$$(X, Y, Z) = \begin{pmatrix} a, & b, & c \\ a', & b', & c' \\ a'', & b'', & c'' \end{pmatrix}(x, y, z),$$

and the consideration of such a system of equations leads to most of the fundamental notions in the theory of matrices. It will be seen that matrices (attending only to those of the same order) comport themselves as single quantities; they may be added, multiplied or compounded together, etc.: the law of the addition of matrices is precisely similar to that for the addition of ordinary algebraical quantities; as regards their multiplication (or composition), there is the peculiarity that matrices are not in general convertible; it is nevertheless possible to form the powers (positive or negative, integral or fractional) of a matrix, and thence to arrive at the notion of a rational and integral function, or generally of any algebraical function, of a matrix. I obtain the remarkable theorem that any matrix whatever satisfies an algebraical equation of its own order, the coefficient of the highest power being unity, and those of the other powers functions of the terms of the matrix, the last coefficient being in fact the determinant; the rule for the formation of this equation may be stated in the following condensed form, which will be intelligible after a perusal of the memoir, viz. the determinant, formed out of the matrix diminished by the matrix considered as a single quantity involving the matrix unity, will be equal to zero. The theorem shows that every rational and integral function (or indeed every rational function) of a matrix may be considered as a rational and integral function, the degree of which is at most equal to that of the matrix, less unity; it even shows that in a sense, the same is true with respect to any algebraical function whatever of a matrix. One of the applications of the theorem is the finding of the general expression of the matrices which are convertible with a given matrix. The theory of rectangular matrices appears much less important than that of square matrices, and I have not entered into it further than by showing how some of the notions applicable to these may be extended to rectangular matrices.

1. For conciseness, the matrices written down at full length will

in general be of the order 3, but it is to be understood that the definitions, reasonings and conclusions apply to matrices of any degree whatever. And when two or more matrices are spoken of in connexion with each other, it is always implied (unless the contrary is expressed) that the matrices are of the same order.

2. The notation

$$\left(\begin{matrix} a, & b, & c \\ a', & b', & c' \\ a'', & b'', & c'' \end{matrix} \right)(x, y, z)$$

represents the set of linear functions

$((a, b, c\,(x, y, z), (a', b', c'\,(x, y, z), (a'', b'', c''\,(x, y, z)),$

so that calling these (X, Y, Z), we have

$$(X, Y, Z) = \left(\begin{matrix} a, & b, & c \\ a', & b', & c' \\ a'', & b'', & c'' \end{matrix} \right)(x, y, z)$$

and, as remarked above, this formula leads to most of the fundamental notions in the theory.

3. The quantities (X, Y, Z) will be identically zero, if all the terms of the matrix are zero, and we may say that

$$\left(\begin{matrix} 0, & 0, & 0 \\ 0, & 0, & 0 \\ 0, & 0, & 0 \end{matrix} \right)$$

is the matrix zero.

Again, (X, Y, Z) will be identically equal to (x, y, z), if the matrix is

$$\left(\begin{matrix} 1, & 0, & 0 \\ 0, & 1, & 0 \\ 0, & 0, & 1 \end{matrix} \right)$$

and this is said to be the matrix unity. We may of course, when for distinctness it is required, say, the matrix zero, or (as the case may be) the matrix unity *of such an order*. The matrix zero may for the most part be represented simply by 0, and the matrix unity by 1.

4. The equations

$$(X, Y, Z) = \left(\begin{matrix} a, & b, & c \\ a', & b', & c' \\ a'', & b'', & c'' \end{matrix} \right)(x, y, z), \quad (X', Y', Z') = \left(\begin{matrix} \alpha, & \beta, & \gamma \\ \alpha', & \beta', & \gamma' \\ \alpha'', & \beta'', & \gamma'' \end{matrix} \right)(x, y, z)$$

give

$$(X+X', \ Y+Y', \ Z+Z')=\begin{pmatrix} a+\alpha, & b+\beta, & c+\gamma \\ a'+\alpha', & b'+\beta', & c'+\gamma' \\ a''+\alpha'', & b''+\beta'', & c''+\gamma'' \end{pmatrix}\!(x, y, z)$$

and this leads to

$$\begin{pmatrix} a+\alpha, & b+\beta, & c+\gamma \\ a'+\alpha', & b'+\beta', & c'+\gamma' \\ a''+\alpha'', & b''+\beta'', & c''+\gamma'' \end{pmatrix}=\begin{pmatrix} a, & b, & c \\ a', & b', & c' \\ a'', & b'', & c'' \end{pmatrix}+\begin{pmatrix} \alpha, & \beta, & \gamma \\ \alpha', & \beta', & \gamma' \\ \alpha'', & \beta'', & \gamma'' \end{pmatrix}$$

as a rule for the addition of matrices; that for their subtraction is of course similar to it.

5. A matrix is not altered by the addition or subtraction of the matrix zero, that is, we have $M\pm0=M$.

The equation $L=M$, which expresses that the matrices L, M are equal, may also be written in the form $L-M=0$, i.e. the difference of two equal matrices is the matrix zero.

6. The equation $L=-M$, written in the form $L+M=0$, expresses that the sum of the matrices L, M is equal to the matrix zero, the matrices so related are said to be *opposite* to each other; in other words, a matrix the terms of which are equal but opposite in sign to the terms of a given matrix, is said to be opposite to the given matrix.

7. It is clear that we have $L+M=M+L$, that is, the operation of addition is commutative, and moreover that $(L+M)+N=L+(M+N)=L+N$, that is, the operation of addition is also associative.

8. The equation

$$(X, \ Y, \ Z)=\begin{pmatrix} a, & b, & c \\ a', & b', & c' \\ a'', & b'', & c'' \end{pmatrix}\!(mx, my, mz)$$

written under the forms

$$(X, \ Y, \ Z)=m\begin{pmatrix} a, & b, & c \\ a', & b', & c' \\ a'', & b'', & c'' \end{pmatrix}\!(x, y, z)=\begin{pmatrix} ma, & mb, & mc \\ ma', & mb', & mc' \\ ma'', & mb'', & mc'' \end{pmatrix}\!(x, y, z)$$

gives

$$m\begin{pmatrix} a, & b, & c \\ a', & b', & c' \\ a'', & b'', & c'' \end{pmatrix}=\begin{pmatrix} ma, & mb, & mc \\ ma', & mb', & mc' \\ ma'', & mb'', & mc'' \end{pmatrix}$$

N

as the rule for the multiplication of a matrix by a single quantity. The multiplier m may be written either before or after the matrix, and the operation is therefore commutative. We have it is clear $m(L+M)=mL+mM$, or the operation is distributive.

9. The matrices L and mL may be said to be similar to each other; in particular, if $m=1$, they are equal, and if $m=-1$, they are opposite.

10. We have, in particular,

$$m\begin{pmatrix} 1, & 0, & 0 \\ 0, & 1, & 0 \\ 0, & 0, & 1 \end{pmatrix}=\begin{pmatrix} m, & 0, & 0 \\ 0, & m, & 0 \\ 0, & 0, & m \end{pmatrix}$$

or replacing the matrix on the left-hand side by unity, we may write

$$m=\begin{pmatrix} m, & 0, & 0 \\ 0, & m, & 0 \\ 0, & 0, & m \end{pmatrix};$$

the matrix on the right-hand side is said to be the single quantity m considered as *involving the matrix unity*.

11. The equations

$$(X, Y, Z)=\begin{pmatrix} a, & b, & c \\ a', & b', & c' \\ a'', & b'', & c'' \end{pmatrix}(x, y, z), \quad (x, y, z)=\begin{pmatrix} \alpha, & \beta, & \gamma \\ \alpha', & \beta', & \gamma' \\ \alpha'', & \beta'', & \gamma'' \end{pmatrix}(\xi, \eta, \zeta),$$

give

$$(X, Y, Z)=\begin{pmatrix} A, & B, & C \\ A', & B', & C' \\ A'', & B'', & C'' \end{pmatrix}(\xi, \eta, \zeta)=\begin{pmatrix} a, & b, & c \\ a', & b', & c' \\ a'', & b'', & c'' \end{pmatrix}\begin{pmatrix} \alpha, & \beta, & \gamma \\ \alpha', & \beta', & \gamma' \\ \alpha'', & \beta'', & \gamma'' \end{pmatrix}(\xi, \eta, \zeta),$$

and thence, substituting for the matrix

$$\begin{pmatrix} A, & B, & C \\ A', & B', & C' \\ A'', & B'', & C'' \end{pmatrix}$$

its value, we obtain

$$\begin{pmatrix} ((a, & b, & c)(\alpha, \alpha', \alpha''), & (a, & b, & c)(\beta, \beta', \beta''), & (a, & b, & c)(\gamma, \gamma', \gamma'')) \\ (a', & b', & c')(\alpha, \alpha', \alpha''), & (a', & b', & c')(\beta, \beta', \beta''), & (a', & b', & c')(\gamma, \gamma', \gamma'') \\ (a'', & b'', & c'')(\alpha, \alpha', \alpha''), & (a'', & b'', & c'')(\beta, \beta', \beta''), & (a'', & b'', & c'')(\gamma, \gamma', \gamma'') \end{pmatrix}$$

$$=\begin{pmatrix} a, & b, & c \\ a', & b', & c' \\ a'', & b'', & c'' \end{pmatrix}\begin{pmatrix} \alpha, & \beta, & \gamma \\ \alpha', & \beta', & \gamma' \\ \alpha'', & \beta'', & \gamma'' \end{pmatrix}$$

as the rule for the multiplication or composition of two matrices. It is to be observed, that the operation is not a commutative one; the component matrices may be distinguished as the first or further component matrix, and the second or nearer component matrix, and the rule of composition is as follows, viz. any *line* of the compound matrix is obtained by combining the corresponding *line* of the first or further component matrix successively with the several *columns* of the second or nearer compound matrix.

[We may conveniently write

$$
\begin{array}{c|ccc}
 & (\alpha, \alpha', \alpha''), & (\beta, \beta', \beta''), & (\gamma, \gamma', \gamma'') \\
\hline
(a, \ b, \ c\ ) & ,, & ,, & ,, \\
(a', \ b', \ c') & ,, & ,, & ,, \\
(a'', \ b'', \ c'') & ,, & ,, & ,, \\
\end{array}
$$

to denote the left-hand side of the last preceding equation.]

12. A matrix compounded, either as first or second component matrix, with the matrix zero, gives the matrix zero. The case where any of the terms of the given matrix are infinite is of course excluded.

13. A matrix is not altered by its composition, either as first or second component matrix, with the matrix unity. It is compounded either as first or second component matrix, with the single quantity m considered as involving the matrix unity, by multiplication of all its terms by the quantity m: this is in fact the before-mentioned rule for the multiplication of a matrix by a single quantity, which rule is thus seen to be a particular case of that for the multiplication of two matrices.

14. We may in like manner multiply or compound together three or more matrices: the order of arrangement of the factors is of course material, and we may distinguish them as the first or furthest, second, third, etc., and last or nearest component matrices: any two consecutive factors may be compounded together and replaced by a single matrix, and so on until all the matrices are compounded together, the result being independent of the particular mode in which the composition is effected; that is we have $L.MN=LM.N=LMN$, $LM.NP=L.MN.P$, etc., or

the operation of multiplication, although, as already remarked, not commutative, is associative.

15. We thus arrive at the notion of a positive and integer power L^p of a matrix L, and it is to be observed that the different powers of the same matrix are convertible. It is clear also that p and q being positive integers, we have $L^p.L^q=L^{p+q}$, which is the theorem of indices for positive integer powers of a matrix.

16. The last-mentioned equation, $L^p.L^q=L^{p+q}$, assumed to be true for all values whatever of the indices p and q, leads to the notion of the powers of a matrix for any form whatever of the index. In particular, $L^p.L^0=L^p$ or $L^0=1$, that is, the 0th power of a matrix is the matrix unity. And then putting $p=1$, $q=-1$, or $p=-1$, $q=1$, we have $L.L^{-1}=L^{-1}.L=1$; that is, L^{-1}, or as it may be termed the inverse or reciprocal matrix, is a matrix which, compounded either as first or second component matrix with the original matrix, gives the matrix unity.

17. We may arrive at the notion of the inverse or reciprocal matrix, directly from the equation

$$(X,\ Y,\ Z)=\begin{pmatrix} a, & b, & c \\ a', & b', & c' \\ a'', & b'', & c'' \end{pmatrix}(x,\ y,\ z),$$

in fact this equation gives

$$(x,\ y,\ z)=\begin{pmatrix} A, & A', & A'' \\ B, & B', & B'' \\ C, & C', & C'' \end{pmatrix}(X,\ Y,\ Z)=\begin{pmatrix} a, & b, & c \\ a', & b', & c' \\ a'', & b'', & c'' \end{pmatrix}^{-1}(X,\ Y,\ Z),$$

and we have, for the determination of the coefficients of the inverse or reciprocal matrix, the equations

$$\begin{pmatrix} A, & A', & A'' \\ B, & B', & B'' \\ C, & C', & C'' \end{pmatrix}\begin{pmatrix} a, & b, & c \\ a', & b', & c' \\ a'', & b'', & c'' \end{pmatrix}=\begin{pmatrix} 1, & 0, & 0 \\ 0, & 1, & 0 \\ 0, & 0, & 1 \end{pmatrix},$$

$$\begin{pmatrix} a, & b, & c \\ a', & b', & c' \\ a'', & b'', & c'' \end{pmatrix}\begin{pmatrix} A, & A', & A'' \\ B, & B', & B'' \\ C, & C', & C'' \end{pmatrix}=\begin{pmatrix} 1, & 0, & 0 \\ 0, & 1, & 0 \\ 0, & 0, & 1 \end{pmatrix},$$

which are equivalent to each other, and either of them is by itself sufficient for the complete determination of the inverse or reciprocal matrix. It is well known that if ∇ denote the determinant, that is, if

$$\nabla = \begin{vmatrix} a, & b, & c \\ a', & b', & c' \\ a'', & b'', & c'' \end{vmatrix}$$

then the terms of the inverse or reciprocal matrix are given by the equations

$$A = \frac{1}{\nabla} \begin{vmatrix} 1, & 0, & 0 \\ 0, & b', & c' \\ 0, & b'', & c'' \end{vmatrix}, \qquad B = \frac{1}{\nabla} \begin{vmatrix} 0, & 1, & 0 \\ a', & 0, & c' \\ a'', & 0, & c'' \end{vmatrix}, \text{ etc.,}$$

or what is the same thing, the inverse or reciprocal matrix is given by the equation

$$\begin{vmatrix} a, & b, & c \\ a', & b', & c' \\ a'', & b'', & c'' \end{vmatrix}^{-1} = \frac{1}{\nabla} \begin{vmatrix} \partial_a \nabla, & \partial_{a'} \nabla, & \partial_{a''} \nabla \\ \partial_b \nabla, & \partial_{b'} \nabla, & \partial_{b''} \nabla \\ \partial_c \nabla, & \partial_{c'} \nabla, & \partial_{c''} \nabla \end{vmatrix}$$

where of course the differentiations must in every case be performed as if the terms a, b, etc., were all of them independent arbitrary quantities.

18. The formula shows, what is indeed clear *a priori*, that the notion of the inverse or reciprocal matrix fails altogether when the determinant vanishes: the matrix is in this case said to be indeterminate, and it must be understood that in the absence of express mention, the particular case in question is frequently excluded from consideration. It may be added that the matrix zero is indeterminate; and that the product of two matrices may be zero, without either of the factors being zero, if only the matrices are one or both of them indeterminate.

19. The notion of the inverse or reciprocal matrix once established, the other negative integer powers of the original matrix are positive integer powers of the inverse or reciprocal matrix, and the theory of such negative integer powers may be taken to be known. The theory of the fractional powers of a matrix will be further discussed in the sequel.

20. The positive integer power L^m of the matrix L may of course be multiplied by any matrix of the same degree: such multiplier however, is not in general convertible with L; and to preserve as

far as possible the analogy with ordinary algebraical functions we may restrict the attention to the case where the multiplier is a single quantity, and such convertibility consequently exists. We have in this manner a matrix cL^m, and by the addition of any number of such terms we obtain a rational and integral function of the matrix L.

21. The general theorem before referred to will be best understood by a complete development of a particular case. Imagine a matrix

$$M = \begin{pmatrix} a, & b \\ c, & d \end{pmatrix},$$

and form the determinant

$$\begin{vmatrix} a-M, & b \\ c & , & d-M \end{vmatrix},$$

the developed expression of this determinant is

$$M^2 - (a+d)M^1 + (ad-bc)M^0;$$

the values of M^2, M^1, M^0 are

$$\begin{pmatrix} a^2+bc, & b(a+d) \\ c(a+d), & d^2+bc \end{pmatrix}, \begin{pmatrix} a, & b \\ c, & d \end{pmatrix}, \begin{pmatrix} 1, & 0 \\ 0, & 1 \end{pmatrix}$$

and substituting these values the determinant becomes equal to the matrix zero, viz. we have

$$\begin{vmatrix} a-M, b \\ c & , d-M \end{vmatrix} = \begin{pmatrix} a^2+bc, & b(a+d) \\ c(a+d), & d^2+bc \end{pmatrix} - (a+d) \begin{pmatrix} a, & b \\ c, & d \end{pmatrix} + (ad-bc) \begin{pmatrix} 1, & 0 \\ 0, & 1 \end{pmatrix}$$

$$= \begin{pmatrix} (a^2+bc)-(a+d)a+(ad-bc), & b(a+d)-(a+d)b \\ c(a+d)-(a+d)c & , & d^2+bc-(a+d)d+ad-bc \end{pmatrix} = \begin{pmatrix} 0, & 0 \\ 0, & 0 \end{pmatrix};$$

that is

$$\begin{vmatrix} a-M, & b \\ c & , & d-M \end{vmatrix} = 0,$$

where the matrix of the determinant is

$$\begin{pmatrix} a, & b \\ c, & d \end{pmatrix} - M \begin{pmatrix} 1, & 0 \\ 0, & 1 \end{pmatrix}$$

that is, it is the original matrix, diminished by the same matrix as a single quantity involving the matrix unity. And this is the general theorem, viz. the determinant, having for its matrix a given matrix less the same matrix considered as a single quantity involving the matrix unity, is equal to zero.

22. The following symbolical representation of the theorem is, I think, worth noticing: let the matrix M, considered as a single quantity, be represented by $\widetilde{M}$, then writing 1 to denote the matrix unity, $\widetilde{M}.1$ will represent the matrix M, considered as a single quantity involving the matrix unity. Upon the like principles of notation, $\widetilde{1}.M$ will represent, or may be considered as representing, simply the matrix M, and the theorem is

$$\text{Det. } (\widetilde{1}.M - \widetilde{M}.1) = 0.$$

23. I have verified the theorem, in the next simplest case of a matrix of the order 3, viz. if M be such a matrix, suppose

$$M = \left(\begin{array}{ccc} a, & b, & c \\ d, & e, & f \\ g, & h, & i \end{array}\right),$$

then the derived determinant vanishes, or we have

$$\begin{vmatrix} a-M, & b & , c \\ d & , e-M, & f \\ g & , h & , i-M \end{vmatrix} = 0,$$

or expanding

$$M^3 - (a+e+i)M^2 + (ei+ia+ae-fh-cg-bd)M$$
$$- (aei+bfg+cdh-afh-bdi-ceg) = 0;$$

but I have not thought it necessary to undertake the labour of a formal proof of the theorem in the general case of a matrix of any degree.

24. If we attend only to the general form of the result, we see that any matrix whatever satisfies an algebraical equation of its own order, which is in many cases the material part of the theorem.

25. It follows at once that every rational and integral function, or indeed every rational function of a matrix, can be expressed as a rational and integral function of an order at most equal to that of the matrix, less unity. But it is important to consider how far or in what sense the like theorem is true with respect to irrational functions of a matrix. If we had only the equation satisfied by the matrix itself, such extension could not be made; but we have besides the equation of the same order satisfied by the irrational function of the matrix, and by means of these two equations, and

the equation by which the irrational function of the matrix is determined, we may express the irrational function as a rational and integral function of the matrix, of an order equal at most to that of the matrix, less unity; such expression will however involve *the coefficients of the equation satisfied by the irrational function*, which are functions (in number equal to the order of the matrix) of the terms, assumed to be unknown, of the irrational function itself. The transformation is nevertheless an important one, as reducing the number of unknown quantities from n^2 (if n be the order of the matrix) down to n. To complete the solution, it is necessary to compare the value obtained as above, with the assumed value of the irrational function, which will lead to equations for the determination of the n unknown quantities.

36. Two matrices such as

$$\begin{pmatrix} a, & b \\ c, & d \end{pmatrix}, \begin{pmatrix} a, & c \\ b, & d \end{pmatrix},$$

are said to be formed one from the other by transposition, and this may be denoted by the symbol tr.; thus we may write

$$\begin{pmatrix} a, & c \\ b, & d \end{pmatrix} = \text{tr.} \begin{pmatrix} a, & b \\ c, & d \end{pmatrix}.$$

The effect of two successive transpositions is of course to reproduce the original matrix.

37. It is easy to see that if M be any matrix, then

$$(\text{tr.}M)^p = \text{tr.}(M^p),$$

and in particular,

$$(\text{tr.}M)^{-1} = \text{tr.}(M^{-1}).$$

38. If L, M be any two matrices,

$$\text{tr.}(LM) = \text{tr.}M \cdot \text{tr.}L,$$

and similarly for three or more matrices, L, M, N, etc.,

$$\text{tr.}(LMN) = \text{tr.}N \cdot \text{tr.}M \cdot \text{tr.}L, \text{ etc.}$$

40. A matrix such as

$$\begin{pmatrix} a, & h, & g \\ h, & b, & f \\ g, & f, & c \end{pmatrix}$$

which is not altered by transposition, is said to be symmetrical.

41. A matrix such as

$$\begin{pmatrix} 0, & v, & -\mu \\ -v, & 0, & \lambda \\ \mu, & -\lambda, & 0 \end{pmatrix}$$

which by transposition is changed into its opposite, is said to be skew symmetrical.

42. It is easy to see that any matrix whatever may be expressed as the sum of a symmetrical matrix, and a skew symmetrical matrix; thus the form

$$\begin{pmatrix} a, & h+v, & g-\mu \\ h-v, & b, & f+\lambda \\ g+\mu, & f-\lambda, & c \end{pmatrix}$$

which may obviously represent any matrix whatever of the order 3, is the sum of the two matrices last before mentioned.

43. The following formulae, although little more than examples of the composition of transposed matrices, may be noticed, viz.

$$\begin{pmatrix} a, & b \\ c, & d \end{pmatrix} \begin{pmatrix} a, & c \\ b, & d \end{pmatrix} = \begin{pmatrix} a^2+b^2, & ac+bd \\ ac+bd, & c^2+d^2 \end{pmatrix}$$

which shows that a matrix compounded with the transposed matrix gives rise to a symmetrical matrix. It does not however follow, nor is it the fact, that the matrix and transposed matrix are convertible. And also

$$\begin{pmatrix} a, & c \\ b, & d \end{pmatrix} \begin{pmatrix} a, & b \\ c, & d \end{pmatrix} \begin{pmatrix} a, & c \\ b, & d \end{pmatrix} = \begin{pmatrix} a^3+bcd+a(b^2+c^2), & c^3+abd+c(a^2+d^2) \\ b^3+acd+b(a^2+d^2), & d^3+abc+d(b^2+c^2) \end{pmatrix}$$

which is a remarkably symmetrical form. It is needless to proceed further, since it is clear that

$$\begin{pmatrix} a, & c \\ b, & d \end{pmatrix} \begin{pmatrix} a, & b \\ c, & d \end{pmatrix} \begin{pmatrix} a, & c \\ b, & d \end{pmatrix} \begin{pmatrix} a, & b \\ c, & d \end{pmatrix} = \left\{ \begin{pmatrix} a, & c \\ b, & d \end{pmatrix} \begin{pmatrix} a, & b \\ c, & d \end{pmatrix} \right\}^2.$$

44. In all that precedes, the matrix of the order 2 has frequently been considered, but chiefly by way of illustration of the general theory; but it is worth while to develop more particularly the theory of such a matrix. I call to mind the fundamental properties which have been obtained, viz. it was shown that the matrix

$$M = \begin{pmatrix} a, & b \\ c, & d \end{pmatrix},$$

satisfies the equation

$$M^2-(a+d)M+ad-bc=0,$$

and that the two matrices

$$\begin{pmatrix} a, & b \\ c, & d \end{pmatrix}, \quad \begin{pmatrix} a', & b' \\ c', & d' \end{pmatrix},$$

will be convertible if

$$a'-d':b':c'=a-d:b:c,$$

and that they will be skew convertible if

$$a+d=0, \quad a'+d'=0, \quad aa'+bc'+b'c+dd'=0,$$

the first two of these equations being the conditions in order that the two matrices may be respectively periodic of the second order to a factor *près*.

45. It may be noticed in passing, that if L, M are skew convertible matrices of the order 2, and if these matrices are also such that $L^2=-1$, $M^2=-1$, then putting $N=LM=-ML$, we obtain

$$L^2=-1, \qquad M^2=-1, \qquad N^2=-1,$$
$$L=MN=-NM, \quad M=NL=-LN, \quad N=LM=-ML$$

54

GEORG CANTOR
(1845–1918)

GEORG FERDINAND LUDWIG PHILIPP CANTOR, whose theory of transfinite numbers and whose contributions to the theory of sets was to usher in a new epoch in the development of mathematics, was born in St Petersburg (Leningrad), Russia, in 1845. His father, Georg Waldemar Cantor, had, as a youth, emigrated from Copenhagen, Denmark, to St Petersburg, where he had become a successful merchant. His mother, Maria Bohm Cantor, came of a family well known in the fields of music and art. In 1856, the father contracted a pulmonary illness and, in search of a less inclement climate, the family moved to Germany. At about this time, young Georg's mathematical talents became evident, but the boy's desire to devote himself to mathematics ran counter to his father's wish that he study engineering. The tensions generated by this opposition of aims were relieved at last when the elder Cantor recognized the depth and sincerity of his son's choice of a career. Just before entering upon his university training, Georg received permission to pursue his mathematical studies and he replied to his father in a highly emotional letter. The letter contained more than words of thanks. It painted a picture in broad strokes of a devoted son and a conscientious personality, modest and at the same time hopeful and very ardently desirous of proving himself worthy.

Cantor studied at Zurich, Berlin and Gottingen, where he was greatly influenced by Kummer, by Kronecker and especially by Weierstrass. In 1867, he received his degree at Berlin, and as his father, who had died in 1863, had left him well provided for, he was ready to embark on what he undoubtedly believed was to be a tranquil life-work in the calm and peaceful atmosphere of university halls. It was his destiny, however, to be led through his quiet and erudite researches to a field of mathematics so new and so strange to his contemporaries that its immediate effect was completely disturbing. Instead of sedate honours, his innovations brought him misunderstanding, disappointment, controversy, discouragement and illness. The post Cantor wanted most, an appointment at the University of Berlin, was never to be his. He taught for a time at a girl's school. He was examined for and received a licence to teach children. In 1869, he was appointed Privat-dozent at the University of Halle, and this assignment was the beginning of a teaching career at Halle which was to last for more than forty years.

The basic concepts of Cantor's theory of transfinite numbers began to be clear to him as early as 1871. In 1874, he published his first paper on the theory of sets. In 1883, his famous *Grundlagen einer allgemeinen Mannigfaltigkeitslehre* appeared, in which he distinguished between the 'improper' and 'proper' mathematical infinite, developing the theory of the latter, the transfinite numbers. A violent storm of protest against his work was led by Kronecker and Poincaré. Their criticisms effectively discouraged many mathematicians from even attempting to understand Cantor's novel concepts. However, the support given him by Dedekind, Mittag-Leffler and others, though it was less noisy and aggressive, was the more understanding and the more fruitful. In the early twentieth century, academic honours were awarded to Cantor in Italy, England and Denmark as well as in Germany. However, academic honours and recognition could not stem the nervous breakdown which first beset him in 1884 and recurred from time to time to the end of his life. Cantor died in the psychiatric clinic at Halle in 1918.

A new generation of mathematicians, turning their attention to subtle points which Cantor had not settled, have given themselves to the study and development of his theories. Now widely accepted, Cantor's novel concepts lie firmly imbedded in the bases of mathematical analysis.

———

CONTRIBUTIONS TO THE FOUNDING OF THE THEORY OF TRANSFINITE NUMBERS

translated by Philip E. B. Jourdain

§ 1

THE CONCEPTION OF POWER OR CARDINAL NUMBER

By an 'aggregate' (*Menge*) we are to understand any collection into a whole (*Zusammenfassung zu einem Ganzem*) M of definite and separate objects m of our intuition or our thought. These objects are called the 'elements' of M.

In signs we express this thus:

(1) $$M = \{m\}.$$

We denote the uniting of many aggregates $M, N, P, \ldots$, which have no common elements, into a single aggregate by,

(2) $$(M, N, P, \ldots).$$

The elements of this aggregate are, therefore, the elements of M, of N, of P, $\ldots$, taken together.

We will call by the name 'part' or 'partial aggregate' of an aggregate M any other aggregate M_1 whose elements are also elements of M.

If M_2 is a part of M_1 and M_1 is a part of M, then M_2 is a part of M.

Every aggregate M has a definite 'power', which we will also call its 'cardinal number'.

We will call by the name 'power' or 'cardinal number' of M the general concept which, by means of our active faculty of thought, arises from the aggregate M when we make abstraction of the nature of its various elements m and of the order in which they are given.

[482] We denote the result of this double act of abstraction, the cardinal number or power of M, by

(3) $$\overline{\overline{M}}.$$

Since every single element m, if we abstract from its nature, becomes a 'unit', the cardinal number $\overline{\overline{M}}$ is a definite aggregate composed of units, and this number has existence in our mind as an intellectual image or projection of the given aggregate M.

We say that two aggregates M and N are 'equivalent', in signs

(4) $$M \sim N \quad \text{or} \quad N \sim M,$$

if it is possible to put them, by some law, in such a relation to one another that to every element of each one of them corresponds one and only one element of the other. To every part M_1 of M there corresponds, then, a definite equivalent part N_1 of N, and inversely.

If we have such a law of coordination of two equivalent

aggregates, then, apart from the case when each of them consists only of one element, we can modify this law in many ways. We can, for instance, always take care that to a special element m_0 of M a special element n_0 of N corresponds. For if, according to the original law the elements m_0 and n_0 do not correspond to one another, but to the element m_0 of M the element n_1 of N corresponds, and to the element n_0 of N the element m_1 of M corresponds, we take the modified law according to which m_0 corresponds to n_0 and m_1 to n_1 and for the other elements the original law remains unaltered. By this means the end is attained.

Every aggregate is equivalent to itself:

(5) $$M \sim M.$$

If two aggregates are equivalent to a third, they are equivalent to one another; that is to say:

(6) from $M \sim P$ and $N \sim P$ follows $M \sim N$.

Of fundamental importance is the theorem that two aggregates M and N have the same cardinal number if, and only if, they are equivalent: thus,

(7) from $M \sim N$ we get $\overline{\overline{M}} = \overline{\overline{N}}$,

and

(8) from $\overline{\overline{M}} = \overline{\overline{N}}$ we get $M \sim N$.

Thus the equivalence of aggregates forms the necessary and sufficient condition for the equality of their cardinal numbers.

[483] In fact, according to the above definition of power, the cardinal number $\overline{\overline{M}}$ remains unaltered if in the place of each of one or many or even all elements m of M other things are substituted. If, now, $M \sim N$, there is a law of coordination by means of which M and N are uniquely and reciprocally referred to one another; and by it to the element m of M corresponds the element n of N. Then we can imagine, in the place of every element m of M, the corresponding element n of N substituted, and, in this way, M transforms into N without alteration of cardinal number. Consequently

$$\overline{\overline{M}} = \overline{\overline{N}}.$$

The converse of the theorem results from the remark that between the elements of M and the different units of its cardinal number

$\bar{M}$ a reciprocally univocal (or biunivocal) relation of correspondence subsists. For, as we saw, $\bar{M}$ grows, so to speak, out of M in such a way that from every element m of M a special unit of M arises. Thus we can say that

(9) $$M \sim \bar{M}.$$

In the same way $N \sim \bar{N}$. If then $\bar{\bar{M}} = \bar{\bar{N}}$, we have, by (6), $M \sim N$.

We will mention the following theorem, which results immediately from the conception of equivalence. If $M, N, P, \ldots$ are aggregates which have no common elements, $M', N', P', \ldots$ are also aggregates with the same property, and if

$$M \sim M', \ N \sim N', \ P \sim P', \ldots,$$

then we always have

$$(M, N, P, \ldots) \sim (M', N', P', \ldots).$$

§ 2

'GREATER' AND 'LESS' WITH POWERS

If for two aggregates M and N with the cardinal numbers $a = \bar{\bar{M}}$ and $b = \bar{\bar{N}}$, both the conditions:

(a) There is no part of M which is equivalent to N,

(b) There is a part N_1 of N, such that $N_1 \sim M$,

are fulfilled, it is obvious that these conditions still hold if in them M and N are replaced by two equivalent aggregates M' and N'. Thus they express a definite relation of the cardinal numbers a and b to one another.

[484] Further, the equivalence of M and N, and thus the equality of a and b, is excluded; for if we had $M \sim N$, we would have, because $N_1 \sim M$, the equivalence $N_1 \sim N$, and then, because $M \sim N$, there would exist a part M_1 of M such that $M_1 \sim M$, and therefore we should have $M_1 \sim N$; and this contradicts the condition (a).

Thirdly, the relation of a to b is such that it makes impossible the same relation of b to a; for if in (a) and (b) the parts played by M and N are interchanged, two conditions arise, contradictory to the former ones.

We express the relation of a to b characterized by (a) and (b) by saying: a is 'less' than b or b is 'greater' than a; in signs

(1) $$a < b \ \text{ or } \ b > a.$$

We can easily prove that,

(2)   if $a<b$ and $b<c$, then we always have $a<c$.

Similarly, from the definition, it follows at once that, if P_1 is part of an aggregate P, from $a<\bar{\bar{P}}_1$ follows $a<\bar{\bar{P}}$ and from $\bar{\bar{P}}<b$ follows $\bar{\bar{P}}_1<b$.

We have seen that, of the three relations

$$a=b,\ a<b,\ b<a,$$

each one excludes the two others. On the other hand, the theorem that, with any two cardinal numbers a and b, one of those three relations must necessarily be realized, is by no means self-evident and can hardly be proved at this stage.

Not until later, when we shall have gained a survey over the ascending sequence of the transfinite cardinal numbers and an insight into their connexion, will result the truth of the theorem:

A. If a and b are any two cardinal numbers, then either $a=b$ or $a<b$ or $a>b$.

From this theorem the following theorems, of which, however, we will here make no use, can be very simply derived:

B. If two aggregates M and N are such that M is equivalent to a part N_1 of N and N to a part M_1 of M, then M and N are equivalent;

C. If M_1 is a part of an aggregate M, M_2 is a part of the aggregate M_1, and if the aggregates M and M_2 are equivalent, then M_1 is equivalent to both M and M_2;

D. If, with two aggregates M and N, N is equivalent neither to M nor to a part of M, there is a part N_1 of N that is equivalent to M;

E. If two aggregates M and N are not equivalent, and there is a part N_1 of N that is equivalent to M, then no part of M is equivalent to N.

[485]   § 3

THE ADDITION AND MULTIPLICATION OF POWERS

The union of two aggregates M and N which have no common elements was denoted in § 1, (2), by (M, N). We call it the 'union-aggregate (*Vereinigungsmenge*) of M and N'.

If M' and N' are two other aggregates without common elements, and if $M \sim M'$ and $N \sim N'$, we saw that we have

$$(M, N) \sim (M', N').$$

Hence the cardinal number of (M, N) only depends upon the cardinal numbers $\overline{M} = a$ and $\overline{N} = b$.

This leads to the definition of the sum of a and b. We put

(1) $a + b = (\overline{\overline{M}}, \overline{\overline{N}})$.

Since in the conception of power, we abstract from the order of the elements, we conclude at once that

(2) $a + b = b + a$;

and, for any three cardinal numbers a, b, c, we have

(3) $a + (b + c) = (a + b) + c$.

We now come to multiplication. Any element m of an aggregate M can be thought to be bound up with any element n of another aggregate N so as to form a new element (m, n); we denote by $(M.N)$ the aggregate of all these bindings (m, n), and call it the 'aggregate of bindings (*Verbindungsmenge*) of M and N'. Thus

(4) $(M.N) = \{(m, n)\}$.

We see that the power of $(M.N)$ only depends on the powers $\overline{M} = a$ and $\overline{N} = b$; for, if we replace the aggregates M and N by the aggregates

$$M' = \{m'\} \quad \text{and} \quad N' = \{n'\}$$

respectively equivalent to them, and consider m, m' and n, n' as corresponding elements, then the aggregate

$$(M'.N') = \{(m', n')\}$$

is brought into a reciprocal and univocal correspondence with $(M.N)$ by regarding (m, n) and (m', n') as corresponding elements. Thus

(5) $(M'.N') \sim (M.N)$.

We now define the product $a.b$ by the equation

(6) $a.b = (M.N)$.

[486] An aggregate with the cardinal number $a.b$ may also be made up out of two aggregates M and N with the cardinal

numbers a and b according to the following rule: We start from the aggregate N and replace in it every element n by an aggregate $M_n...M$; if, then, we collect the elements of all these aggregates M_n to a whole S, we see that

$$(7) \qquad\qquad S \sim (M.N),$$

and consequently

$$\bar{\bar{S}} = a.b.$$

For, if, with any given law of correspondence of the two equivalent aggregates M and M_n, we denote by m the element of M which corresponds to the element m_n of M_n, we have

$$(8) \qquad\qquad S = \{m^n\};$$

and thus the aggregates S and $(M.N)$ can be referred reciprocally and univocally to one another by regarding m_n and (m, n) as corresponding elements.

From our definitions result readily the theorems:

$$(9) \qquad\qquad a.b = b.a,$$

$$(10) \qquad\qquad a.(b.c) = (a.b).c,$$

$$(11) \qquad\qquad a(b+c) = ab + ac;$$

because

$$(M.N) \sim (N.M),$$

$$(M.(N.P)) \sim ((M.N).P),$$

$$(M.(N, P)) \sim ((M.N), (M.P)).$$

Addition and multiplication of powers are subject, therefore, to the commutative, associative, and distributive laws.

§ 4

THE EXPONENTIATION OF POWERS

By a 'covering of the aggregate N with elements of the aggregate M', or, more simply, by a 'covering of N with M', we understand a law by which with every element n of N a definite element of M is bound up, where one and the same element of M can come repeatedly into application. The element of M bound up with n is, in a way, a one-valued function of n, and may be denoted by $f(n)$; it is called a 'covering function of n'. The corresponding covering of N will be called $f(N)$.

[487] Two coverings $f_1(N)$ and $f_2(N)$ are said to be equal if, and only if, for all elements n of N the equation

(1) $$f_1(n) = f_2(n)$$

is fulfilled, so that if this equation does not subsist for even a single element $n = n_0$, $f_1(N)$ and $f_2(N)$ are characterized as different coverings of N. For example, if m_0 is a particular element of M, we may fix that, for all n's

$$f(n) = m_0;$$

this law constitutes a particular covering of N with M. Another kind of covering results if m_0 and m_1 are two different particular elements of M and n_0 a particular element of N, from fixing that

$$f(n_0) = m_0$$
$$f(n) = m_1,$$

for all n's which are different from n_0.

The totality of different coverings of N with M forms a definite aggregate with the elements $f(N)$; we call it the 'covering-aggregate (*Belegungsmenge*) of N with M' and denote it by $(N \mid M)$. Thus:

(2) $$(N \mid M) = \{f(N)\}.$$

If $M \sim M'$ and $N \sim N'$, we easily find that

(3) $$(N \mid M) \sim (N' \mid M').$$

Thus the cardinal number of $(N \mid M)$ depends only on the cardinal numbers $\overline{\overline{M}} = a$ and $\overline{\overline{N}} = b$; it serves us for the definition of a^b:

(4) $$a^b = (\overline{\overline{N}} \mid \overline{\overline{M}}).$$

For any three aggregates, M, N, P, we easily prove the theorems:

(5) $$((N \mid M) . (P \mid M)) \sim ((N, P) \mid M),$$

(6) $$((P \mid M) . (P \mid N)) \sim (P \mid M . N)),$$

(7) $$(P \mid (N \mid M)) \sim ((P . N) \mid M),$$

from which, if we put $\overline{\overline{P}} = c$, we have, by (4) and by paying attention to § 3, the theorems for any three cardinal numbers, a, b and c:

(8) $$a^b . a^c = a^{b+c}_c,$$

(9) $$a^c . b^c = (a . b)^c,$$

(10) $$(a^b)^c = a^{b . c}.$$

[488] We see how pregnant and far-reaching these simple formulae extended to powers are by the following example. If we denote the power of the linear continuum X (that is, the totality X of real numbers x such that $x \geq 0$ and ≤ 1) by $\mathfrak{o}$, we easily see that it may be represented by, amongst others, the formula:

$$(11) \qquad \mathfrak{o} \sim 2^{\aleph_0},$$

where § 6 gives the meaning of $\aleph_0$. In fact, by (4), $2^{\aleph_0}$ is the power of all representations

$$(12) \qquad x = \frac{f(1)}{2} + \frac{f(2)}{2^2} + \ldots + \frac{f(v)}{2^v} + \ldots$$

$$\text{(where } f(v) = 0 \text{ or } 1)$$

of the numbers x in the binary system. If we pay attention to the fact that every number x is only represented once, with the exception of the numbers $x = \dfrac{2v+1}{2^\mu} < 1$, which are represented twice over, we have, if we denote the 'enumerable' totality of the latter by $\{s_v\}$,

$$2^{\aleph_0} = (\overline{\{s_v\}, X}).$$

If we take away from X any 'enumerable' aggregate $\{t_v\}$ and denote the remainder by X_1, we have:

$$X = (\{t_v\}, X_1) = (\{t_{2v-1}\}, \{t_{2v}\}, X_1),$$
$$(\{s_v''\}, X) = (\{s_v\}, \{t_v\}, X_1),$$
$$\{t_{2v-1}\} \sim \{s_v\}, \ \{t_{2v}\} \sim \{t_v\}, \ X_1 \sim X_1;$$
$$X \sim (\{s_v\}, X),$$

and thus (§ 1)

$$2^{\aleph_0} = \overline{\overline{X}} = \mathfrak{o}.$$

From (11) follows by squaring (by § 6, (6))

$$\mathfrak{o} \cdot \mathfrak{o} = 2^{\aleph_0} \cdot 2^{\aleph_0} = 2^{\aleph_0 + \aleph_0} = 2^{\aleph_0} = \mathfrak{o},$$

and hence, by continued multiplication by $\mathfrak{o}$,

$$(13) \qquad \mathfrak{o}^v = \mathfrak{o},$$

where v is any finite cardinal number.

If we raise both sides of (11) to the power[1] $\aleph_0$ we get

$$\mathfrak{o}^{\aleph_0} = (2^{\aleph_0})^{\aleph_0} = 2^{\aleph_0 \cdot \aleph_0}.$$

But since, by § 6, (8), $\aleph_0 \cdot \aleph_0 = \aleph_0$, we have

$$(14) \qquad \mathfrak{o}^{\aleph_0} = \mathfrak{o}.$$

---

1. In English there is an ambiguity.

The formulae (13) and (14) mean that both the ν-dimensional and the $\aleph_0$-dimensional continuum have the power of the one-dimensional continuum. Thus the whole contents of my paper in Crelle's *Journal*, vol. lxxxiv, 1878,[1] are derived purely algebraically with these few strokes of the pen from the fundamental formula of the calculation with cardinal numbers.

<h1 style="text-align:center">§ 6</h1>

THE SMALLEST TRANSFINITE CARDINAL NUMBER ALEPH-ZERO

Aggregates with finite cardinal numbers are called 'finite aggregates', all others we will call 'transfinite aggregates' and their cardinal numbers 'transfinite cardinal numbers'.

The first example of a transfinite aggregate is given by the totality of finite cardinal numbers ν; we call its cardinal number (§ 1) 'Aleph-zero' and denote it by $\aleph_0$; thus we define

$$(1) \qquad \aleph_0 = \{\bar{\nu}\}.$$

That $\aleph_0$ is a *transfinite* number, that is to say, is not equal to any finite number μ, follows from the simple fact that, if to the aggregate $\{\nu\}$ is added a new element e_0, the union-aggregate $(\{\nu\}, e_0)$ is equivalent to the original aggregate $\{\nu\}$. For we can think of this reciprocally univocal correspondence between them: to the element e_0 of the first corresponds the element 1 of the second, and to the element ν of the first corresponds the element $\nu+1$ of the other. By § 3 we thus have

$$(2) \qquad \aleph_0 + 1 = \aleph_0.$$

But we showed in § 5 that $\mu+1$ is always different from μ, and therefore $\aleph_0$ is not equal to any finite number μ:

$$(3) \qquad \aleph_0 > \mu.$$

[493] This follows, if we pay attention to § 3, from the three facts that $\mu = (\overline{1, 2, 3, \ldots, \mu})$, that no part of the aggregate $(1, 2, 3, \ldots \mu)$ is equivalent to the aggregate $\{\nu\}$, and that $(1, 2, 3, \ldots, \mu)$ is itself a part of $\{\nu\}$.

On the other hand, $\aleph_0$ is the least transfinite cardinal number.

1. See Section V of the Introduction.

If a is any transfinite cardinal number different from $\aleph_0$, then

(4) $$\aleph_0 < a.$$

This rests on the following theorems:

A. Every transfinite aggregate T has parts with the cardinal number $\aleph_0$.

Proof: If, by any rule, we have taken away a finite number of elements t_1, t_2, . . . , $t_{\nu-1}$, there always remains the possibility of taking away a further element t_ν. The aggregate $\{t_\nu\}$, where ν denotes any finite cardinal number, is a part of T with the cardinal number $\aleph_0$, because $\{t_\nu\} \sim \{\nu\}$ (§ 1).

B. If S is a transfinite aggregate with the cardinal number $\aleph_0$, and S_1 is any transfinite part of S, then $S_1 = \aleph_0$.

From A and B the formula (4) results, if we have regard to § 2.

From (2) we conclude, by adding 1 to both sides,

$$\aleph_0 + 2 = \aleph_0 + 1 = \aleph_0,$$

and, by repeating this

(5) $$\aleph_0 + \nu = \aleph_0.$$

We have also

(6) $$\aleph_0 + \aleph_0 = \aleph_0.$$

The equation (6) can also be written

$$\aleph_0 . 2 = \aleph_0;$$

and, by adding $\aleph_0$ repeatedly to both sides, we find that

(7) $$\aleph_0 . \nu = \nu . \aleph_0 = \aleph_0.$$

We also have

(8) $$\aleph_0 . \aleph_0 = \aleph_0.$$

Proof: By (6) of § 3, $\aleph_0 . \aleph_0$ is the cardinal number of the aggregate of bindings

$$\{(\mu, \nu)\}$$

where μ and ν are any finite cardinal numbers which are independent of one another. If also λ represents any finite cardinal number, so that $\{\lambda\}$, $\{\mu\}$, and $\{\nu\}$ are only different notations for the same aggregate of all finite numbers, we have to show that

$$\{(\mu, \nu)\} \sim \{\lambda\}.$$

Let us denote $\mu + \nu$ by p; then p takes all the numerical values

2, 3, 4, . . . , and there are in all $p-1$ elements (μ, ν) for which $\mu+\nu=p$, namely:

$$(1, p-1), (2, p-2), \ldots, (p-1, 1).$$

In this sequence imagine first the element $(1, 1)$, for which $p=2$, put, then the two elements for which $p=3$, then the three elements for which $p=4$, and so on. Thus we get all the elements (μ, ν) in a simple series:

$$(1, 1); (1, 2), (2, 1); (1, 3), (2, 2), (3, 1); (1, 4), 2, 3), \ldots,$$

and here, as we easily see, the element (μ, ν) comes at the λth place, where

(9) $$\lambda=\mu+\frac{(\mu+\nu-1)(\mu+\nu-2)}{2}.$$

The variable λ takes every numerical value $1, 2, 3, \ldots,$ once. Consequently, by means of (9), a reciprocally univocal relation subsists between the aggregates $\{\nu\}$ and $\{(\mu, \nu)\}$.

[495] If both sides of the equation (8) are multiplied by $\aleph_0$, we get $\aleph_0^3=\aleph_0^2=\aleph_0$, and, by repeated multiplications by $\aleph_0$, we get the equation, valid for every finite cardinal number ν:

(10) $$\aleph_0\nu=\aleph_0.$$

The theorems E and A of § 5 lead to this theorem on finite aggregates:

c. Every finite aggregate E is such that it is equivalent to none of its parts.

This theorem stands sharply opposed to the following one for transfinite aggregates:

d. Every transfinite aggregate T is such that it has parts T_1 which are equivalent to it.

Proof: By theorem A of this paragraph there is a part $S=\{t_\nu\}$ of T with the cardinal number $\aleph_0$. Let $T=(S, U)$, so that U is composed of those elements of T which are different from the elements t_ν. Let us put $S_1=\{t_{\nu+1}\}$, $T_1=(S_1, U)$; then T_1 is a part of T, and, in fact, that part which arises out of T if we leave out the single element t_1. Since $S\sim S_1$ by theorem B of this paragraph, and $U\sim U$, we have, by § 1, $T\sim T_1$.

In these theorems c and d the essential difference between finite and transfinite aggregates, to which I referred in the year

1877, in volume lxxxiv [1878] of Crelle's *Journal*, p. 242, appears in the clearest way.

After we have introduced the least transfinite cardinal number $\aleph_0$ and derived its properties that lie the most readily to hand, the question arises as to the higher cardinal numbers and how they proceed from $\aleph_0$. We shall show that the transfinite cardinal numbers can be arranged according to their magnitude, and, in this order, form, like the finite numbers, a 'well-ordered aggregate' in an extended sense of the words. Out of $\aleph_0$ proceeds, by a definite law, the next greater cardinal number $\aleph_1$, out of this by the same law the next greater $\aleph_2$, and so on. But even the unlimited sequence of cardinal numbers

$$\aleph_0, \aleph_1, \aleph_2, \ldots, \aleph_\nu, \ldots$$

does not exhaust the conception of transfinite cardinal number. We will prove the existence of a cardinal number which we denote by $\aleph_\omega$ and which shows itself to be the next greater to all the numbers $\aleph_\nu$; out of it proceeds in the same way as $\aleph_1$, out of $\aleph_\omega$ a next greater $\aleph_{\omega+1}$ and so on without end.

[496] To every transfinite cardinal number a there is a next greater proceeding out of it according to a unitary law, and also to every unlimitedly ascending well-ordered aggregate of transfinite cardinal number (a) there is a next greater proceeding out of that aggregate in a unitary way.

For the rigorous foundation of this matter, discovered in 1882 and exposed in the pamphlet *Grundlagen einer allgemeinen Mannichfaltigkeitslehre* (Leipsig, 1883) and in volume xxi of the *Mathematische Annalen*, we make use of the so-called 'ordinal types'.

CHRONOLOGICAL TABLE

B.C.

3rd millennium onwards	The Babylonians
c. 1850	The Moscow Papyrus
c. 1650	The Rhind Mathematical Papyrus
12th century	Wan Wang
c. 1100	Chou Kung
5th century	Hippocrates of Chios
c. 428–*c.* 347	Plato
c. 300	Euclid of Alexandria
3rd century	Apollonius of Perga
c. 287–212	Archimedes
d. 152	Chang Tsang

A.D.

c. 100	Nicomachus of Gerasa
c. 150	R. Nehemiah
3rd century	Liu Hui
3rd century	Sun-Tsu
c. 250	Diophantus
c. 300	Pappus
4th century	The Bakhshali Manuscript
4th–9th centuries	The Maya Civilization, Classical Period
410–485	Proclus
b. 476	Āryabhata, the Elder
c. 500	Metrodorus
b. 598	Brahmagupta
6th century	Hsai-Hou Yang
7th century	Wang Hs'Iao-T'Ung
c. 825	Mohammed Ben Musa Al-Khowarizmi
c. 1044–1123	Omar Khayyám
1114–*c.* 1185	Bhāscara Āchārya

c. 1178–*c.* 1265	Li Yeh
c. 1250	Ch'in Chiu-Shao
c. 1275	Yang Hui
c. 1300	Chu Chi-Chieh
1340(?)–1400	Geoffrey Chaucer
c. 1350	Immanuel Ben Jacob Bonfils
15th and 16th centuries	The Peruvian Quipu (highest development)
1471–1528	Albrecht Dürer
c. 1510–1558	Robert Recorde
1548–1620	Simon Stevin
1596–1650	René Descartes
1630–1677	Isaac Barrow
1642–1727	Sir Isaac Newton
1646–1716	Gottfried Wilhelm v. Leibniz
1704–1752	Gabriel Cramer
1745–1818	Caspar Wessel
1777–1855	Carl Friedrich Gauss
1806–1871	Augustus De Morgan
1809–1880	Benjamin Peirce
1814–1897	James Joseph Sylvester
1815–1864	George Boole
1821–1895	Arthur Cayley
1834–1923	John Venn
1839–1914	Charles Sanders Peirce
1845–1918	Georg Cantor
1848–1925	Gottlob Frege

INDEX

TO VOLUMES ONE AND TWO

411

INDEX